GERM PROOF YOUR KIDS

The Complete Guide to Protecting (without Overprotecting) Your Family from Infections

Also by Harley Rotbart

Human Enterovirus Infections (ASM Press, 1995)

The On Deck Circle of Life: 101 Lessons from the Dugout
(iUniverse, Inc., 2007)

For timely updates and insightful perspectives on today's scary germ headlines, visit www.germproofyourkids.com and access Dr. Rotbart's up-to-the minute GERMBlog

The author has no financial interests, including stock ownership, stock options, or royalities, in any product or service mentioned in this book. The author receives no financial support, consultancy fees, speakers' honoraria, or gifts from any pharmaceutical industry, consumer products company, or government source.

GERM PROOF YOUR KIDS

The Complete Guide to Protecting (without Overprotecting) Your Family from Infections

Harley A. Rotbart, M.D.
University of Colorado School of Medicine
The Children's Hospital of Denver
Denver, Colorado

Washington, DC

Address editorial correspondence to ASM Press, 1752 N St. NW, Washington, DC 20036-2904, USA

Send orders to ASM Press, P.O. Box 605, Herndon, VA 20172, USA
Phone: (800) 546-2416 or (703) 661-1593
Fax: (703) 661-1501
E-mail: books@asmusa.org
Online: estore.asm.org

Library of Congress Cataloging-in-Publication Data

Rotbart, Harley A.
Germ proof your kids : the complete guide to protecting (without over-protecting) your family from infections / Harley A. Rotbart.
p. cm.
Includes index.
ISBN 978-1-55581-427-4 (alk. paper)
1. Microbiology—Popular works. 2. Infection—Popular works.
3. Hygiene. I. Title.
QR56.R598 2008
616.9′041—dc22

2007035694

10 9 8 7 6 5 4 3 2 1

To Matt, who was spared a spinal tap for his first fever but had to endure the "bulb syringe torture technique" for his first cold

To Emily, whose "severe colic" was cured as soon as her ear started draining pus and whose first rotavirus infection left its mark all over the house

To Sam, who brought resistant staphylococcus home from school and had a plantar wart so big that he named it "Pumbaa"

And especially to Sara, who got them through it all despite being married to a pediatric infectious diseases doctor.

Science and technology revolutionize our lives, but memory, tradition and myth frame our response.

Arthur M. Schlesinger, historian (1917-2007)

Contents

Introduction: the Germ Theory versus My Mom's Theory

In the mid-19th century, a revolution took place in science and medicine. With the discoveries of microorganisms and their role in causing disease, the research giants Ignaz Semmelweis, Louis Pasteur, Joseph Lister, and Robert Koch defined the germ theory of disease. According to this new paradigm, microscopic living organisms could spread from person to person, invade our bodies, and cause severe illness and even death. In the ensuing 150 years, the germ theory has proven to be the single most important contribution of science to the practice of medicine.

My mother, however, didn't get the memo.

I grew up certain that illness came from being underdressed in cold weather or overdressed in warm weather. If my hair was wet when I went outdoors, guaranteed sickness. If my feet got wet in the snow or rain, sick again. Food and drink, of course, were also critical determinants of health. Hot was healing—chicken soup and oatmeal—but heaven help the poor soul who drank something too cold on a cold day or, for that matter, on a hot day. Ice was a public health menace. Hot tea with honey was curative; iced tea—are you kidding? Menthol rubs, "gogle mogels" (a concoction of hot milk, egg, and honey), six layers of blankets, and hot steam vaporizers treated the mysterious causes of most illnesses that occurred above the belt. For the illnesses below the belt, a variety of binding substances would be ingested, each of which tasted chalkier and was more nauseating than the last. Our cuts and scrapes were painted with mercurochrome, a burning antiseptic that today is banned because of the risk of mercury poisoning! If we complained that the treatment made us feel worse than the disease, we'd be reminded of what our grandparents did to our poor parents when they were sick as kids: "bankes" (a torture device using a wax candle in a glass that was applied to the chest), cod-liver oil, and schnapps. We would also be reminded that this was the price we paid for not wearing our boots in the rain. We caught a chill!

As with most diametrically opposed theories, the truth about infections rests somewhere between Mom and Pasteur. Recognizing the germs that cause infections has certainly given us powerful tools to combat them. Long before today's worries over epidemics like bird flu and SARS (severe acute respiratory syndrome) crossing onto our shores, quarantines and importation restrictions prevented the spread of yellow fever and cholera from contaminated European ships docked in our ports. The implementation of basic community sanitation innovations, like sewage treatment and water purification systems, by the early 1930s had eliminated homegrown waterborne epidemics of cholera and typhoid fever. Coincident with these innovations, but not coincidental to them, the infant mortality rate fell from 110/1,000 live births in 1900 to 29/1,000 by 1950, and U.S. life expectancy rose from 47 years to 68 years in the same period. However, a modern-day scientific "miracle" also contributed mightily to the improved quality and duration of life in America in the first half of the 20th century. The "miracle of antibiotics" began with the discovery of penicillin in 1928. Its mass production, just in time to provide enough medicine for all our troops by D-Day in 1944, was followed by the introduction of other antibiotics, including the first to be effective against tuberculosis. Contrast the dramatic life-saving impact of penicillin on our troops in World War II with the stark statistic that two-thirds of the 360,000 Union soldiers who perished in the preantibiotic-era Civil War died not from bullets but from infectious diseases.

Our germ conquests continue. From 1950 to the present, infant mortality in the United States has decreased even further, to fewer than 7 deaths per 1,000 live births, contemporaneous with, and largely attributable to, the "miracle of vaccines." This miracle began in 1950, when a grassroots campaign to eradicate polio emerged in Phoenix, Arizona, with the slogan, "Turn on your porch light, fight polio tonight." In the ensuing few years, the Mother's March of Dimes raised a fortune in door-to-door contributions from around the country, directly leading to the development and administration of effective polio vaccinations. As a result, there were no cases of polio in the United States between 1999 and 2006 (when a case of imported polio was diagnosed), compared with 21,000 U.S. cases in 1951 just before vaccines were introduced. More vaccine miracles followed: there are 200,000 fewer cases of diphtheria each year in the United States, 250,000 fewer pertussis (whooping cough) cases, and 900,000 fewer cases of measles than there were in early and mid-20th century America.

Okay, so germs are important causes of disease and fighting them with sophisticated scientific advances is effective. Does that mean that our mothers and grandmothers were snake oil saleswomen? Quite the opposite. The highly focused and skeptical lens of today's science has proven that the simple interventions that generations of moms have embraced are also effective in protecting kids against infections—good hygiene, healthy nutrition, sleep, exercise, and yes, maybe even those rubber boots in the rain.

Neither the science nor the folklore of infectious diseases stands still. In the past 30 years, no fewer than 25 "new" germs have been identified and established as causing significant human disease; more properly, 25 "old" germs have been newly identified, since there is evidence that all of these infections have been around, albeit unnamed and unstudied, for decades or centuries. Some, such as Legionnaires' disease and peptic ulcers, can be cured with our current armamentarium. Others, such as AIDS and hepatitis C, can be managed but not cured. Still others, like West Nile virus and Ebola virus infections, can at best be contained. Novel mechanisms of infection have also been identified, such as the recently recognized toxin-mediated damage due to a very old and familiar bacterium, *Staphylococcus aureus;* the resulting disease is toxic shock syndrome. Another such example is the bizarrely abnormal protein molecule that defies all definitions of infection as it alters brain structure and causes mad cow disease.

If the past is at all predictive of the future, as it almost always is, new germ challenges to our health will continue to emerge with alacrity and regularity. Challenges will also continue to arise from nongerm sources—the societal, economic, and political realities we face in our war on germs. Examples are the recent trends affecting the very same infection-fighting weapons hailed just yesterday as miracles: antibiotics and vaccines. Both have now come under widespread scrutiny and criticism. Zealous marketing and excessive use of antibiotics have contributed to the emergence of supergerms that are more difficult to treat. Public amnesia about past plagues that have been controlled by successful vaccination programs has opened the door to questions about vaccine safety; such questions never surfaced in earlier eras when children were dying of infections that can now be prevented by vaccines. Predictably, immunization rates have dropped in the face of uncertainty regarding safety, and increased numbers of cases of previously controlled diseases are being observed.

For an infectious-disease specialist like me, these perpetual challenges amount to job security, but I'd happily forgo that uneasy peace of mind for a world in which infections could be prevented before they had to be cured. We can already prevent many, if not most, infections with a combination of individual common sense and societal common purpose. It sounds like a mantra for a new movement, but by coupling our cumulative scientific ingenuity with our grandparents' wisdom, your children could become the healthiest, most infection-free generation in history. This book is your first step toward joining the new movement.

Now, a word about how to read this book. I have written it in four parts, and it can be read, logically and I hope seamlessly, cover to cover. However, I suspect most of you will find yourselves frequently flipping back and forth rather than reading straight through. That's because I've also written the book as a home bookshelf reference tool that you can use to look up specific germs and diseases that you are concerned about, that you read about in the morning newspaper, or that your child's doctor has diagnosed. You can also look up specific treatments, both prescribed and home brew, as well as specific preventives, those that are recognized by the medical establishment and those that were touted by your mom and her mom before her.

"Part I: Worthy Enemies" introduces you to the germs by name and reputation, as well as to the ways in which they travel in and out of your child. This part of the book is also where you should turn when you want to know about AIDS, bird flu, the common cold, diabetes (yes, it might be caused by germs), ehrlichiosis, and the rest of the alphabet of infections through zoonoses (that's not pronounced "zoo-noses" but rather "zoa-no-sees," and it means infections caught from animals). Rather than being presented in alphabetical order, though, the diseases are organized into those that you most hate, those that you most fear, and those that you used to hate and fear but that are now mostly gone.

"Part II: Weapons in the War" is the good news to counter all the bad news in Part I. These chapters describe your children's natural immunity to infections, as well as the armamentarium that doctors and scientists have amassed to combat germs: antibiotics, antiviral medicines, antifungal medicines, antiparasitics, insecticides, and vaccines. In this part of the book, you'll find answers to your questions about the safety of these life-saving weapons, as well.

"Part III: Wear Your Boots in the Rain" contains the reconciliation chapters. How do Mom's ideas of infection prevention jive with those we have been taught by modern medicine? Is there science to support Mom's intuition? You'll be surprised.

Finally, "Part IV: Wisdom of the Ages" contains a single but important chapter that puts it all in perspective. How should you balance phobia with prudence in protecting your kids? How much prevention is too much? How should you interpret each new germ threat and health warning in the daily news?

Germ Proof Your Kids guides you in protecting, without overprotecting, the ones you love from the invisible enemies all around them.

Acknowledgments

With gratitude to:

Our parents and grandparents (Helen and Max, Ruth and Gene, Miriam and Lou, Judith and Lou, Devorah and Zvi, and Vivian and Leo), who understood everything about infections except the part about germs.

My colleagues at The Children's Hospital of Denver, who inspire by their good care and good hearts.

Marianne Neifert, Bart Schmitt, Jim Todd, and Gail Reichlin who encouraged me toward this mid-life crisis; and Nancy Love, who heard a "voice" in my writing that I hadn't heard.

The good folks at ASM Press, especially Jeff Holtmeier, who had the courage to push the envelope; Pamela Lacey, who understood how to fold and seal the envelope; and Jennifer Adelman, Lindsay Williams, Alaina Scalercio, Scott Hanson, and the rest of the staff who stamped the envelope and put it in the mail.

I. WORTHY ENEMIES

1

Germinology: Know Your Enemy

The microscopic world of germs far exceeds, in numbers and perhaps even in diversity, the living species that we can see with the naked eye. Within each human being, there are more bacteria than there are people on earth: 100 billion bacteria in our mouths and 100 trillion bacteria in our gastrointestinal tracts. Add to this the 1 trillion bacteria living on each person's skin, and it's easy to see why infectious diseases are ubiquitous. In the latest of many such attempts to profile the germs that dwell in and on us, a 2007 study found that many of the bacteria on normal human skin display a unique genetic makeup, unlike any germs previously recognized (see chapter 4). It even appears that one human's resident germs are distinct from those of the next.

Yet, remarkably, despite the fact that we live in intimate contact with nearly infinite multitudes and arrays of germs, most of them never cause us harm—in fact, many are beneficial. The most harmful germs often come from the outside, invading our bodies through breaches in our protective defenses. One gram of garden soil, for example, may contain as many as 2.5 billion bacteria.

The following paragraphs name names, introducing you to the germs and the disease syndromes with which they are most closely associated. The organization of this chapter, a prologue of sorts to the world of infections, is based on the type of germ and the major body site where it causes infection. The goal of this chapter is to put germs into context—what kinds they are, which children they infect, and what the general manifestations of those infections are in children. More complete descriptions of the diseases caused by the germs most likely to actually infect your kids (as well as those you're most worried about but which are actually highly unlikely to infect your kids) follow in chapter 3, as does a discussion of each disease's contagiousness, prevention, and treatment. After reading this first chapter, however, I hope you'll understand generally how the germ world is organized; refer back to this chapter for a refresher on "germinology" as you encounter the names and classifications of germs throughout the remainder of the book.

Like the visible world, in which an impossibly complex cornucopia of living species can be classified into a relatively few simple categories—vertebrates, invertebrates, mammals, reptiles, etc.—germs can also be catalogued into user-friendly groups for the purposes of general discussion: viruses, bacteria, fungi, parasites, and the other harder to classify infectious agents.

Although there are now even simpler germs that have been identified—and I will discuss those at the very end of this chapter—it is convenient to begin a germ tutorial with viruses.

Viruses—What Are They, and How Do They Make Your Kids Sick?

Viruses are by far the most common causes of human infections, resulting in many millions of sick children (and adults) every year. Viruses are composed of nothing more than a protein, or a protein and lipid (fat), coat surrounding a tiny piece of genetic material, either RNA (ribonucleic acid) or DNA (deoxyribonucleic acid). The genes of a virus are spartan, containing only enough information for the survival and reproduction of the virus—no redundancy, no excess, no fluff. In fact, the viral genes are only smart enough to allow the virus' survival and reproduction within the well-equipped cell of a host organism, a human, for example. Viruses cannot survive for more than a short time, and cannot reproduce at all, outside of a living host cell. This will be important when we discuss the transmissibility of viruses from host to host (as opposed to from toilet seat to host) in chapter 2. Despite being such elementary germs, viruses are responsible for a vast repertoire of diseases ranging from the barely banal to the uniformly fatal.

The outer surface of the virus has deviously evolved to contain a docking mechanism for landing on the target host cell. Every virus has a binding site on its coat that matches a site on the host, like two interlocking pieces of a puzzle. The host cell port in which the virus docks is a wonderful example of the efficient ingenuity of germ evolution and the comparatively lumbering development of host defenses. Viruses usurp host cell surface receptors that are meant for something else; certainly the host doesn't evolve sites for the purpose of facilitating virus infection! Rather, the virus identifies reliable surface structures on a host cell, structures that are stable over time, and the virus adapts its own surface to fit. You can think of this as something akin to skateboarders using ramps built for handicapped access. Once bound to the outer surface of the host cell, the virus finds a way to penetrate into the cell, often helped by the relatively dimwitted host cell that thinks it's bringing something useful into its inner self. This Trojan horse approach makes viruses the wiliest of all invading germs. After the virus gets inside, it ejects its genes and takes over. Within minutes to hours, the virus genes take command of host cell machinery to result in the production of more viruses and sometimes, depending on the specific type of virus, the death of the cell. Whether the host cell dies or is merely damaged, the virus infection triggers an immune response by the host; it is the combination of the cell damage caused by the virus and the resultant immune response that causes the symptoms of viral infection.

Virus Varieties

Common Causes of Respiratory Illnesses. Most respiratory illness is caused by viruses, and there are hundreds of different viruses involved, with enough seasonal diversity to cover the entire year in facial tissues. Infections of the nose, throat, sinuses, and ears are termed upper respiratory tract infections (or URIs), whereas infections of the lungs and airways leading to the lungs are called lower respiratory tract infections (or LRIs). The term URI is also frequently used synonymously with the common cold even though, as noted above, the upper respiratory tract is more complicated than that.

More than 100 rhinoviruses alone cause one-third to one-half of all cases of the common cold, clustered into fall and early-winter epidemics every year, with a smaller second peak occurrence in the spring. Coronaviruses cause between 10 and 20% of all colds, also largely in the winter and spring months. Summer colds, the hard-to-shake, nagging variety, are largely due to the enteroviruses, more benign cousins of the polioviruses (which were also summer denizens in the prevaccine era); other members of the enterovirus group of viruses include the coxsackieviruses and echoviruses.

The more serious respiratory infections, LRIs like croup, bronchiolitis, and pneumonia, are caused by numerous virus types, including household names such as influenza viruses (winter) and adenoviruses (year round) and the more esoteric parainfluenza viruses (spring and fall) and respiratory syncytial virus (winter); all of these viruses, but particularly influenza viruses, can also cause a flu-like syndrome of high fever, chills, and body aches. To add to the complexity of treating viral respiratory tract infections, the viruses that typically cause the more serious LRIs may also cause only mild, common-cold-like symptoms in many infected children, and all of the respiratory viruses can cause ear infections that can be difficult to distinguish from those due to bacteria (see below).

Precise diagnostic tests are available, but impractical or unnecessary, for many of the viruses that cause respiratory infections: impractical because the tests may be expensive or technically difficult and unnecessary because we do not have specific antiviral medicines available to treat most respiratory-virus infections (see chapter 6). Most viral respiratory infections are treated similarly, with a tincture of time, and sometimes with medicines that provide symptomatic relief (see chapter 8).

In the absence of specific diagnostic tests, these germs are distinguished only imprecisely by their seasonality, as noted above, and by the location in the respiratory tract (upper or lower) of the major symptoms. Although most children and adults recover from respiratory viruses, wherever they are located in the respiratory tract, newborns and children with compromised immune systems (see chapter 4) may develop severe, even fatal, disease.

Common Causes of Skin Infections. Many viruses are associated with rashes resulting from the widespread distribution of the viruses through the bloodstream and the host's response to those infections. Some viral rashes, like those due to measles, German measles (rubella), and chicken pox (varicella) viruses, are distinctive in appearance and can be diagnosed by grandmother and doctor alike; however, as vaccines (see chapter 7) have almost eliminated measles and German measles, and are about to eliminate chicken pox, grandmother's diagnosis based on personal experience may be more accurate than that of the doctor who began practice in the vaccine era! Parvovirus B-19 causes a curious slapped-cheek rash, also known as "fifth disease" because it was the fifth of the rash diseases identified by doctors early in the 20th century.

Other system-wide virus infections cause very nondescript rashes that can be confused for one another and confused with allergic reactions, as well. These generic rash viruses include the enteroviruses, adenoviruses, human herpesvirus 6, and human herpesvirus 7. The last two germs are the most common causes of roseola (also once known as "sixth disease"—can you guess why?), an otherwise nondescript rash that appears after a period of high fever, just as the fever breaks.

There are also viruses that cause local eruptions only, not the diffuse whole-body rashes described above. Herpes simplex virus type 1 (HSV-1) causes the distinctive cold sores/fever blisters/sun blisters on your kids' lips, and its close cousin HSV-2 causes the same type of blisters on sexually active adolescents' genital areas. HSV-1 and HSV-2 are both more severe when they're first acquired and then lie dormant in your kids' bodies until they are awakened in their recurrent form by stress (such as another infection, sun exposure, or menstruation). Papillomaviruses cause warts. Depending on the specific strain of the infecting virus, the resulting wart may be a common skin wart (including the uncomfortable plantar wart on the bottom of your kids' feet) or a sexually transmitted genital wart. Molluscum contagiosum is a benign but very contagious skin rash caused by a molluscipoxvirus. This infection may linger for months and spread from one part of the body to another as children scratch, but the infection eventually resolves.

In the fetus, and in newborns and children with compromised immune systems, many of these skin and rash viruses can be extremely dangerous, proving that they are really not skin viruses at all; rather, they are system-wide internal infections that manifest themselves on the outside.

Common Causes of Gastrointestinal Illness. Despite its name, the stomach flu, or viral gastroenteritis, is not related to the respiratory flu or to the influenza viruses. Rather, the stomach flu is caused by a unique set of viruses that limit their mayhem to the gastrointestinal tract, specializing in causing vomiting and diarrhea. The first to be identified and the leading cause of viral gastroenteritis in the United States and worldwide is rotavirus. This highly contagious winter bug is particularly potent in young infants, resulting in 55,000 hospitalizations in the United States each year, mostly for dehydration; worldwide, rotaviruses cause more than 600,000 deaths each year, also from dehydration. Although they are also a significant problem in children, much viral gastroenteritis in adults is due to the noroviruses (formerly known as Norwalk viruses and sometimes called caliciviruses), which are responsible for more than half of all food-borne outbreaks of gastroenteritis, including those on cruise ships, which have made some think twice as they enter the buffet lines. Rounding out the players in viral gastroenteritis are the astroviruses (named for their star-shaped appearance under the electron microscope) and the enteric strains of adenoviruses. These germs—astroviruses and adenoviruses—are numerically far less significant than the rotaviruses and noroviruses but cause indistinguishable disease and similarly high risks of dehydration in children.

Common Causes of Life-Threatening Disease. For simple little germs, viruses often pack a wallop—a *fatal* wallop, in fact. As noted in the sections above, the everyday respiratory, skin, and gastrointestinal viruses can cause severe illness and death in newborns and immune-compromised children. Even in otherwise healthy older kids, severe cases of measles, chicken pox, herpes, and influenza are well known. However, there are certain viruses for which severity and even lethality are the rule rather than the exception.

Atop the list of life-threatening virus infections, of course, is AIDS, caused by human immunodeficiency virus (HIV). This virus' unique modus operandi is that it infects the body's own immune system, rendering the host incapable of fighting many infections, including HIV itself. HIV also distinguishes itself as the only significant human-infecting germ that runs its genetic operation backwards—rather than the usual path of turning DNA into RNA, HIV is a retrovirus that turns RNA into DNA. This trait has made it particularly resilient, as well as refractory to many of our usual strategies for antiviral treatments.

Viral infections of the liver can also be deadly. Fatal hepatitis can occur following infection with hepatitis B virus, either via a sudden, fulminating infection or by destruction of the liver over many years. The latter pattern, and only rarely the former, can also develop after hepatitis C virus infection. When hepatitis D virus infects an individual already chronically infected with hepatitis B virus, the combination is frequently fatal. Rounding out the hepatitis virus alphabet soup, hepatitis A virus and hepatitis E virus are rarely fatal but can cause a miserable and prolonged illness typified by jaundice, gastrointestinal discomfort, and fatigue. Epstein-Barr virus (EBV) and cytomegalovirus (CMV) both cause nonfatal hepatitis but are often accompanied by a long-lasting fatigue and weakness syndrome called mononucleosis, or "mono." EBV and CMV are members of the extended herpes family of germs, and they cause a variety of nonliver illnesses as well, particularly in immune-compromised patients.

Viral infections of the brain (encephalitis), spinal cord (myelitis), and the membranes lining the brain and spinal cord (meningitis) are frequently severe because of the irreplaceable nature of nerve cells. Numerous viruses mentioned above as causes of respiratory and skin infections can also infect the nervous system. The enteroviruses and herpes simplex viruses are the leading causes of meningitis and encephalitis, respectively. HIV also causes serious neurological disease, as can measles virus, chicken pox virus, EBV, and CMV. Mumps virus, another germ on the run because of vaccines (see chapter 7), can cause both meningitis and encephalitis, in addition to the better-known infection of the saliva-forming parotid glands in the cheeks that gives mumps patients their distinctive chipmunk-like appearance. A disparate group of viruses colloquially called arboviruses also cause both meningitis and encephalitis; included in this group are West Nile virus, St. Louis encephalitis virus, Japanese encephalitis virus, and the California encephalitis group. The unifying trait of the arboviruses is their transmission by blood-feeding insects (e.g., mosquitoes and ticks; see chapter 2). The most famous of all viruses targeting the spinal cord, with the resultant myelitis, is poliovirus, now only rarely seen in our hemisphere because of the vaccine miracle of the 1950s (see chapter 7); the three strains of polioviruses are members of the enterovirus group, several other varieties of which can also cause a poliomyelitis syndrome of paralysis. The dread rabies virus causes a rapidly progressive and almost universally fatal form of encephalitis.

Bacteria—What Are They, and How Do They Make Your Kids Sick?

Bacteria were the only life form on earth for more than 2 billion years, and with this head start, they represent a great leap forward from viruses in germ sophistication. Each bacterium is a self-contained, often independently living single-cell organism. The cell is the basic unit of all living organisms (with the exception of the questionably living viruses described above); humans are made up of billions of highly differentiated cells, each functioning as part of an organ system, e.g., liver cells, kidney cells, nerve cells, and blood cells. In contrast, bacterial cells are each stand-alone organisms rather than differentiated components of a larger living thing. This is critical to understanding the survival and spread of bacteria. Most of these germs can exist somewhat independently for long periods of time, feeding off of whatever substance, surface, or structure they find themselves on. From our own skin, throat, and intestines to the depths of Yellowstone's geysers and the peaks of Antarctica's glaciers, bacteria colonize our world. Bacteria can accommodate not only the extremes of hot and cold, but also the extremes of oxygen-rich and oxygen-poor environments. While most of the bacteria that cause infections in humans prefer to grow where oxygen is plentiful, like our skin and mucous-membrane surfaces, our lungs, and even our blood, other bacterial species can grow and cause serious disease in the absence of oxygen. These anaerobic bacteria are particularly plentiful in the gastrointestinal tract of humans, but also in many locales in nature, such as soil and water.

Bacteria grow very quickly in most environments, and their rapid growth results in adaptive mutations. The ability of bacteria to efficiently mutate explains their ubiquity and their prowess in developing resistance to each new antibiotic we engineer (see chapter 5).

Although some bacteria enter cells in a way reminiscent of viruses, these intracellular bacterial infections are far less common than infections in which the bacteria cause their damage from the outer surfaces of tissues and cells. When bacteria are introduced into the body through wounds or other unnatural openings, they dissect their way through our tissues using unique structures of motility called flagella or cilia. Infections can also occur without a breach in the body's fortress walls. Bacteria that are already residing within the body, at sites where they are harmless (e.g., the mouth, the gastrointestinal tract, and the vagina), can sometimes manipulate themselves into deeper tissue or blood venues. The invasive movement of bacteria is facilitated by proteins they secrete, called enzymes or toxins, which further damage neighboring tissues, creating fertile growth environments for these nutrient-seeking germs.

If bacteria gain access to our bloodstream, their spread through the body rapidly accelerates. Bacteria may reach the blood either by their own nefarious means or by the body's well-intentioned efforts to use the blood to clear the germs. Once in the blood, bacteria may, indeed, be successfully cleared by the host's immune organs—lymph nodes, liver, and spleen (see chapter 4). However, often the clearance mechanisms are insufficient and blood-borne germs gain the upper hand, resulting in severe and life-threatening diseases, as discussed briefly below and in greater detail in chapter 3. The previously mentioned bacterial toxins, in addition to causing local damage, may themselves spread in the blood, causing system-wide illness that is well out of proportion to the seemingly well-contained and localized germs producing them.

Bacterial Breeds

Common Causes of Respiratory Illnesses. The two most common bacterial respiratory pathogens are cousins in the genus *Streptococcus*. *Streptococcus pyogenes* (also called group A strep) causes the enormous societal nuisance known as strep throat. Besides the discomfort and fever, strep throat is most worrisome for its rare complications, like rheumatic fever, scarlet fever, kidney disease, and necrotizing fasciitis. *Streptococcus pneumoniae* (pneumococcus), as the name suggests, is the most common bacterial cause of pneumonia in children and adults but is much more common, and therefore more troublesome, as a cause of ear and sinus infections. Both of these bugs are hearty, able not only to cause us harm, but also to colonize our bodies without causing infection, an important consideration in treatment decisions. A former major cause of pneumonia in children, *Haemophilus influenzae* type b, has been almost completely eradicated by a vaccine introduced in the late 1980s (see chapter 7). *Haemophilus* species other than *H. influenzae* type B continue to cause respiratory infections, such as pinkeye (conjunctivitis) and, in adults, bronchitis. Fortunately somewhat less common, but often more dangerous, are other pneumonia-causing bacteria, like *Mycobacterium tuberculosis,* the cause of tuberculosis; *Bordetella pertussis,* the cause of infant whooping cough (and a successful vaccine target [see chapter 7]); *Legionella pneumophila,* the cause of Legionnaire's disease and Pontiac fever; and *Staphylococcus aureus,* more typically a cause of skin infection (see below), which can wreak havoc when it gets into the lungs.

Another very common cause of pneumonia, *Mycoplasma pneumoniae,* is not quite as sophisticated as are typical bacteria—these germs lack an outer coat and the ability to survive as free-living germs, but they otherwise behave like bacteria and are treated as such. *M. pneumoniae* and *Chlamydophila pneumoniae* (formerly known as *Chlamydia pneumoniae;* there is no end to the renaming and reclassifying of germs) are the leading causes of "walking pneumonia."

Common Causes of Skin Infections. Far and away the most common bacterial skin germs are *S. aureus* (staph) and *S. pyogenes* (strep). These two bacteria cause or contribute to almost all cases of bacterial skin infections, ranging from annoyances, like impetigo, paronychia (nail bed infections), and folliculitis (hair follicle infections), to more worrisome conditions, like cellulitis, carbuncles, and furuncles, and to potentially devastating infections of the skin and deeper tissue layers. These deeper infections include erysipelas, ecthyma, and necrotizing fasciitis. Although these are also usually due to the common skin staph and strep bacteria, anaerobic germs, like *Clostridium perfringens,* can also thrive in deep layers of tissue where oxygen is scarce.

Acne, a component of which is due to infection, can involve numerous bacteria that creep into skin pores and hair follicles; among the most commonly associated germs are *Propionibacterium acnes.*

Common Causes of Gastrointestinal Illness. It is hard to imagine anyplace where the quantity and diversity of bacteria are greater than in the gastrointestinal tract of humans. Most of the bacteria in the human intestines are benign and helpful; they produce enzymes that help digest and process the food we eat, and they fill a void that otherwise would be overtaken by more malicious germs. Evidence for the goodness of our intestinal tenants is what often happens when our children are treated with antibiotics—the diarrhea that results reflects the damaged ability of the intestines to absorb food because the beneficial bacteria have been killed along with the bad (see chapter 5). The antibiotic toll on good intestinal germs also allows the opportunistic bad ones that survive the antibiotics to take over and overgrow our children's intestines, making recovery from the diarrhea more prolonged and complicated.

However, diverse bacterial germs, acting by diverse mechanisms, also are important causes of gastrointestinal disease. Bacteria can harm the intestinal-lining cells directly or by releasing toxins, or by both strategies. Most harmful bacteria come from food contaminated at the original source (e.g., cows, chickens, and pigs) or from human food handlers who have themselves become infected (see chapter 2). *Campylobacter jejuni* is the most commonly identified bacterium causing diarrhea in North America and is usually acquired by eating undercooked poultry or drinking unpasteurized milk; it acts by directly invading intestinal cells. Food-borne origins are also found for most cases of *Salmonella* and *Yersinia* infections, although these two types of organisms are more prone to causing serious system-wide disease than are the *Campylobacter*. In contrast, *Shigella* species have no known animal reservoir, spreading from person to person via direct contact or by human contamination of food; shigellae combine a direct hit on our kids' intestinal cells with a toxin released by the germ for a double whammy. *Escherichia coli,* a normally benign and important good bacterium, can acquire *Shigella*-like traits and cause severe direct and toxin-mediated dysentery; *E. coli* is the most frequently blamed germ in traveler's diarrhea. Other bacteria that normally are benign intestinal residents but can acquire dangerous toxin-producing abilities include the anaerobic *Clostridium* species of bacteria; these organisms are among those that can take advantage of the void created by antibiotic treatment to grow into large and menacing factories of diarrhea-causing toxins. *Vibrio parahaemolyticus* travels to us in undercooked seafood and is a close cousin of *Vibrio cholerae,* the infamous cause of the cholera outbreaks that are now limited to developing countries. *Vibrio* species also act by a combination of direct and toxin effects.

Common Causes of Urinary Tract Disease. Although viruses and fungi can occasionally cause urinary tract infections, bacteria are the major players. At the risk of sounding indelicate, it is the proximity of the anus to the urethral meatus (the portal of exit for urine) that defines the germs causing bladder and kidney infections; almost all such infections are caused by gastrointestinal bacteria. Kids' bathroom hygiene (see chapter 9), as well as their intrinsic anatomy, contributes to the ability of intestinal germs to find their way into the urinary structures; anatomy makes little girls more susceptible than little boys. The most prolific of these germs is *E. coli,* which causes 70 to 90% of children's urinary tract infections, followed by fellow intestinal bacteria including *Enterococcus, Klebsiella, Proteus,* and *Pseudomonas* species.

Common Causes of Life-Threatening Disease. Many bacteria, including those named in the sections above, have the capacity to cause great harm and death to those infected. The factors that determine whether the impact of a bacterial infection will be limited to minor morbidity or will cause potentially life-threatening illness are manifold and not often clear cut. A complicated interplay between the potency of the germ and the competency of the immune system (see chapter 4) accounts for most variations in the spectrum of infection severity due to any germ, and certainly that is true for bacteria. However, other factors are also important: the inoculum of the bacteria, i.e., how many bacteria get into us; contemporaneous events, such as concurrent nonbacterial infections (e.g., coexisting viral infections that predispose to worse illness when bacteria arrive) or tissue damage (e.g., accidental traumatic injury or surgery); and hormonal influences, such as adolescence, menstruation, and pregnancy.

Sepsis is a condition wherein germs (usually, but not always, bacteria) cause widespread infection throughout the body with failure of one or more organ systems (kidneys, heart, lungs, liver, or brain) (see chapter 3). The most common causes of this syndrome are germs mentioned above in the context of their more localized infections: *S. aureus, S. pyogenes,* and *E. coli* (and other usually benign organisms from the gastrointestinal tract). These germs and others also have the capacity to stay localized but to release chemical toxins into the bloodstream or into nearby tissues, or both, causing toxic shock syndrome, necrotizing fasciitis ("flesh-eating" infections), and hemolytic uremic syndrome (associated with fast food and other food contamination outbreaks of *E. coli*). *Neisseria meningitidis* causes both severe sepsis and meningitis. This is often the organism responsible for terrifying headaches and headlines when meningitis outbreaks spread through a classroom, school building, or college dormitory.

Bacteria cause important sexually transmitted diseases that have defied our educational and pharmaceutical attempts at eradication. The most important bacteria among those causing venereal infections are *Neisseria gonorrhoeae* and *Treponema pallidum,* which cause gonorrhea and syphilis, respectively. Both are highly contagious, putting an adolescent at risk not only for infertility and other complications of these bacteria, but also for infections with other bad actors with which they travel, including HIV (AIDS) and *Chlamydia trachomatis.* Infection by the latter, chlamydia, is due to a bacterium that acts somewhat like a virus, requiring a living cell to reproduce itself and do damage to the host. Chlamydia genital infections are often without symptoms and may result in infertility, particularly in women.

The cause of syphilis, *T. pallidum,* as noted above, is an unusual spiral-shaped bacterium; a similar spiral bacterium, *Borrelia burgdorferi,* causes Lyme disease, an infection that can involve multiple organ systems, including the brain. Rickettsiae are another unusual form of bacteria; these germs can live only inside of cells and can't be grown on artificial media (culture plates) like other bacteria; in this sense, rickettsiae also behave more like viruses than like bacteria. Rickettsiae cause severe and potentially life-threatening human diseases, including Rocky Mountain spotted fever, Q fever, ehrlichiosis, and typhus.

Fungi—What Are They, and How Do They Make Your Kids Sick?

With fungi, we again move up the evolutionary ladder from viruses and bacteria. Whereas some fungi, like bacteria, are single-cell organisms, fungi are eukaryotic and bacteria are prokaryotic. The difference is in the compartmentalization of the structures within the cell: fungi organize their major components within membranes, whereas bacterial parts are a jumble within the cell soup. One-cell fungi are called yeasts, and more organized, multicellular fungi are called molds. Molds are conglomerations of cells working together to form sophisticated lattices, scaffolding that facilitates invasion and infection of the host. Some of the most important human-infecting fungi can grow in both yeast and mold forms and are called dimorphic (having two shapes).

Like bacteria, fungi are everywhere and can survive for long periods on inanimate objects and in the environment. Three patterns of fungal invasiveness occur, differentiated by the virulence (potency) of the particular fungus. Certain fungi are very potent and capable of causing infections in people with normal immune systems and without other underlying risk factors for infection. Other fungi are less potent and are able to invade only a host compromised by any of a variety of factors, such as an abnormal immune system (see chapter 4), open wounds or traumatized tissues, the presence of artificial material inside the body (e.g., intravenous lines in a hospital setting), diabetes, intravenous-drug abuse, malnutrition, or prolonged antibiotic use. Finally, a third pattern of fungal infection is that of skin and superficial fungi that crawl into minor skin openings and hair follicles or simply proliferate in moist and raw areas of the skin of otherwise healthy people.

Fungal Flavors

Common Causes of Respiratory Illnesses. Respiratory fungi in otherwise healthy children (those without any underlying predisposing condition for infection) have unique geographic distributions but common pathways of infection. All are acquired by inhaling the fungus from the air, and all can spread from the lungs to other parts of the body. The resulting illnesses are generally mild, with brief episodes of fever, cough, and chest pain. *Histoplasma capsulatum,* the cause of histoplasmosis, is the most commonly occurring of these infectious fungi in the United States but is primarily focused in the Mississippi and Ohio River valleys. Coccidioidomycosis, caused by *Coccidioides immitis,* is found almost exclusively in the southwestern United States, Mexico, and Central and South America. Blastomycosis is caused by *Blastomyces dermatitidis,* a fungus primarily found in western New York, eastern Ontario, along the St. Lawrence River, and in areas overlapping with histoplasmosis in the Mississippi and Ohio River valleys. Of the three lung diseases caused by fungi, blastomycosis is typically the most severe, with a higher risk of complications.

Many other fungi can cause severe, life-threatening lung infections, but only in otherwise compromised hosts. Indeed, fungal infections are among the greatest threats to patients with abnormal immune systems and other underlying, predisposing conditions. Some of these opportunistic fungi are noted below in the section on life-threatening fungal infections.

Common Causes of Skin (and Mucous-Membrane) Infections. The most common fungal skin infections are due to *Candida* species and *Trichophyton* species. *Candida* infections, colloquially referred to by most people as "yeast infections" (which they are, but not all yeast infections are due to *Candida* species), occur on healthy babies' bottoms, causing a nasty diaper rash. *Candida* species also infect the mucous membranes in babies' mouths, causing a condition called thrush. Older children, adolescents, and adults can get

Candida infections of hair follicles (folliculitis); moist skin fold regions, like the groin and underarms (intertrigo); the skin around fingernails and toenails (paronychia); and the nails themselves (onychomycosis). *Candida* can also cause infection of vaginal mucous membranes (vaginitis) in adolescents on antibiotics or oral contraceptives. Much more severe complications of *Candida* infection occur in compromised patients, a common theme for almost all infections, but particularly for fungi.

Trichophyton and several related fungi cause ringworm, jock itch, athlete's foot, scalp infections, and the really ugly, discoloring fingernail and toenail infections that can progress to loss of the nail. The good news is that this infection stays local, without the potential for dissemination to other parts of the body.

A less common fungal skin infection occurring in otherwise healthy individuals is called sporotrichosis and is caused by *Sporothrix schenckii*. In children, this germ is introduced into the skin of the hands and fingers through minor cuts and scrapes, especially from wood splinters and plant thorns. The fungus spreads to the glands (lymph nodes) that drain the affected area, resulting in a chain-like appearance on the arm, but it rarely spreads beyond.

Common Causes of Life-Threatening Disease. Although fungal infections like athlete's foot, ringworm, and baby's diaper rash are everyday nuisances in otherwise healthy patients, almost all fungi are very opportunistic germs, patiently lying in wait for us to become somehow compromised and more susceptible to their languid style of attack. Given an opening when our immune systems are weakened, fungi meander into us and grow into masses that slowly bulldoze everything in their path or disseminate through the blood to infect multiple vital organ systems. Indeed, patients compromised by abnormalities in their immune systems, diabetes, tissue damage, malnutrition, intravenous-drug abuse, and even prolonged antibiotic use may develop life-threatening fungal infections. No organ system is spared, and no nook or cranny of the body is invulnerable to invasive fungal disease. To name but a few, the soil and plant fungi *Aspergillus, Fusarium,* and *Alternaria* can all cause severe infections of the lungs, sinuses, brain, heart, liver, and spleen. The molds named Zygomycetes grow in decaying organic material, such as spoiled food, and cause mucormycosis, a condition that typically begins in the sinuses and spreads from there directly to the lungs and brain. *Cryptococcus neoformans* causes cryptococcosis, which begins with inhalation of the germ into the lungs but can quickly spread to any organ system; spread to the brain is the most vicious and most difficult to treat. Finally, *Pneumocystis carinii* is an organism that is especially famous for its horrific pneumonia in AIDS patients and other immune-compromised children. Unlike other fungi, this germ rarely spreads beyond the lungs; *Pneumocystis* is physically unlike other fungi as well, with some microscopic characteristics of bacteria.

Parasites—What Are They, and How Do They Make Your Kids Sick?

In a sense, viruses, bacteria, and fungi are all parasites, feeding on and growing in us for their own gain. The term parasites, when used to describe a specific group of germs, however, is reserved for protozoa (one-celled germs that can move on their own) and helminths (worms). This eclectic collection of human-infecting germs and worms is the most exotic pathogen group of all, because of their species diversity, their dramatic presentation (worms crawling in and out of a person *are* dramatic!), and their intriguing geographic locales.

Parasites get into us by one of two routes—either we eat them (via contaminated food) or they penetrate our skin; in the latter model, some parasites are able to directly puncture our skin or crawl through an available opening; others require the help of mosquitoes or biting flies (see chapter 2). The more advanced worm parasites (classified as nematodes, trematodes, and cestodes for you worm term trivia buffs) cycle through several stages (egg, larva, and adult) and several hosts (fish, mollusks, pigs, cows, or others) and may also contaminate water and soil during their fascinating lives. Some have both male and female reproductive organs and can even self-fertilize. Fascinating lives, indeed. Most clinical disease reflects the route of entry—ingested parasites cause primarily gastrointestinal illness and inoculated parasites cause disease where the germ is distributed after inoculation, e.g., the skin, bloodstream, and other organ systems.

Parasite Potpourri

Common Domestic (U.S.) Parasitic Infestations. There are few significant domestic parasites infesting children. The happiest (microscopic pictures of this germ actually look like a smiling face), most prevalent, and most acclimated parasite in the United States is *Giardia lamblia,* the cause of giardiasis. Although its ability to be transmitted from person to person is rare among parasites, most transmission still occurs from ingesting contaminated water or food. *Giardia* is famous for causing day care center outbreaks of diarrhea and for the diarrhea that hikers get from drinking stream water (where feces from wild animals have contaminated the pristine-looking stream with *Giardia* cysts) (see chapter 2). Cryptosporidiosis, caused by *Cryptosporidium parvum,* causes a similar but somewhat longer-lasting diarrheal illness and can also be transmitted from person to person, especially in day care centers; again, waterborne transmission, including swimming pools, is the most common route. *Toxoplasma gondii* is a protozoan parasite of cats and other animals that has its main impact on the unborn fetus when pregnant women are infected. Mothers themselves develop a usually mild flu-like illness or no symptoms at all, but the parasite can cross the placenta and cause eye and other neurological damage to the baby. Although toxoplasmosis can be acquired by eating undercooked meat, the most common route in the United

States is via the litter box, where contaminated cat feces meet human hands. Trichomoniasis (also known as "trich") is a sexually transmitted disease of 3 to 4 million people every year in the United States caused by the parasite *Trichomonas vaginalis.* Vaginitis (soreness, irritation, discharge, and painful urination) is the manifestation in adolescent girls; boys get urethritis, a syndrome of painful urination and penile discharge. A domestic variety of a malaria-type parasite, *Babesia* species, causes babesiosis, transmitted by tick bites, mostly along the east coast of the United States. A flu-like illness accompanies anemia resulting from the parasite's destruction of red blood cells, a condition from which most patients recover uneventfully.

Giardia, Cryptosporidium, Toxoplasma, Trichomonas, and *Babesia* are all protozoans, the simple, one-cell, motile parasites. In contrast, worm parasites are much less common in the United States, but the most common of them is, well, *very* common: pinworm. Pinworms, *Enterobius vermicularis,* are ingested (again, day care centers and schools are the hot zones) after hand-to-hand or hand-to-elsewhere contact among kids who then put their hands in their mouths. The ingested pinworm eggs hatch in kids' intestines, larvae descend through the gut, and adult worms (thankfully still tiny) migrate out onto the skin around the anus, causing itching—lots of itching—but nothing more serious than that.

Common Foreign and Travel Parasite Infestations. The diversity and multiplicity of foreign parasites are as profound as the diversity and multiplicity of the people they infest. The few parasites found commonly in the United States are found even more commonly abroad. Added to that is a panoply of simple and complex parasites that brazenly crawl in and out of their hosts. Starting with the simple (protozoa) category, *Entamoeba histolytica* is the cause of amebiasis in travelers to (and natives of) Central and South America, India, and Africa. Severe colitis (dysentery, bloody stools, and abdominal cramping) may be complicated by development of abscesses in the colon or liver. Leishmaniasis is caused by sandfly bites transmitting *Leishmania* species in India, China, Africa, and Central and South America. The disease is severe, with high death rates, and includes weakness and wasting, skin discoloration, anemia, involvement of multiple organs, and superimposed infections with other germs taking advantage of the debilitation caused by this parasite.

Caused by four distinct *Plasmodium* species, each of which results in different clinical presentations, malaria is transmitted by mosquitoes in tropical and equatorial climes. As many as 2 million people die worldwide every year from malaria, an infection of red blood cells. It is the effect on the red blood cells that causes all of the symptoms of the disease, including high fevers, anemia, lung failure, seizures and severe neurological deterioration, rupture of the spleen, and kidney failure. Malaria has literally altered the course of world affairs for centuries, with historical descriptions dating back to 2700 BC making it one of the oldest recognized diseases of humankind.

Moving on from the simple protozoan infestations to the much more complex and sophisticated worm infestations, the most common, causing more than 1 billion infections worldwide (100 to 400 million of them are in children), is *Ascaris lumbricoides,* the cause of ascariasis. The fascinating lifestyle of parasitic worms is nicely illustrated by this worm. Worm eggs are ingested from contaminated soil; the eggs hatch in the small intestine, and the resulting larvae burrow their way through the gut wall, getting into the blood vessels and making their way to the lungs. Once there, the larvae personally cause damage and elicit more harm by the immune response to their presence. These intrepid little larvae then make their way up the airway tree into the mouth, where they are swallowed and get back to the intestines; there, they mature into adult worms that can grow to more than a foot in length. Yes, you read that correctly—*more than a foot* in length. As you might imagine, this is not good for the intestines, and they become obstructed and may rupture. Adult worms can migrate from the intestines to the liver, gall bladder, and pancreas.

A similar pathway exists for hookworm infestation, caused by *Ancylostoma* species, also occurring in numbers approaching 1 billion people worldwide. These adult worms are distinguished by the hook-like apparatus on their mouths that allows them to attach to the intestine wall and feed—*on us!* Bleeding and sometimes severe anemia result in children, causing delays in growth, development, and learning ability. Before sanitation innovations (see chapter 9), hookworm infections were regularly seen in the southeastern United States; now they are limited to China, India, Latin America, and Southeast Asia. An estimated 800 million people in the tropics and subtropics are infected with *Trichuris trichiura,* or whipworm. This little fellow stays in the intestines, causing dysentery, bleeding, and obstruction. Still other worms cause visceral larval migrans, in which larvae migrate from the intestine throughout the body, invading almost all organs, including the brain; trichinosis, a disease of heart and skeletal muscle; elephantiasis, a worm blockade of glands (lymph nodes) that causes massive swelling of the legs, arms, and genitals; malnutrition syndromes; brain cysts and abscesses; and a variety of other colorful and potentially lethal diseases. In our increasingly mobile world, a single plane ride is all that separates us from the exotic world of these parasites.

Germs That Are Harder To Classify

There are also free-living insects that infest kids. Head lice, *Pediculus humanus capitis,* are very contagious, spreading from the scalp of one child to the scalp of another and feeding on tiny blood meals from both scalps along the way. Scabies is caused by a mite, *Sarcoptes scabiei* subsp. *hominis.* This little guy burrows into your kids' skin, causing a very itchy, red, raised, sometimes blistery rash. Yet another unpleasant bug on parents' "I can't believe my kids have (fill in the blank)" list is *Cimex lectularius,* popularly known as bed bugs. These hide in the warm, cozy spots that your kids frequent, feeding by biting and taking a sip of your kids' blood.

The newest, and weirdest, of germ classes to be described are the prions. In contrast to all other known causes of infection, which contain genetic material, either RNA or DNA, prions appear to be nothing more than simple proteins. These proteins are present in a normal form in all living things, at least all advanced living things, like people and animals. When a normal prion protein assumes an abnormal shape, it is able to accumulate within the brains of infected hosts and cause profound damage: mad cow disease in both bovine and human forms, chronic wasting disease in deer and elk, and similar syndromes in other animal species. Most intriguing of all, prions appear to be contagious despite not having genetic material and therefore not being able to reproduce. The contagiousness of prions results from protein-to-protein contact—when a normal prion protein is in contact with an abnormal prion, the normal protein changes shape and becomes abnormal, perpetuating the germ within the host and between hosts.

Take a Deep Breath and Relax; It's Going To Be Okay

After that seemingly endless parade of grim and gruesome germs, it's understandable if you are left somewhat anxious and afraid. Fortunately, most of the bugs you just discovered will never discover your kids. Those that might are discussed, along with the diseases they cause, in chapter 3. After chapter 3 comes the good news—the rest of the book is devoted to our natural defenses against germs and strategies for prevention and treatment of infections.

But first, chapter 2 presents the travel plans for germs trying to get in and out of your kids. Knowing how to interrupt those plans can be important for protecting your kids.

All right, now breathe, and read on.

2

Germ Roots and Germ Routes: Where They Come From, Where They Go, and How They Get from Here to There

Chapter 1 introduced you to the names and categories of dozens of germs and to the microscopic reality that germs are everywhere. They are in us, on us, and around us, a rich and complex invisible world of organisms that call us home. That said, the sources of human infection, i.e., where the germs that cause our kids to get sick come from, can be classified into just four simple categories that read like a Jeopardy TV show quiz board: Other People, Animals and Birds, Food and Drinking Water, and the Environment.

I'll use those game board categories to explain the roots and routes of germs as they move in and out of your kids' bodies (but I'll try not to phrase my answers in the form of questions!). After appreciating the tactics that germs use in spreading disease, the strategies for preventing that spread will make more sense; those strategies are the essence of chapters 4 to 11.

At the conclusion of this chapter, we'll also take a look at infection amplifiers, those hot zones in your kids' world where infections spread more rapidly and the risks for getting sick are higher—day care centers, school classrooms, petting zoos, doctors' offices, and college dormitories.

How Germs Spread

Other People

Far and away the most common and important source of human infectious diseases is other humans. From Typhoid Mary Mallon to Patient Zero Gaetan Dugard, the person thought to have first introduced AIDS to the United States, humans have always prolifically infected other humans. We are, after all, a social species, living in intimate contact with others who, like ourselves, harbor billions of germs per person—and we find ample ways to share them among ourselves.

The underlying theme of human-to-human transmission is straightforward: contagious body fluids from one person are acquired by another. Disgusting, but true. Every episode of your kids' common cold, flu, chicken pox, strep throat, lice, and warts comes from their contact with someone else who has the malady. The same holds true for potentially lethal diseases, like AIDS, hepatitis, and tuberculosis. Whoever gives an infection to your kids first got it from someone they were in contact with, and so forth back far enough to raise chicken-and-egg questions about this apparently endless perpetuation of original illness.

Humans have developed numerous approaches to sharing germs: inhaling, impaling, ingesting, injecting, and impacting (I needed a word for touching that started with an "i" to complete the alliteration, hence, impacting; sorry). Each germ, in turn, has developed a preferred route for infecting humans, although some enlightened microbes have adapted to take advantage of more than one path to their daily meal. The following sections of our Other People category review the germs' options for human-to-human transmission: direct contact, droplets, airborne transmission, blood, breast milk, and indirect contact.

Direct Contact. The shortest distance between two points is a straight line. Direct contact, including touching, kissing, and sex, is the most common and most important path that germs take from victim to victim. The hands of a karate or judo expert can be lethal weapons, but as you'll read in chapter 9, your kids' hands may also be lethal weapons even if they've never practiced the martial arts. Hands not only carry the germs that are expected to be found on the skin, they also carry the germs that kids have coughed into them, sneezed into them, and wiped onto them after going to the bathroom. And hands aren't the only perpetrators; kids' lips, tongues, and other mucous membranes are often in direct contact with their counterparts in other kids.

Most germs that live natively on the skin are bacteria. Of those that cause infection, staphylococci (staph) and streptococci (strep) are the most common. As you remember from chapter 1, most staph and strep are harmless, but some acquire genes that give them greater potency, allowing them to produce toxins or enzymes that facilitate a bacterium's ability to penetrate beneath the skin and invade deeper tissues or get into the bloodstream and spread throughout the body. Below the belt, gastrointestinal bacteria, like *Escherichia coli,* live on the skin around the anus. When kids wipe and don't wash, those germs are ready for hand-to-hand combat (see chapter 9). In addition to the ubiquitous *E. coli,* all the bacteria that are shed from kids with gastroenteritis and dysentery (severe gastroenteritis) can travel on unwashed hands—salmonella, shigella, and campylobacter to name but three. This makes day care centers, where pre-potty-trained kids roam, unique amplifiers of gastroenteritis (see below).

The mucous membranes of the mouth and nose are also heavily colonized with common bacteria that usually cause no harm and whose identities are too esoteric to review here. However, these surfaces are also the preferred landing sites of more dangerous bacteria, like invasive *Streptococcus pneumoniae* (the noninvasive variety is present normally in many children's mouths and noses) and *Neisseria meningitidis*, both of which can cause sepsis and meningitis, as well as *Streptococcus pyogenes*, which causes strep throat. It's also on the mucous membranes of the nose and mouth that bacteria that cause pneumonia, ear infections, pinkeye (conjunctivitis), sinus infections, and infections of the gums and teeth gain a foothold and from which they can spread to the noses and mouths of others by direct contact. Mucous membranes below the waist may be the source of sexually transmitted bacterial diseases, like syphilis, gonorrhea, and chlamydia.

There are viruses that live for long periods of time, or recur frequently, on skin and mucous-membrane surfaces and are readily spread by direct contact. They include the herpesviruses that cause cold sores (fever blisters) on your kids' lips and cause genital sores in sexually active adolescents. These herpesviruses live dormantly in nerve roots and reactivate frequently (i.e., recurrent herpes) to again become contagious. Molluscum contagiosum is caused by a contagious (hence the name) skin virus (molluscipoxvirus) that can hang around for months, causing a distinctive rash that is spread by direct contact or indirectly via inanimate objects such as towels and clothing (see below). Papillomaviruses, the germs that cause skin and plantar warts, also pass from the skin or mucous membranes of one child to another. Warts may be present and infectious for long periods and tend to recur in the same location or spread to other spots in the same child. Genital warts due to other strains of papillomaviruses spread by direct contact among sexually active teens. That high-risk group also shares infectious secretions from mucous membranes that may contain another long-term virus inhabitant, the human immunodeficiency virus (HIV), which causes AIDS.

Whereas herpesviruses, molluscipoxviruses, papillomaviruses, and HIV are chronic or recurrent denizens of the skin and mucous membranes, many other viruses infect mucous membranes only fleetingly but can be transmitted by direct contact while they're around. These include all the respiratory viruses that cause the common cold, the flu, viral ear infections, pinkeye, and infections of the airways and lungs, as well as all the gastrointestinal viruses that cause gastroenteritis. All of these viruses make only cameo appearances, hit-and-run infections that don't establish a foothold beyond the few days that they are causing symptoms. In a devious evolutionary tactic, however, these mucous-membrane-infecting viruses are frequently present and contagious shortly *before* symptoms develop, making them harder to avoid and guaranteeing their successful passage to the next victims.

Intermediate between the long-term or recurrent viruses of the skin and mucous membranes and the fleeting viruses of the nose, mouth, and gastrointestinal tract are Epstein-Barr virus (EBV), which causes infectious mononucleosis, and hepatitis A virus. "Mono" immediately conjures up images of kissing teenagers, the perfect illustration of direct-contact transmission. EBV may be in the nose and throat of an infected patient for weeks, but eventually it is cleared from the mucous membranes and the patient with "mono" is no longer contagious. Similarly, hepatitis A virus may be shed in the feces of infected patients for days or weeks after the symptoms resolve, but the virus also eventually clears.

Fungi called dermatophytes may spread by direct skin-to-skin contact, causing ringworm and athlete's foot. *Trichomonas vaginalis* causes the sexually transmitted infection known as "trich," a parasitic infection that spreads by sexual contact. Other parasites that spread by direct contact include giardiae and cryptosporidia, causes of gastroenteritis; these germs travel from child to child by the same feces-to-hand-to-mouth (fecal-oral) route as bacteria and viruses that don't get washed away after wiping. The same route is taken by pinworms, a common parasite of children that causes an itchy rash on your child's bottom. Finally, head lice and scabies mites are little tiny insects that cause significant angst and embarrassment and are transmitted by your kids' direct skin contact with their comrades.

The common sexually transmitted diseases (STDs) are discussed fully in chapter 3, and the germs' access routes to your kids by direct contact are no mystery. Sexual intercourse is the most efficient route of spread for the STDs, of course, but many STDs, including syphilis, gonorrhea, herpes, and HIV infection, can also be caught by oral sex. That may be of help in counseling your adolescent kids about sex (see chapter 9).

Droplets. There are two types of flying objects that can spread infection. Droplets are the secretions that fly out of your kids' noses and mouths when they cough, sneeze, spit, or even talk moistly. The second type of flying object is a germ that can actually travel suspended in the air; that route of infection spread, called airborne transmission, is discussed as a separate category below.

Your kids have different names for collections of droplets that come from their noses or mouths—flugies, boogers, drool, dribble, slobber, spittle, snot; you get the idea. Droplets don't need large mucus collections to travel; they may be tiny, even microscopic (as small as 5/1,000 of a millimeter, although most are larger than that) emissions of secretions. Droplets can fly only a short distance, 3 feet or less, since they are weighed down by their fluid content, but if another child is close enough, the droplet may land on the mucous membranes of his or her eye, nose, or mouth, efficiently spreading germs that were stowed away on the magical mucus carpet ride.

Almost all of the bacterial germs discussed above that can be spread by direct contact with mucous membranes (but not those bacteria spread only by skin contact) can also be spread by droplets. They include the bacterial causes of respiratory tract infections, like pneumonia, ear infections, and strep throat, as well as those that cause meningitis and sepsis. Droplets are also a very significant route of transmission for all the viruses discussed above that cause respiratory infections; in contrast, viruses like herpesviruses, molluscipoxviruses, papillomaviruses, and HIV, which cause longer-term infections, are not spread by droplets, nor is EBV. Fungi, parasites, and insects are also not droplet borne.

Generally speaking, germs that can be spread by droplets are more contagious than those that require direct contact. Another way of saying this is that when direct contact is required to spread infection, the chances of another person getting the infection are lower. When your child can catch an infection by just being near someone (whose secretions have taken wing), that germ is more contagious.

Airborne Transmission. When a germ can float suspended in air, traveling distances longer than the 3-foot range that limits droplets, that germ is said to have airborne transmission. Extending the argument about the relative contagiousness of germs with different routes of human-to-human spread, if droplet-spread germs are more contagious than those that require direct contact, airborne germs should be the most contagious of all—and they are.

Germs that go airborne don't actually float on their own. They attach either to very tiny evaporated droplets (less than 5/1,000 of a millimeter, and often much less than that) of moisture or to floating dust particles. These germs can travel in the air across a room, down a hallway, and even from building to building, where they are then inhaled by a susceptible individual. There are not many organisms that can manage airborne transmission, but those that do are particularly difficult to control and have very high rates of spread. For example, the varicella-zoster virus, which causes chicken pox, and the rubeola virus, which causes measles, can be spread by airborne mechanisms, and the result is a greater-than-90% rate of infection among household members breathing the same air as an infected patient, regardless of the meticulous household hygiene measures (see chapter 9) that might be undertaken. Similarly, schoolrooms, doctors' offices, emergency room waiting areas, and any other airspace containing a patient with chicken pox or measles put everyone else in the room at risk. Airborne viruses access ventilation systems and can spread throughout a building, between floors, and out into the nearby community. The more dilute the virus is in the air, e.g., outdoors, the less likely is a passerby to contract the infection.

Other germs that are able to leave an infected person and leap tall buildings in a single bound include the bacteria that cause tuberculosis and the feared weaponized spores of the anthrax bacterium; recall the spread from floor to floor within contaminated buildings during the anthrax bioterrorist attack of October 2001 (see chapter 3).

I have mentioned only those airborne germs that arise in one person and fly to another (from the Other People category, as the Jeopardy format permits). Additional airborne hazards are discussed below under "Environment."

Blood. There was a time not long ago in this country when blood transfusions posed considerable risk for transmission of serious disease. Modern screening methods have substantially reduced that risk, although some germs may still escape detection.

Before describing the blood-borne germs, I'll say a word about blood products. Whole blood is what comes out of the vein; that blood may then be fractionated into blood components, like red blood cells, white blood cells, platelets, and the liquid plasma that remains after all of those particles are removed. Cells and plasma may be pooled from many donors, and pooled plasma may be further fractionated into proteins like gamma globulin and albumin. Whereas whole blood is collected entirely from volunteers, plasma often comes from paid donors at commercial facilities.

Volunteer donors are less likely to carry infectious diseases than are paid donors, and the screening of volunteer donors through interviews is extensive, as is the laboratory testing of their blood for known infectious agents. In contrast, the combination of paying donors and pooling their plasma greatly increases the risk of infectious agents getting into the plasma supply; however, the sterilization techniques that can be applied to plasma to kill viruses and other germs are much more vigorous and effective than the sterilization methods that can be used on whole blood (which has to be handled gently enough to preserve the important cell components). The result is that both whole-blood and plasma products in the United States are very safe.

Any germ that gets into the blood of an infected person is a potential contaminant of blood products. The determination by the Food and Drug Administration, which regulates blood donations, of which germs to screen blood for is based on a number of factors, including the risk of a germ causing life-threatening disease (see chapter 3), the prevalence of the germ in the population, the cost of screening, and the effectiveness of screening tests.

All blood donations in the United States are currently screened for five viruses: HIV (the cause of AIDS), hepatitis B virus, hepatitis C virus, and two retroviruses (viruses with a backwards genetic strategy of going from RNA to DNA in their reproductive cycle), human T-cell lymphotropic virus type 1 (HTLV-1) and HTLV-2. The last two germs travel in white blood cells and have been linked to leukemia and to a serious neurological disease; importantly, they are very commonly found in the blood of intravenous-drug abusers. Blood donations are also routinely screened for the bacterium that causes syphilis, and many blood centers also screen for West Nile virus.

As a result of universal screening for the germs mentioned above, the risks of contracting them by blood transfusion have been reduced to 1 in 2 million for HIV and hepatitis C virus, less than 1 in 60,000 for hepatitis B virus, and less than 1 in 600,000 for HTLV-1 and HTLV-2. Why is there any risk at all, though, if all blood is screened? Because screening depends on the presence in the blood of detectable markers of the viruses, and for some viruses, there is an incubation period of many days to weeks during which the donor is infectious but his or her blood is negative by screening. There is much work being done to detect other markers that are present earlier in the incubation period to further reduce the already very small risk of transfusion-associated infection with these viruses.

There are many other germs that reach the bloodstream of infected patients and therefore pose a transfusion risk if those infected patients unknowingly donate blood. Among them is cytomegalovirus (CMV), a member of the family of herpesviruses, which can pose a risk for severe disease in certain blood product recipients, such as premature newborns and people with abnormal immune systems. When blood products are required by these patients, as they frequently are, the specific units of blood designated for those at-risk patients are selectively screened for CMV by most hospitals and doctors. For other blood recipients, CMV screening is not performed because the virus is so common in both the donor and recipient populations—screening out all CMV-positive donors would drastically shrink the available blood supply, and most recipients are already immune, or partially immune, from their own past infection with CMV.

The transfusion risk of contracting parvovirus B19, the cause of childhood fifth disease (see chapter 3) and a rare cause of stillbirth when pregnant women are infected (see below), is about 1 in 10,000; hepatitis A virus contaminates less than 1 per million units of donated blood. Neither is screened for routinely. Other germs that have a blood phase but for which the exact risk of transfusion acquisition is not known and for which screening is not routinely performed are the rickettsia bacteria that cause Rocky Mountain spotted fever, ehrlichiosis, and Lyme disease; the parasites that cause malaria and babesiosis; and the viruses that cause Colorado tick fever and West Nile virus infections. There is also a theoretical risk of transmission of prion diseases, like mad cow disease (see chapter 3), by blood transfusion; no screening method is currently available. Finally, very rarely blood products, like almost anything else, become contaminated with bacteria from the environment. When that occurs, the recipient is at risk for sepsis and other infections, as well as for transfusion reactions.

Blood is a dangerous body fluid—many of the germs that can be transmitted by blood cause life-threatening disease. The foregoing discussion of blood-borne germs focused on human-to-human transmission by transfusion, the intentional sharing of blood between people for the good of the recipient. Each of the germs mentioned, however, can also be transmitted by blood via needle sharing in drug abuse, and some of the germs can be transmitted by risky sexual behavior that results in even minimal bleeding. Rarely, household or other contacts with sharp objects other than needles, e.g., razors, have been implicated in the transmission of dangerous viruses; even toothbrushes from infected patients with bleeding gums pose a theoretical risk. These risks are the basis for the important "sharing is bad" principle detailed in chapter 9.

Wherever blood is spilled, whether in the hospital, on the playground, or in the bathroom at home, it needs to be promptly cleaned up, and children must be taught to avoid contact with it (see chapter 9). No matter how much your kids would like to help their friends with scrapes or nosebleeds, teach them that the best way to help a bleeding friend is to find an adult who knows what to do (see chapter 9). It goes without saying that kids should not become blood brothers or blood sisters with their friends.

Womb and Birth Canal. Although the concept of the womb as a protective sanctuary for the unborn is fundamentally accurate (see chapter 4), germs often find a way in. Guarded by the placenta, the bag of waters, and the cervix and cervical mucus, the fetus inside the womb is indeed sheltered from the harsh world outside. Sometimes, however, the world *inside* of Mom can be harsh as well. There are two routes by which germs infect the fetus and the newborn exiting the womb: the mother's blood and the mother's vaginal fluids.

The germs discussed above as potentially transmitted by blood all apply to the fetus and the pregnant woman; whatever gets into Mom's blood can sometimes find its way across the placenta. The placenta serves as the way station between mother and baby, the tissue through which the fetus' nutrition is provided and the fetus' waste is cleared; both of those processes occur via an exchange between the mother's blood and the fetus' blood. Although the two blood supplies, Mom's and the fetus', remain mostly separated, their contents are exchanged through an elaborate network of capillaries and cells that comprise the placenta. It is through this exchange, vital to the survival of the fetus, that blood-borne germs may slip through from Mom to baby. With the exception of CMV, which can be present in the white blood cells of many people under normal circumstances, any germ in the bloodstream of a pregnant woman is abnormal. Furthermore, any germ, including CMV, in the bloodstream of a pregnant woman poses a significant risk to the baby.

The second route that germs take from the mother to the fetus is via the birth canal, the vaginal space through which the baby will descend from the womb. Vaginal germs come from several sources. First, the vagina is not a closed space; bacteria from the outside skin have access, as do bacteria from the nearby anus. As a result, the normal vaginal flora, those bacteria that are found in the vaginas of healthy women, includes skin germs, like staphylococcus and streptococcus; gastrointestinal germs, like *E. coli,* group B streptococcus, enterococcus, and klebsiella; and gastrointestinal anaerobic bacteria, like *Clostridium, Lactobacillus,* and *Bacteroides* species. The vagina may also harbor germs that are not present in normal situations, including herpes simplex virus from genital-herpes sores and other sexually transmitted diseases (see above).

Finally, as with almost everything else in medicine that we try to put into neat-and-clean categories, there is overlap between the two routes of germ transmission from pregnant woman to baby—some germs can take both routes, because they may be in the birth canal and also gain access to Mom's bloodstream; conversely, blood-borne germs may also be found in the bloody secretions of the birth canal.

Infections of the fetus are called congenital infections, a term also used to describe some infections acquired by the newborn baby on the way out of the womb. Another term for infections acquired on the way out of the womb is peripartum, meaning around the time of birth.

Table 2.1 summarizes the most important germs causing congenital and peripartum infections and the maternal sources (blood, birth canal, or both) of those germs. Many of these infections can cause stillbirth, poor growth of the fetus in the womb, and prematurity; all of the germs listed in Table 2.1 can cause significant disease in the baby. Descriptions of the diseases caused

Table 2.1 Germs causing congenital and peripartum infections

Germ	Route[a]		Disease in baby[b]	
	Blood	Birth canal	Birth defects	Sepsis
Viruses				
AIDS (HIV)	X	X	X	
Chicken pox (varicella) virus	X		X	X
Cytomegalovirus	X	X	X	
Enteroviruses	X	X	(X)	X
Fifth disease virus (parvovirus B19)	X		X[c]	
German measles (rubella) virus	X		X	
Hepatitis B virus	X	X	Liver	
Herpes simplex virus	X	X	X	X
Bacteria				
Chlamydia trachomatis		X	Eye, lung	
E. coli	(X)	X		X
Gonorrhea (*Neisseria gonorrhoeae*)		X	Eye	
Group B strep (*Streptococcus agalactiae*)	(X)	X		X
Listeria monocytogenes	X	X		X
Lyme disease (*Borrelia burgdorferi*)	(X)		(X)	(X)
Syphilis (*Treponema pallidum*)	X	X	X	(X)
Tuberculosis (*M. tuberculosis*)	X	X	X	
Parasites				
Malaria parasite	X			X
T. gondii	X		X	

[a]X, route used; (X), rarely reported.
[b]X, causes disease; (X), rarely reported.
[c]Parvoviruses cause fetal defects with resultant stillbirth; live-born babies with birth defects are not reported.

by most of these germs are found in chapter 3 and elsewhere in this book. Generally speaking, the diseases in the infected fetus or the newborn baby infected on the way out of the womb fall into two clinical categories: birth defects and sepsis. Sepsis (see chapter 3) refers to a life-threatening whole-body illness, often including shock and inadequate circulation to vital organs; neonatal sepsis has a high fatality rate.

Breast Milk. As you'll read extensively in chapter 10, breast milk is a wonderful thing for babies. Among its many qualities is its unrivaled ability to protect babies from a bevy of infections. Like many wonderful things, however, there is a potential dark side—several infections can actually be transmitted from mother to baby via breast milk. Most of the germs of concern from breast milk are viruses, and for many of them, the benefits of breast-feeding still outweigh the risks, especially in parts of the world where the risks of any other infant nutrition source are much greater. Hepatitis C virus, hepatitis B virus, HIV, CMV, and HTLV-1 and HTLV-2, all of which were discussed above as potential blood-borne pathogens, have also been detected in human milk. Of those, only HIV, HTLV-1, and HTLV-2 are enough of a risk for breast milk transmission to the baby that U.S. women with known infections with those viruses should not breast-feed. Premature babies are at higher risk for CMV disease, and mothers known to have had infection with CMV (as measured by the presence of antibodies to the virus in their blood) probably should not breast-feed their premature babies, or their milk should be expressed and pasteurized. Routine treatment with vaccine and hepatitis gamma globulin of babies born to hepatitis B virus-infected mothers (see chapter 7) protects those babies from the virus in their mother's milk. Hepatitis C virus has not been proven to be transmitted by milk, but infected mothers may choose not to nurse their babies if they are concerned about the theoretical risk.

Sores on the breast of a nursing woman may transmit germs to babies; these include bacteria (e.g., from a breast abscess or mastitis) and herpesvirus (e.g., from herpesvirus breast lesions). Women with breast abscesses or herpes lesions on the breast should not breast-feed until the sores are gone.

Breast milk banks function much like blood banks. Through these facilities, such as those belonging to the Human Milk Banking Association of North America, breast milk can be provided to babies who cannot receive their own mother's milk at the time it is available (e.g., a premature baby who is not yet feeding or cannot suckle) and to babies whose mothers cannot provide milk, in which case milk from a lactating woman other than their own mother is made available. For a mother banking milk for her own child, proper collection and storage of the milk reduce the risk of contamination with environmental germs. Donor mothers who provide milk for babies other than their own must be screened in almost exactly the same way that blood donors are screened at blood banks, and for the same germs. Also, like blood banks, milk banks belonging to the Human Milk Banking Association of North America follow the guidelines of the Food and Drug Administration, which include not only the aforementioned screening, but also a heat treatment that kills bacteria and HIV.

Indirect Contact. We're well into chapter 2 of this book with nary a mention of toilet seats. Now it's time to lift the lid on this subject. When nonliving intermediaries, also known as inanimate objects, are required for the spread of infection, the contact is said to be indirect. The success of a germ in getting from person to person via a toilet seat or other nonliving thing depends on two factors: how long the germ can live on the thing and how long it takes for a susceptible person to come in contact with the thing. A 2006 study of rhinovirus-infected adults staying overnight in hotel rooms found that one-third of the objects they touched in the room became contaminated and that rhinoviruses could be recovered for up to 24 hours from light switches, telephones, and TV remote control devices, objects that might not receive routine cleaning.

Free-living germs, like bacteria and fungi, survive on inanimate objects longer than viruses because viruses require living cells to survive for prolonged periods (see chapter 1). Moisture helps all germs survive longer; even some viruses can hang around in a moist environment for several days without drying up and dying off.

The most dangerous inanimate objects have already been discussed—those sharp and even not-so-sharp things that can be contaminated by blood (see above). Although the diseases transmitted by blood-contaminated objects are potentially life threatening, such transmission fortunately occurs only very rarely among children. Much more commonly serving as vectors of infection are everyday things, like drinking glasses, toothbrushes (not blood-tinged, just full of saliva germs), hats, hair brushes, bed sheets and other laundry, and, yes, occasionally even the much maligned toilet seat (see chapter 9).

Table 2.2 summarizes some of the germs and diseases that can be spread from person to person via inanimate intermediaries; most require sharing—that most evil of all evils perpetrated on kids by well-meaning adults. Sharing is bad (see chapter 9 for a full and scathing indictment of sharing).

One of the diseases listed in Table 2.1 that you may not recognize from elsewhere in this book is herpes gladiatorum, a curious malady of wrestlers. The scrapes and scratches on wrestlers' skin expose them to infection with herpes simplex viruses from the respiratory secretions of their opponents and from secretions left behind on the wrestling mats. An outbreak of herpes gladiatorum in Minnesota among 24 wrestlers representing 10 different schools in a tournament forced the suspension of all high school wrestling in the state in January and February 2007. When the virus gets into the abraded skin, it causes a painful eruption of skin blisters. Kids who don't wrestle but have eczema or suck their thumbs can get a similar disease from their own oral secretions during a herpes infection or from another child with herpes, but those are by direct contact, not involving the novel inanimate object at fault with wrestlers.

Animals and Birds

Transmission of germs from animals to humans can occur either by direct contact between a person and an animal (or an animal's secretions) or via an intermediary insect, a mosquito or a tick under most circumstances. Although many germs are species specific, meaning they can infect only one or a few closely related species (see chapter 3), other germs are able to cross species barriers and pass from animals or birds to humans; some can even go from humans to animals, but I'll leave that for a veterinarian's book (*Germ Proof Your Pets?*).

In general, viruses and fungi are less likely to be transmitted from animals to humans, whereas bacteria and parasites that are normally found in animals find humans to be suitable hosts as well; thus, many of the animal-derived infections discussed below are bacterial or parasitic. Among the viruses that do spread from animals to humans, most (but not all) accomplish the leap via an insect intermediary.

Table 2.2 Everyday "things" that can spread infections

Thing(s)	Germ(s)	Disease(s)/condition(s)
Drinking glasses; water bottles; toothbrushes; eating utensils	Respiratory bacteria Respiratory viruses	Strep throat; pertussis (whooping cough); common cold; flu; ear infections; pinkeye; sinus infections; pneumonia
	Herpesviruses	Cold sores
	Epstein-Barr virus	"Mono"
Sinks; toilets; toilet seats	Gastrointestinal bacteria Gastrointestinal viruses	Gastroenteritis (stomach flu)
Shower stall	Viruses	Plantar warts
	Fungi	Athlete's foot; jock itch
Hair brushes; combs; hats; sport and bike helmets	Fungi	Ringworm of the scalp
	Insects	Head lice
Wrestling mats	Herpesviruses	Herpes gladiatorum
Laundry	Molluscipoxvirus	Molluscum contagiosum skin rash
	Skin bacteria	Skin infections (impetigo, others)
	Genital bacteria	Syphilis, gonorrhea
	Gastrointestinal bacteria Gastrointestinal viruses	Gastroenteritis (stomach flu)
	Fungi	Ringworm
	Parasites	Pinworms
	Insects	Scabies; bed bugs
Razor blades (face or leg)	Viruses	HIV infection/AIDS; hepatitis B; hepatitis C

Diseases caused in humans by animal germs are called zoonoses (pronounced "zoa-no-sees," not "zoo noses"), and there are hundreds of them. I believe that the most useful way to present this information to you is with an animal-by-animal detailing of the most common germs each animal can give to your kids; that way, you can just scan down the page to the pets or wildlife most likely to have contact with your kids and see the risks. Once you find the animal of interest, the discussion focuses on how the animal's germs get into your child (e.g., via bites, scratches, or contact with secretions or droppings) and what germs and diseases are of greatest concern. This list is by no means exhaustive, but it should give you an idea of the most important beasts and their bugs.

For the purposes of discussion, and hopefully clarity, I've divided the animal kingdom into three subcategories: household pets, farm animals, and wild animals. Most of the animal infections that travel to humans via insect intermediaries (mosquitoes, ticks, and fleas) are found among the wild animals, and those diseases are presented in "Wild Animals." It's in "Farm Animals," though, that you'll read about houseflies, because farms have so many more attractions for those pesky pests.

Prevention strategies for the zoonoses are detailed in chapter 9.

Household Pets. *Cats.* You either love 'em or you hate 'em (it is said that dogs have masters, but cats have slaves. . .), but there's no doubt that cats are a significant source of human infection. It is estimated that there are 400,000 cat bites every year in the United States. Cats' teeth are sharp and pointy, and they can inject germs deep into a bite wound, explaining the very high rate of infection (50%) following cat bites. Any of the bacteria in a cat's mouth can potentially cause infection, but the most dangerous is *Pasteurella multocida,* an anaerobic (see chapter 1) bacterium found in the mouths of almost all cats. This germ rapidly causes redness and swelling around the bite site, which may progress to involve a large area of skin and deeper layers of soft tissue. Glands (lymph nodes) near the bite may become infected, and the germs may even extend down to bone. Infections of hand joints can be particularly severe. These bacteria can also spread to the bloodstream of the bite victim, causing sepsis, meningitis, and infections of virtually all organ systems. *P. multocida* can also be transmitted by cat scratches (the germs on the cat's paw include those from the mouth because of the way the animals groom themselves).

Another worrisome infection associated with cat scratches, especially from adorable little kittens, is called, appropriately, cat scratch disease. This infection, caused by the bacterium *Bartonella henselae,* is associated with swollen, red, and tender glands (lymph nodes) in the region of the bite, as well as a fever and flu-like illness. The germ can also spread via the blood to involve many organs, including the brain, liver, lungs, and heart.

In recent years, rabies (see chapter 3) has been diagnosed twice as often in cats as in dogs in the United States, but there have been no proven cases of rabies transmitted from a cat to a human in many years in this country. Cats acquire rabies by contact with infected wildlife—raccoons, bats, squirrels, etc.

Pregnant women who change the kitty litter are at risk for acquiring *Toxoplasma gondii* parasites, shed in the feces of many pet cats. Congenital toxoplasmosis results from infection of the fetus, manifesting as a rash, swollen glands (lymph nodes), and enlarged liver and spleen at birth; other babies may have even more dramatic abnormalities at birth, including infections of the brain and eyes. Many of these babies, as well as those that may have no outward signs of infection at birth, go on to develop blindness, deafness, and mental retardation, which are diagnosed as they grow older.

A similar-sounding parasitic disease of kittens and cats is toxocariasis (also known as visceral larval migrans or ocular larval migrans), caused by *Toxicara cati,* a roundworm excreted in the feces of cats. Kids who eat soil (pica) are particularly at risk because the eggs from infected animals may remain in soils in hot and humid climates for prolonged periods of time. The illness in children is a fever and flu-like disease with abnormalities in blood counts, enlargement of the liver, and occasionally involvement of other organ systems, including the brain. Still another feces-transmitted infection from cats is gastroenteritis (see chapter 3) caused by *Campylobacter jejuni* bacteria, shed by both cats and dogs.

Ringworm is caused by several species of fungi and is one of the few fungal infections that can pass from pets (cats and dogs) to people.

Dogs. There are 10 times as many dog bites every year in the United States as there are cat bites, but the risk of infection is somewhat lower. The flatter teeth and larger jaws of dogs result in more crush-type injuries, rather than the deep penetration wounds seen with cat bites; 10 to 15% of dog bites result in infection. *P. multocida* is also present in dogs' mouths and can cause the same complications as when acquired from cats. Dog bites frequently become infected with other bacterial germs, as well, including common staph and strep bacteria and anaerobic bacteria (see chapter 1). Infected dog bites may cause a range of findings, from localized redness and tenderness to more widespread involvement of deeper tissues, including bone, and even spread via the bloodstream to internal organs. *Capnocytophaga canimorsus* is a unique dog bite bacterium that can cause severe human infection, including sepsis and meningitis (see chapter 3); fortunately, these are rare, with only a few dozen proven cases over the past 3 decades.

The almost universal immunization of U.S. dogs against rabies (see chapter 3) has greatly reduced the canine role in transmitting the infection to humans. Although dog bites remain the leading cause of human rabies worldwide, there have been no proven cases of human rabies in the United States associated with a dog bite (by a U.S. dog) in many years; there were five cases of dog bite rabies diagnosed in the United States between 1990 and 2001, but all of the bites occurred abroad. During that period, there were two cases of human rabies in Texas in which the exposure was unknown and a dog or coyote strain of the rabies virus was identified; they may have been dog bites that were not recalled by the victims or their families.

Dog tapeworm, *Dipylidium caninum,* is carried by fleas. Kids who accidentally swallow dog fleas from close contact with the pets may become infested with the worm, which can be diagnosed clinically as gastroenteritis and in the laboratory by seeing worm eggs or little swimming seeds in specimens from the child's bowel movements. Another dog tapeworm of less concern for kids and more native to foxes than to domestic dogs, *Echinococcus multilocularis,* causes cysts in human organs, especially the liver and lungs; the infection is contracted by ingesting soil or other matter contaminated by animal feces. Dog hookworm, *Ancyclostoma caninum,* can get from the feces of pet dogs under the skin of children, where the parasite burrows and causes an extremely itchy, red rash. Like cats, dogs (especially puppies) shed their own species of roundworm parasite (*Toxacara canis*) in the feces, which causes toxacariasis. The disease is nearly identical to that described for cats above, and it also gets into kids who eat soil contaminated with dog excrement. Also, as noted above for cats, dogs shed *C. jejuni* bacteria in their feces, a cause of human gastroenteritis (see chapter 3). Rarely, the gastroenteritis-causing parasites *Cryptosporidium parvum* and *Giardia intestinalis* may be caught by kids in contact with dog feces; more common routes for these germs are discussed above (see "Other People") and below (see "Environment"). Dogs can carry ticks into the home; the ticks may be carriers of the Lyme disease bacterium (see "Wild animals" below).

As noted above for cats, ringworm is caused by fungi and can also pass from dogs to people.

Birds. Pet birds shed the bacterium *Chlamydophila psittaci* in their droppings, which causes an illness in people called psittacosis, a respiratory tract infection including fever, cough, and flu-like symptoms that may progress to pneumonia; involvement of other organ systems, like the liver and brain, has also been seen occasionally. Humans acquire the infection by inhaling the airborne dust from the droppings of the birds. This is one of the few infectious diseases for which kids seem to be at lower risk than adults, although the reason isn't known. Also shed in bird droppings is the fungus *Cryptococcus neoformans,* a fungal cause of severe pneumonia, meningitis, and other complications that mainly affects people with abnormal immune systems.

Rodents (mice, rats, hamsters, gerbils, guinea pigs, and prairie dogs). There are dozens of rodent-related human diseases, but most occur as a result of contact with wild rodents; some of those infections will be discussed in "Wild animals" below. In this section, the germs from household rodent pets that can infect your kids are discussed.

It's hard to understand why people would keep rats as pets, but it's hard to understand why people do a lot of things. Rat bite fever is caused by two different bacteria, *Streptobacillus moniliformis* and *Spirillum minus.* The first germ causes most cases of rat bite fever in this country; symptoms include fever, muscle aches, headache, vomiting, joint pain and swelling, and a rash; involvement of major organ systems occurs occasionally. The infection is transmitted, as the name suggests, by the bite or scratch of a rat; mice can also harbor the germ and may pass it on to kids.

The house mouse and pet hamsters shed lymphocytic choriomeningitis virus in their urine and droppings. Rarely, this germ is passed on to humans, where it causes a flu-like illness, often accompanied by meningitis and/or encephalitis (see chapter 3). When pregnant women become infected with lymphocytic choriomeningitis virus, spontaneous abortion may result, and live-born babies may have abnormalities of their brains, including blindness and mental retardation. The infection is caught by inhaling the dust of rodent excrement or by ingesting food contaminated with rodent droppings or urine.

Rodents may be shedders of salmonella bacteria, a common cause of human gastroenteritis; complications may include spread through the blood to other organs, sepsis, and meningitis (see chapter 3). Although much more strongly associated with reptile pets (see below), rodent-caused salmonella infections in children are well known to occur by ingestion of the germ, which is shed in rodent feces; rodent fur may also be contaminated by fecal salmonella.

In 2003, 37 cases of monkeypox were reported in the United States, associated with pet prairie dogs. This was the first occurrence of the infection in this country. Monkeypox in humans causes a milder version of smallpox (see chapter 3), but milder is not very reassuring, since smallpox is such a devastating disease. Fever, flu-like illness, total-body blistering rash, and swollen glands (lymph nodes) are some of the symptoms; among the U.S. cases, two children required intensive care unit treatment, one for encephalitis (see chapter 3) and the other for severe and painful swelling of the glands (lymph nodes). In Africa, where the disease is native to rodents of the rain forest, the fatality rate among infected humans is between 1 and 10%. Spread from person to person occurs by contact with respiratory secretions or with the pox rash sores on the skin. The pet prairie dogs linked to the U.S. outbreak were apparently infected by other prairie dogs and rodents imported from Ghana. Laws restricting such imports have since been passed by the U.S. government, but it is not yet known how far the monkeypox virus may have spread within the U.S. rodent population.

Reptiles (turtles, lizards, salamanders, snakes, and iguanas). The CDC estimates that each year 70,000 people in the United States develop salmonella bacteria infections from pet reptiles (the CDC also estimates that 3% of all U.S. homes have a pet reptile!). Kids under 5 years old, and people of any age with abnormal immune systems, are at particular risk for severe disease: gastroenteritis, with the risk of spread to other organ systems, sepsis, and meningitis (see chapter 3). The risk of salmonella disease from pet turtles is so great that the sale of turtles smaller than 4 inches has been banned in the United States since 1975. In 2007, legislation is being considered to reinstate turtles as legal pets in the United States; at the time of writing, an amendment permitting the sale of small turtles had passed the U.S. Senate and was being considered by the House. It is not completely clear why the senators now consider turtles to be safer than in 1975, when hundreds of thousands of cases of salmonella infections occurred in the United States; the occurrence rate dropped dramatically after the ban was instituted. I suspect that there must be a strong (but slow) turtle lobby in this country that is just now coming out of its shell.

Farm Animals. *Cows and sheep.* Since the first reported human cases of mad cow disease, it has been hard to think of anything else when we worry about cow-related infections in humans, but there have been fewer than 200 cases of mad cow prion disease in humans (see chapter 3). A "mad sheep disease" called scrapie is not transmitted to humans; "mad deer and elk" disease, also known as chronic wasting disease, has been a scourge on deer and elk farms, but it also has not been proven to be transmissible to people.

Cows pass numerous other infections to people, and in far greater numbers than mad cow disease. The *E. coli* strain O157:H7 causes severe bloody gastroenteritis in children (also fully described in chapter 3), which may advance to kidney failure and death. Associated with food outbreaks, this germ comes from the feces of infected cows and contaminates unpasteurized milk, unpasteurized juices (from fruit grown in pastures where cows graze), undercooked meat, and vegetables (also grown in fields where cows graze). Direct contact by children with the feces of calves and cows can also result in infection. Brucellosis is caused by the bacteria *Brucella abortus* and *Brucella melitensis*. The disease occurs in three stages: an early flu-like illness, a waxing and waning fever over months, and a chronic form of arthritis and fatigue. The infection is caught from cows and sheep (and goats) by drinking unpasteruized milk or by contamination of open wounds with cow feces. Cow and sheep feces may also transmit numerous germs that cause gastroenteritis (see chapter 3), including campylobacter, salmonella, and cryptosporidium (see above and chapter 3).

Q fever is caused by the bacterium *Coxiella burnetii*. The disease in humans starts out looking like a severe case of the flu but may advance to serious lung and liver disease. It is acquired by inhaling barnyard dust, particularly where calves or lambs have recently been born. The bacterium is in high concentration in the placenta and birth products of cows and sheep, as well as in their urine and feces.

Cows can get tuberculosis from the bacterium *Mycobacterium bovis*, a cousin of the strain that causes human disease (see chapter 3). Kids drinking unpasteurized milk can contract tuberculosis by this route, with abdominal symptoms predominating over the respiratory symptoms associated with the human strain.

Pigs. Gastroenteritis caused by the bacterium *Yersinia enterocolitica* can be acquired from contact with pig feces or ingestion of pig feces-contaminated food. In some patients, this infection can mimic appendicitis. Eating undercooked meat from pigs can result in transmission of *Trichinella spiralis*, a roundworm parasite that initially causes a gastroenteritis-type illness in humans but may progress to involve the lungs and muscles in the torso and heart.

Horses. Most infections acquired by humans from horses are transmitted by ticks or mosquitoes that first bite the horse and then the person (see below).

Houseflies. Of course, houseflies are not animals, but they spend so much time on the farm that they might as well be. The formal genus-species name for the housefly is *Musca domestica*, and these pests don't limit themselves to farmhouses. Houseflies may carry as many as 4 million bacteria on their little legs and bodies and 28 million in their stomachs (don't ask me how those studies are conducted). During the Spanish-American War at the turn of the 20th century, typhoid fever (a severe salmonella infection) killed more soldiers than bullets did, and common houseflies were deemed to be the 2nd most likely route of transmission (after direct human-to-human contact). Flies are attracted to anything organic, including animal feces. Bacteria in feces can be picked up by the flies and carried to food and water, the route of spread of typhoid (salmonella is found in the gastrointestinal tracts of humans and animals [see above and chapter 3]). With modern sanitation and modern farming facilities, housefly-spread infection is far less of a risk than it was 100 years ago, but *E. coli* O157:H7, the cause of severe hemorrhagic gastroenteritis and hemolytic uremic syndrome in children (see chapter 3), has been found on flies at cattle ranches, and campylobacter (another cause of gastroenteritis [see chapter 3]) has been recovered from flies at poultry farms. The potential for fly-transmitted disease is always present, but proving that flies are involved is very difficult. Housefly prevention strategies are discussed in chapter 9.

Wild Animals. There will be a change of format for this section, because the enormous diversity of wild animals makes an animal-by-animal listing less useful than a listing by disease. There are a limited number of important diseases that are acquired from wildlife, but there is significant overlap in the animal species that can harbor the germs. It is also important to note that almost all of the infections acquired from household pets and farm animals can also be transmitted by wild animals, and hence, the divisions into these categories are somewhat arbitrary.

Table 2.3 focuses on the diseases of greatest concern, the primary animal reservoir (a reservoir is the source of the infection, where the germ lives before it gets into humans), and the route by which the germs can get into kids. Most of the clinical conditions associated with each infection are described more fully in chapter 3. It is in this category of wild animals that you'll find many of the tick-borne and mosquito-borne infections—again, there is overlap with animals listed above. Horses, for example, contribute significantly to the life cycles of several of the encephalitis viruses (Table 2.3).

Food and Drinking Water

Contaminated food and drinking water are very common sources of human infection. There is arbitrariness to classifying the infections in this category, because the germs that contaminate food and water have to come from somewhere, and often that somewhere is another person or animal whose contaminated hands or feces slipped or dipped into someone's food and drink. Additionally, most of these infections can be passed on from the first infected person to others without going back through food or water, and hence, person-to-person spread occurs.

As a result of the overlapping routes that many germs take, most of the infections discussed as being spread through food and drinking water have already been discussed in "Other people" or "Animals and birds" above. Table 2.4 lets you look for your child's major symptoms and decide whether they could be the result of something he or she ate. The clinical features of most of these infections are more extensively reviewed in chapter 3, and pre-

Table 2.3 Diseases transmitted to humans by wild animals

Disease	Clinical condition	Germ	Animal reservoir(s)	Route of acquisition
Colorado tick fever	Fever, flu-like illness, meningitis, encephalitis, abnormal blood counts	Virus: Colorado tick fever virus	Rodents	Tick bites infected animal and then bites human
Ehrlichiosis	Fever, flu-like symptoms, rash, stomach flu-like symptoms, nausea, vomiting, weight loss, abnormal blood counts	Bacteria: *Ehrlichia chaffeensis; Anaplasma phagocytophila; Ehrlichia ewingii*	Deer, mice, other rodents	Tick bites infected animal and then bites human
Encephalitis	See chapter 3	Viruses: West Nile, eastern equine, western equine, St. Louis, Japanese, California serogroup viruses, others	Birds, horses, cattle, rodents	Mosquito bites infected animal and then bites human
Hantavirus pulmonary syndrome	Fever, fatigue, muscle aches, severe pneumonia	Hantavirus	Deer mouse, other rodents	Inhaling dust from rodent excretions; rarely by direct contact
Leptospirosis	Fever, flu-like symptoms, rash, pinkeye, severe muscle aches, jaundice, meningitis, sepsis	Bacteria: *Leptospira* species	Many different wild animals, rodents, dogs, farm animals	Contact of mucous membranes or skin wounds with water or soil contaminated by animal urine; ponds and lakes
Lyme disease	See chapter 3	Bacterium: *Borrelia burgdorferi*	Deer	Tick bites infected animal and then bites human
Plague	See chapter 3	Bacterium: *Yersinia pestis*	Rodents	Flea bites infected animal and then bites human; also person-to-person spread of pneumonia form (see chapter 3)

(Continued on following page)

Table 2.3 Diseases transmitted to humans by wild animals *(Continued)*

Disease	Clinical condition	Germ	Animal reservoir(s)	Route of acquisition
Rabies	See chapter 3	Rabies virus	Bats, raccoons, squirrels	Bites; secretions (see chapter 3)
Rocky Mountain spotted fever	Fever, flu-like symptoms, rash, stomach flu-like symptoms, abnormal blood counts, sepsis, encephalitis, bleeding	Bacteria: *Rickettsia rickettsii*	Dogs, rodents	Tick bites infected animal and then bites human
Tularemia	See chapter 3	Bacterium: *Franciscella tularensis*	Rabbits, rodents	Direct contact with infected animal; also may be via a tick that has bitten an infected animal and then bites a human

Table 2.4 Clinical illnesses associated with contaminated food or drinking water

Illness	Clinical condition	Germ(s)	Food or water source(s)
Botulism	Muscle paralysis (see chapter 3)	Bacterial toxin from *Clostridium botulinum;* formed in foods prior to ingestion	Improperly stored or preserved food; commercial canned meat; home canned, bottled, or jarred food; restaurant outbreaks
Botulism, newborn	Constipation, weak cry, decreased appetite, muscle paralysis, cessation of breathing	Bacteria: *C. botulinum* germ ingested by baby, produces toxin in the intestines	Honey; other unidentified sources
Food poisoning	Abrupt (within a few hours of eating) onset of severe nausea, vomiting, abdominal cramps, diarrhea	Bacterial toxins from *Staphylococcus aureus* or *Bacillus cereus*	Food contaminated by infected handlers and then left unrefrigerated; salad dressing, pastries, poultry, deli meat, fried rice
Gastroenteritis	Stomach flu (see chapter 3)	Viruses: noroviruses	Water, ice, salads, shellfish, cruise ship outbreaks
		Bacteria: *Salmonella, E. coli, Shigella, Campylobacter, Yersinia*	Poultry, eggs, dairy, meat, fruits, vegetables, pork, well water
		Parasites: cryptosporidia, giardiae	Water
Hemolytic uremic syndrome	Bloody diarrhea, kidney failure (see chapter 3)	*E. coli* H7:O157	Undercooked meat, unpasteurized dairy products, unpasteurized juices, vegetables grown in fields where cattle have grazed
Hepatitis	See chapter 3	Virus: hepatitis A virus	Sandwiches, salads, pastries; any food handled by infected person not using proper hand hygiene
Listeriosis	Fever, flu-like symptoms, diarrhea, meningitis (see chapter 3); severe in pregnant women, may cause stillbirth, prematurity, sepsis, and meningitis in newborn	Bacteria: *Listeria monocytogenes*	Unpasteurized dairy products, soft cheese (especially imported), undercooked meat, unwashed vegetables
Traveler's diarrhea	Stomach flu (see chapter 3)	Bacteria: *E. coli*	Water, ice, food washed with water, raw produce, salads, undercooked meats and vegetables

vention strategies for food and drinking-water infections are presented in chapter 9.

Environment

Once again, please don't overlook the overlap—many environmentally acquired infections have their original source in animals or, to a lesser extent, other people. Consider, for example, leptospirosis, which is often acquired by bathing in or swallowing pond or lake water contaminated by animal urine; this infection was included among those with wild-animal sources above and will not be repeated here. Rather, this section focuses on germs that are primarily acquired from the air your kids breathe and the water they come in contact with (other than drinking water, which is discussed above). If the original source of the environmental germ was other people or animals, it is probably discussed elsewhere in this chapter.

Air. A number of human fungal infections are acquired from the environment. Airborne fungi usually originate in moist soil, where they produce spores that float in the air until they're inhaled by susceptible kids or adults; different fungi inhabit different regions of the country (see chapter 1). Dry and windy conditions, especially dust storms, facilitate the dissemination of spores in the air, as does excavation and construction on contaminated land. Hospital outbreaks have occurred because of nearby construction with uptake of fungal spores into the ventilation systems. Histoplasmosis, caused by the soil fungus *Histoplasma capsulatum,* is a flu-like illness with chest pain, swollen glands (lymph nodes) in the chest cavity, and a usually mild pneumonia. A more severe variety of the disease can occur in children under 2 years of age, in whom weight loss, lengthy periods of fever, enlargement of the liver and spleen, and meningitis develop. Another soil fungus, *Coccidioides immitis,* causes coccidioidomycosis, clinically very similar to histoplasmosis. Again, flu-like illness predominates; some patients develop a rash. Many patients are left with abnormal, but clinically insignificant, changes on chest X ray. This fungus also occasionally infects through wounds in the skin. Young infants are at risk for more serious infection, with organ systems other than the lung involved. The most potentially dangerous of the inhaled soil fungal diseases is blastomycosis, caused by *Blastomyces dermatitidis.* This fungus causes lung and skin infection, as seen with coccidioidomycosis; additionally, blastomycosis can cause a whole-body form of disease that begins in the lungs and spreads to cause multiorgan system infection, including the brain. Infection in children is usually characterized by pneumonia, but the chest X ray may lead to mistaken diagnoses of bacterial pneumonia, tuberculosis, or a tumor.

Water. Tuberculosis (see chapter 3) is caused by a bacterium called *Mycobacterium tuberculosis* and is spread from person to person. A number of related organisms, collectively called nontuberculous mycobacteria (NTB), are environmentally acquired germs, mostly from water sources, that cause infections in kids. NTB are either ingested from contaminated water or soil or accidentally inoculated into the skin through wounds, again usually by exposure of the wounds to contaminated water. Most infections in children remain local, that is, they do not spread within the child's body unless the child has an abnormal immune system (see chapter 4), in which case severe and even fatal infections can occur. The most common illness caused by NTB in kids is swollen neck glands (lymph nodes), thought to be caused by drinking contaminated water; the germ infects the mouth, with limited spread to the nearby glands. The glands may become so enlarged that surgery is required for their removal, which is usually curative. NTB species can also cause skin infections; particularly common are those acquired from infections of cuts or scratches with aquarium water that harbors the germs. Other species of NTB sneak in and cause infection through ear tubes (placed to treat chronic ear infections [see chapter 3]).

One of the fascinating infectious-disease detective success stories that you'll read about in chapter 3 is how the mystery of Legionnaire's disease was solved more than 30 years ago. The causative bacterium, *Legionella pneumophila,* was finally tracked down in the water system of a hotel where hundreds of American Legion conventioneers were staying for their bicentennial meeting. The germ is inhaled from water vapor in showers or other misty locations if the water temperature isn't high enough to kill the bacteria.

Other water environment risks for infection are swimming pools, hot tubs, and spas. The CDC logged 62 different infection outbreaks from recreational water use between 2003 and 2004; more than two-thirds of the outbreaks resulted from improper water treatment or disinfection. Three-quarters of the nearly 2,700 affected individuals became ill with gastroenteritis due to the viral, bacterial, and parasitic germs discussed above (and in chapter 3). Although these germs are usually passed to your kids by other people or animals, water is sometimes the intermediary. While chlorination of swimming pools and spas inactivates many germs, others can survive the process. Table 2.5 lists the infections *other than* gastroenteritis that your kids can catch from the pool.

Furniture? This brings us to bed bugs (perhaps that was not my smoothest transition). You probably don't think of beds in the same context of environmental risks for your kids as air and water; nevertheless, beds are the major environmental source for bed bugs—otherwise we'd call them something else! For most of the last half-century, bed bugs were absent from the United States, apparently eliminated by the liberal use of insecticides. They remained a problem in most of the rest of the world during those 50 years, however, and have now hitched a ride back to the United States with unsuspecting travelers. These insects (they're not really germs, but where else are you going to read about them?) are formally known as *Cimex lectularius,* and they live by taking tiny and painless blood meals from sleeping kids

Table 2.5 Infections (other than gastroenteritis) from recreational water use

Disease	Clinical condition	Germ(s)	Water source
Cellulitis	Skin infection	Bacteria: *Vibrio parahaemolyticus; Vibrio vulnificus*	These germs usually cause gastroenteritis but can infect an open wound and cause skin infection; exclusively in salt water
Cercarial dermatitis	Swimmer's itch; itchy rash	Parasites: several, from birds, snails, or water animals	Animals lay eggs in the water, which hatch, and the larvae attach to swimmers and burrow under skin
Conjunctivitis	Pinkeye (see chapter 3)	Virus: adenovirus	Swimming pools inadequately chlorinated
Hot tub rash	Rash	Bacteria: *Pseudomonas aeruginosa*	Hot water breaks down chlorine faster; frequent water testing necessary
Meningoencephalitis	Severe infection of the brain and its coverings (see chapter 3)	Parasite: *Naegleria fowleri*	Lakes, rivers, hot springs; rare in United States; parasite enters nose, travels to brain
Molluscum contagiosum	Rash	Virus: *Molluscipoxvirus*	Not certain—swimming pools or shared towels?
Otitis externa	Swimmer's ear; itchy ear canal with pus drainage (see chapter 3)	Bacteria: *Pseudomonas aeruginosa,* others	Swimming pool water lingering in ear canal, especially if canal is irritated

(and adults). The bugs lie in wait for their victims, hiding out in the bed sheets and in cracks between the boards of the bed frame. Bed bug bites are not dangerous, but they can sensitize kids' skin and cause a nasty allergic-type reaction, which can be difficult to diagnose, especially for doctors who have never seen bed bug bites—and that includes most doctors who have practiced for the past 50 years! Chapter 9 presents bed bug prevention tips.

Infection Amplifiers: Hot Zones Where Kids and Germs Gather Together in High Numbers

Day Care Centers

Webster's Revised Unabridged Dictionary, edited by Noah Porter and published by G. & C. Merriam Co. in 1913, has the following definition:

> Cesspool..., A cistern in the course, or the termination, of a drain, to collect sedimentary or superfluous matter; a privy vault; any receptacle of filth.

You get the idea. Day care centers are the proverbial double-edged graham cracker, liberating parents and providing prekindergarten age kids with a stimulating and entertaining environment on the one hand, but. . . Not since the trenches of World War I have germs been as efficiently shared as in today's child care centers. The American Academy of Pediatrics (AAP) *Red Book: Report of the Committee on Infectious Diseases* (2006) lists 2 dozen viruses or categories of viruses, more than a dozen bacteria or groups of bacteria, three parasites, a fungus, and two insects (mites causing scabies and lice) among the germs known to be shared in the microbial smorgasbord at day care. Within those categories and groups of germs, hundreds of individual species are included. Rhinoviruses, for example, cause the common cold (see chapter 3), and they are included as just one of the dozens of germ categories in the AAP *Red Book* list of day care scourges. However, there are more than 100 different species of rhinoviruses (see chapter 1), each of which can cause the common cold while leaving your child still susceptible to the remaining species circulating in the day care center this season or next. Sixty-five distinct species of enteroviruses, the most common causes of meningitis, as well as of the summer colds that your kids can't shake, are also jam-packed into just a single category of day care germs listed by the AAP.

All of the routes of transmission discussed above (remember our Jeopardy categories: Other People, Animals and Birds, Food and Drinking Water, and Environment) apply to day care centers but are amplified because of the close proximity of the kids and, more importantly, because of their potty-training status. Indeed, what makes day care centers so unique to this era as amplifiers is that in previous generations, pre-potty-trained kids rarely gathered in one place. When our kids' grandparents were raising us, by the time we got into a crowded school environment—kindergarten, usually—we were already potty trained and far less likely to be putting our hands in the mouths (and elsewhere) of playmates.

While respiratory infections have always been efficiently spread in schools—kids cough and sneeze and hold hands regardless of their age (see below)—the fecal-oral (feces to mouth, usually via the hands [see above]) avenue of transmission has been uniquely opened by day care centers. Among day-care-age kids, unwashed hands travel everywhere on a child's body and then everywhere on every other child's face and body and then everywhere else in the room—the chairs, tables, water fountains, stuffed toys, puzzles, and crayon boxes. Germs spread by the fecal-oral route can cause gastroenteritis and hepatitis. Some of the germs, like giardia and hepatitis A virus, rarely if ever infected pre-school age kids in the past, but in the day care era, those germs have established that demographic as the major reservoir for infection in the United States. Germs that pass from Johnnie to Janie go on to infect Janie's older siblings and parents, changing the entire epidemiology of the infection. Hepatitis A is the quintessential example. Day care outbreaks occur among very young kids, who are usually without symptoms. Their infections go unnoticed until their parents at home start falling ill; by then, the day care outbreak is well underway and multiple households are infected. Adults who contract hepatitis A from their toddlers in day care may be debilitated for weeks (see chapter 3).

Rather than relisting all of the germs and routes of transmission discussed above that are magnified by the day care experience, it's easier to simply list those that are *not* a risk for kids in day care. There aren't many. Among those infections transmitted by the Other People route, only sexually transmitted diseases are (thankfully) missing from the day care menu; unfortunately, blood-borne germs are not absent from day care centers, where bloody noses, knee scrapes, and lost teeth are endemic. Many large day care centers have their pet rodents, reptiles, or parakeets, and many in-home facilities have dogs and cats; hence, the only Animals and Birds germs missing from day care are those from farm animals (which your city kids have ample opportunity to contract at petting zoos [see below]) and wild animals. With the possible exception of botulism, all of the Food and Drinking Water infections are even more available to day care kids: caregivers who prepare the kids' snacks may have just changed a diaper—or 10 diapers! Little Susie's unwashed hands are reaching into the same cookie box as little Sam's hands. Finally, the environmental exposures of kids in day care are the same as for kids at home, save perhaps for the swimming pool germs that day care kids wait to catch on weekends with their families.

The determinants of risk for infections in day care are straightforward: the number of kids in the center, their ages, and the hygienic measures undertaken by the caregivers. The single most critical hygienic measure is hand washing (see chapter 9), by the caregivers and, in an ideal world, by the kids on a regular and enforced schedule. Yeah, right.

For all of you parents of day care kids who, after reading the previous paragraphs, are feeling guilty about dropping your kids into the microbial stew, take comfort. It has long been known that kids in day care have more infections each year than kids in home care, but it has also been suggested in numerous studies that things even out. Home care kids get to preschool and kindergarten and discover all the germs they missed in their earlier years, and their rates of infections and school absenteeism are arguably higher for several years after they start school. To some extent, at least, immunity to many of the germs caught in day care apparently protects day care kids when they get to school.

Things to look for when choosing a day care center for your child (from an infectious disease doctor's perspective) can be found in chapter 9.

Schools

There's not much to add here regarding unique routes of transmission; the standard Jeopardy categories apply to schools, as well. The only reason to separate schools from the rest of the discussion is that the concentration of kids is higher, of course, and schools impart a unique seasonality to childhood infections.

By the time kids get to school age, they are hopefully past the excessive fecal-oral transmission tendencies of day care kids, although kindergarten and the early grades may be exceptions. However, older school age kids are especially good at the spread of respiratory infections. The symphony and cacophony of sneezing, sniffling, and coughing in a typical classroom drown out the poor teacher's lessons, especially since the teacher herself probably has infectious laryngitis. The droplets that are spread up to 3 feet away by every cough and sneeze land on desks and drinking fountains, lockers and lunchroom trays, basketballs and bathroom fixtures. They also end up on the kids' hands and then on their friends' hands and then in their friends' eyes, noses, and mouths. It's no surprise, then, that kids get sick within days of returning to school in the fall, returning to school after winter break, and returning to school after spring break. The well-known seasonality of many germs, in fact, may have more to do with school calendars than with the biology of the germ. The Hong Kong flu pandemic of 1968 (see chapter 3) was probably contained to just 34,000 cases in the United States because it struck this country over winter break from school, limiting the amplifier effect of the classroom. The other two flu pandemics of the 20th century were less well timed and much more extensive in their spread (see chapter 3).

Outbreaks of meningitis in schools, although rare, cause widespread concern and panic whenever they occur. The most severe such cases involve *N. meningitidis* (see chapter 3), but other bacterial and viral causes can also spread quickly in schools. Risk factors include how close your child's desk is to that of a child with the infection, because as with other germs carried in the mouth and nose, meningitis germs are spread by respiratory droplets and by direct contact.

Strategies for prevention of infection in schools are presented in chapter 9 (a preview: strategic and well-timed hand washing and up-to-date immunizations).

Petting Zoos

Awww . . . aren't they adorable? Those soft and cuddly farm animals at the local petting zoo, county fair, aviaries, and "day at the farm" exhibits teach kids important things about nature's creatures. Unfortunately, some of those creatures are microscopic, germs that city kids are unlikely to otherwise encounter. A brief perusal of "Farm animals" above will reveal all the germs of concern. The main route of transmission at animal exhibits is fecal-oral—from animal feces into your kids' oral cavities. Animals are only slightly less potty polite than kids in day care centers (above). Animals' fecal matter is on their fur, in their mouths, on the holding pens, on the feed and water buckets your kids are holding, and maybe even on the snacks served nearby.

In the decade between 1990 and 2000, investigators identified more than 2 dozen outbreaks of infection linked to petting zoos and other animal exhibits; most of the infections involved intestinal germs that cause gastroenteritis (see chapter 3) or worse (e.g., hemorrhagic gastroenteritis and hemolytic uremic syndrome [see chapter 3]). In 1996, a salmonella outbreak in Colorado was traced to a zoo exhibit of Komodo dragons. Sixty-five children were infected; although none of the kids touched the reptiles, more than 80% of them touched the wooden barrier surrounding the exhibit. The investigators estimated that more than 300 additional children among the many visitors to the Komodos may have become infected but went undiagnosed.

Other infections, ranging from ringworm to tuberculosis, have also been linked to animal exhibits; animal bites pose additional infectious (and other) risks.

Recognizing the risks posed by well-intentioned petting zoo operators, and well-intentioned parents hoping to teach kids about animals that don't ride in elevators, the National Association of State Public Health Veterinarians, Inc., issued guidelines in 2005 for animal exhibitors. From those guidelines, which are endorsed by the CDC, important infection safety tips for parents visiting petting zoos with their kids can be gleaned, and those safety tips are summarized in chapter 9. As a preview, this sentence is from the conclusions of the National Association of State Public Health Veterinarians report:

> The recommendation to wash hands is the single most important prevention step for reducing the risk for disease transmission. (http://www.cdc.gov/mmwr/preview/mmwrhtml/rr5404a1.htm)

As proof of this point, during the aforementioned Komodo dragon outbreak, investigators compared the sick kids with kids who visited the exhibit but didn't get infected. Hand washing after visiting the exhibit was the most important determinant of which kids got infected and which didn't. Numerous other petting zoo outbreak investigations have come to the same conclusion.

Doctors' Offices

Sixty years ago, astute pediatricians observed a correlation between the development of roseola, a rash illness of children (see chapters 1 and 3), and a recent prior visit to the same doctors' offices. Concluding that this disease may be spread from child to child, for example, among those waiting to see the doctors, those astute pediatricians subsequently proved that a transmissible germ was responsible. It would take another 3 decades before human herpesvirus 6 was identified as the cause of roseola (see chapter 1), but the concept of the waiting room contagion was born.

The infectious hazards of the waiting room at your child's doctor's office are just like those of a day care center, but *worse*—this is the equivalent of a short-stay day care center (although sometimes the wait isn't so short) that specializes in sick kids with contagious germs. Transmission of germs from child to child occurs by direct contact, droplets, the airborne route (for only the few germs noted above), and the indirect route via inanimate objects (also as described above). Many doctors' offices have separate waiting areas for kids who are sick and those who are there for routine well-child visits; that seems like a good idea, but the truth is that there's a paucity of evidence that separation of the two populations of kids reduces the spread of germs in waiting rooms. Perhaps that's because even in the well-child area, kids are shedding germs without feeling ill. In addition to separating sick from well kids, there are many other behind-the-scenes steps that doctors take to protect kids in the office, from careful cleaning to properly discarding needles and other sharp objects. Chapter 9 reviews the steps you can take to protect your child from the other children in the doctor's office waiting room.

College Dormitories

Dormitories are day care centers for big kids. In addition to all of the germs described for day care centers, add sexual transmission (see chapter 3) to the list of routes for catching germs from other people. Furthermore, several infections seem uniquely prevalent among college kids, including mumps (see chapter 3). A national epidemic of this previously vaccine-contained infection originated among college students and was amplified by them (see chapter 3). Between 1 January and 14 October 2006, nearly 6,000 cases of mumps occurred in the United States. Most of those infected were college age, and many were college students. *N. meningitidis* infections (including meningitis and meningococcemia [see chapter 3]) also cluster in dormitories. For that reason, the new meningococcal vaccine (see chapter 7) was released first for college students; it has now been recommended for all children. Tips for disease prevention in college dormitories can be found in chapter 9.

This concludes your orientation to germs and their travel plans. Now, proceed with caution to chapter 3, a compendium of the actual diseases caused by the germs you've just read about. However, before you panic over how long the next chapter is and how many threats to your kids' health are contained therein, glance ahead at the cumulative lengths of chapters 4 to 11—those are the remedies!

3

Plagued: a Glossary of Diseases

Now that we know the names and categories of germs and the routes they take to get in and out of us, it's time to press formal charges against the germs for the diseases they cause. In this chapter, we'll review those infections that are most vexing to you as parents, either because of the disruptive frequency of these illnesses in your children's lives or because of the abject fear of the diseases engendered by the cover stories in *Newsweek* and *Time* magazines.

We'll also take a look back and a look forward. A review of the once-dreaded germs that have been eradicated or are on their way toward eradication will remind you of the progress we made in our war on germs during the 20th century and the further strides already made in the 21st century. Finally, we will explore the mysterious and elusive relationship between germs and chronic illnesses for which the true cause is still unclear—these are maladies that may or may not be due to germs, conditions that offer a preview of where tomorrow's great discoveries may occur.

The concise disease vignettes in this chapter set the stage for, and reference, the detailed discussions of strategies for prevention and treatment in subsequent chapters. In this chapter, diseases are briefly summarized, followed by a bulleted-list description of contagiousness and proven (*proven!!*) strategies for prevention and treatment of each disease; unproven strategies, unfounded claims, unsubstantiated data, and unsuccessful interventions are dismantled in chapters 9 to 11. The germs mentioned in this chapter's disease vignettes are put into context within the germ world in chapter 1—you may well find yourself flipping back and forth between the diseases discussed here and the germs catalogued in chapter 1 to see how it all fits together.

In order to cover the most important diseases your kids may get, as well as those they likely will never get (but that you'll still hear enough about to worry yourself sick), the sections of this chapter are somewhat arbitrarily divided as follows:

- Parents' dirty dozen most *hated* infectious diseases
- Parents' baker's dozen most *feared* infectious diseases
- Parents' half-dozen *former* most dreaded infectious diseases that are now "so last year"

- Grandparents' half-dozen (plus one) *former* most dreaded infectious diseases that are now "so last generation"
- Dr. Rotbart's half-dozen *chronic* diseases for which germs may someday be found
- A half-dozen leftovers

A word of caution is in order before you begin reading these vignettes. Many of the clinical descriptions will sound something like this: "The disease typically begins with fever and flu-like symptoms." It's a problem for parents and doctors alike that countless diseases start out looking the same, and it's often tough to distinguish the banal cold or flu from much more serious diseases early in the course of illness. It's important for you not to read too much into your child's fever and flu-like symptoms as a result of your reading the pages to follow; common things are common (see chapter 12), and your child probably has just a minor and fleeting illness. However, it's also important not to dismiss seemingly harmless symptoms if they don't follow the expected course of quick resolution or if new and unusual features develop.

Parents' Dirty Dozen Most *Hated* Infectious Diseases

Fever

Fever is the single most common reason for kids' sick visits to their doctors. The typical child may have as many as 10 fever illnesses each year in the first years of life; that's known as physician job security. How many infections in the first years of life are too many and should be cause for concern? See chapter 4 for the answer to that common question.

Most fevers are due to infections, and most infections are associated with fevers. Fever may accompany all of the syndromes discussed below—colds, ear infections, the flu, stomach flu, etc. However, when fever is the dominant symptom of a child's illness, perhaps along with trivial symptoms, like a runny nose, a slight cough, or a nondistinctive rash, the cause of the fever is a virus at least 97% of the time. Because of the 3% or less chance of a bacterial disease, particularly in the youngest infants, fever generates substantial attention and activity on the part of parents and physicians alike.

What constitutes a fever? A fever is defined as a temperature elevation greater than or equal to 100.4°F (38°C) measured rectally; other methods of taking body temperature are less accurate. Temperature elevations of less than that are of less concern for serious infection (with the possible exception of babies younger than 2 months of age, for whom even lesser temperature elevations may herald serious infection). There is also a trend toward a higher rate of serious infections (usually bacterial) with very high fevers (greater than or equal to 104.0°F [40°C]); still, 80 to 90% of kids with even very high fevers have benign viral illnesses that resolve on their own.

Fevers are indicators of underlying infection, but fever in and of itself is not harmful and is probably beneficial in helping the body fight infection.

Because kids with fevers are miserable, we tend to use antipyretic medicines, like acetaminophen and ibuprofen (see chapter 8), to bring down elevated temperatures and make kids feel better. The theoretical benefit of fever in helping to fight infection is not so great that it warrants resisting treatments that can make kids less miserable. If your child has fever, and unless otherwise instructed by a doctor, it's okay to bring the fever down.

When a baby under 2 months of age develops a fever, most physicians automatically hospitalize the infant and treat him or her with intravenous or intramuscular antibiotics until bacterial infection can be ruled out by a series of tests on the blood, urine, and spinal fluid. Fever is a "hospitalizable offense" in these very young babies because bacterial infection in general can be more serious and progress more rapidly than in older children. The specific bacteria that cause infections in babies younger than 3 months of age can be more dangerous than the bacteria that cause infections in older kids, and younger babies contain infection less well.

When an older infant or child, between the ages of 2 months and 2 years, develops a fever without clear cause (e.g., without an obvious ear infection), some physicians will do testing—blood and sometimes urine or a spinal tap—to help determine the likelihood of bacterial infection; other physicians will rely on their clinical experience and observation skills to distinguish between benign viral illnesses and more worrisome bacterial infections. If testing is done and raises concerns, antibiotic treatment and/or hospitalization will be considered. Beyond 2 years of age, the clinical appearance of the child is the most important determinant of the physician's actions. Because such a high percentage of fevers are caused by benign viruses that resolve on their own, few children older than 2 years need aggressive diagnosis or specific therapy. Antibiotics don't treat viral infections (see chapter 5) and therefore should not be given to the vast majority of kids with fevers.

How contagious is it?

- Fever itself is not contagious, but many of the germs that cause it are.

Prevention

- Personal and household hygiene (see chapter 9)

Treatment

- Tepid bath, cool washcloth
- Acetaminophen or ibuprofen (see chapter 8)

Common Cold

The cold is caused by more than 200 different viruses, guaranteeing that no matter how many colds we get in our lifetime, we will never become "immune" to catching a cold. It is true, however, that the frequency of colds goes down with age: children have many more colds than adults, and young adults have more colds than seniors. Although sometimes associated with fever, most common colds are dominated by signs of upper respiratory tract involvement—runny nose and eyes, sore throat, and cough.

The common cold is society's greatest medical nuisance, resulting in 23 million days of school missed each year, 25 million days of work missed, 84 million doctor visits, and nearly $4 billion spent on over-the-counter remedies (see chapter 8), all in the United States alone. Additionally, more than half of all common-cold cases presenting to doctors result in antibiotic prescriptions—remember, antibiotics don't treat viruses and shouldn't be given for the common cold.

How contagious is it?

- Very contagious, by personal contact and by contact with inanimate objects contaminated by an infected person (see chapter 2)

Prevention

- Personal and household hygiene (see chapter 9)
- Staying warm and dry (see chapter 11)
- Breast (versus bottle) feeding (see chapter 10)
- Vitamin C (see chapter 10)

Treatment

- Fluids and nutrition (see chapter 10)
- Rest (see chapter 11)

Ear Infections

Ears may become infected behind the eardrum (otitis media) or in the outer canal of the ear (otitis externa). Each of the two types of ear infections arises by a different route, is caused by different germs, and is treated with different kinds of medicines.

Otitis Externa. Also known as "swimmer's ear," otitis externa is caused by a breakdown in the skin lining the outer ear canal; frequent swimming or other moisture in the ear causes the skin to chafe and allows bacteria and fungi to set up shop. Overenthusiastic cotton swabbing (e.g., to remove wax) can also damage the skin of the ear canal. Symptoms include drainage from the ear canal, local pain, and itchiness. Otitis externa is not a dangerous condition.

How contagious is it?

- Not contagious

Prevention

- Dry ear canals (gently, with a towel and by positional drainage) after swimming and bathing.
- Avoid putting objects into the ear canal (including aggressive use of cotton swabs).

Treatment

- Acetaminophen or ibuprofen for pain or fever (see chapter 8)
- Topical (drops in the canal) antibiotics (see chapter 5)
- Oral antibiotics (rarely needed; only in severe cases) (see chapter 5)

Otitis Media. While otitis externa is a nuisance, it is far less common and far less bothersome to kids and parents than otitis media, or middle ear infections. By the time kids reach 3 years of age, most have had at least one, and many kids have had more than one, middle ear infection. The middle ear space is protected from outside bacteria and viruses by the eardrum, but it is not protected from germs that ascend through the eustachian tube, that depressurizing canal that connects our mouth and nose cavities to our middle ear space, allowing us, for example, to pop our ears by yawning. The eustachian tube is shorter, straighter, and more prone to blockage in younger children than in older kids, explaining the predilection of the youngest for ear infections. Bacteria and viruses that live in our kids' noses and mouths creep along the eustachian tube and, if conditions in the middle ear are favorable, cause infection and inflammation. Favorable conditions for infection include fluid buildup and congestion from concomitant viral infection or allergy (e.g., hay fever) and residual fluid from past ear infections; household cigarette smoke also promotes ear infections. Bottle-fed babies are more likely to get otitis media than those who are breast-fed for two reasons: the anatomic differences in sucking and swallowing with the two approaches and the immunologic benefits of breast milk (see chapter 10).

Middle ear infections are painful for babies and children and are often associated with fever and nausea or vomiting. Whereas most middle ear infections are benign and resolve with oral antibiotics (and sometimes with no therapy at all), some children develop complications of these episodes. Rare complications include more serious infections, such as bacteremia and meningitis (see "Parents' Baker's Dozen Most *Feared* Infectious Diseases" below). There has been concern for many years that hearing and speech impairment may result from recurrent middle ear infections and persistent fluid behind the eardrum. It is because of this concern that certain kids have traditionally been referred to ear, nose, and throat specialists for insertion of drainage tubes in the eardrum. A 2007 study, however, followed children who had received drainage tubes and compared them to those who had not and found no long-term differences in speech and language development, calling into question one of the most common reasons for tube placement. Drainage tubes may also be recommended for some young children who have many (six or more) painful middle ear infections per year—tubes may reduce the number of infections and make those that do occur less painful. Most children can avoid the need for placement of tubes, which can be quite expensive, by treatment of each new ear infection as needed and, in some instances, prophylactic (preventive) antibiotics given daily for several months during the cold and flu season, when ear infections are also much more common (see chapter 5).

Certain bacterial causes of otitis media (e.g., pneumococcus and haemophilus) are prevented by current immunizations (see chapter 7).

How contagious is it?

- Not contagious (although the viruses that sometimes cause or facilitate otitis media are contagious and may cause colds or fevers in other kids)

Prevention

- Breast (versus bottle) feeding (see chapter 10)
- Avoiding secondhand-smoke exposure (see chapter 11)
- Daily low-dose oral antibiotic (only for kids with recurrent infections at risk for hearing loss and/or tube placement) (see chapter 5)
- Tympanostomy tubes (only for certain kids with frequent recurrent infections, chronic fluid buildup, and evidence of hearing or speech impairment)
- Routine immunizations (see chapter 7)

Treatment

- Acetaminophen or ibuprofen for pain or fever (see chapter 8)
- Watchful waiting, appropriate for many ear infections due to viruses
- Oral antibiotics, appropriate for many ear infections based on the child's age, the severity of clinical findings, and the risk for recurrences (see chapter 5)
- Tympanostomy tubes (only for certain kids with frequent recurrent infections, chronic fluid buildup, and evidence of hearing or speech impairment)

Strep Throat

In contrast to fever, common colds, and ear infections, each of which is caused by many different germs, including different strains of viruses and bacteria, strep throat is a one-germ disease caused by *Streptococcus pyogenes,* also known as group A strep. (To be completely honest, there are actually more than 100 different strains of *S. pyogenes,* but they all have the same name and modus operandi and are considered clinically as if they were all the same.)

Strep paranoia is perhaps second only to fever paranoia among parents and especially among grandparents. The basis for the anxiety rests with strep's notorious associations with rheumatic fever (an autoimmune-type [see chapter 7] attack by the body on its own heart valves and other tissues, triggered by a preceding strep infection), severe scarlet fever (see "Parents' Baker's Dozen Most *Feared* Infectious Diseases" below), and kidney disease. Although rheumatic fever and severe scarlet fever still occur, and there have been brief resurgences of those diseases in recent years, they are no longer the menace that they were during our kids' grandparents' era. It is hypothesized that the germ has changed over time to make it less likely to cause those complications, although the true reason for the waxing and waning seen with many infectious diseases is unknown. More recently, strep has again jumped into the headlines as a cause of "flesh-eating" disease (see Parents' Baker's Dozen Most *Feared* Infectious Diseases below).

Strep throat has created its own cottage industry of doctor visits, throat cultures, high- and low-tech laboratory tests, antibiotic prescriptions, and school nurse employment opportunities. Untreated strep throat gets better on its own but takes several days longer to resolve than if treated with antibiotics. The risk of rheumatic fever and kidney problems is virtually eliminated by antibiotic treatment during the acute episode of strep throat; as a result, treatment of *proven* strep throat is warranted.

The symptoms of severe sore throat are accompanied by a characteristic appearance in the throat which many doctors feel confident in diagnosing even without testing; the availability of rapid (in minutes) testing in the doctor's office, however, means that most kids with sore throats are tested regardless of what their throats look like to the doctor. Therein lies a big part of strep's vexation for doctors and parents—strep can remain on the throats, tonsils, and adenoids of kids long after the actual infection is gone; the germ establishes a carrier, or colonization, state (see chapter 1), living harmlessly in low numbers. But, when the next *viral* infection occurs and causes a viral sore throat, the test in the doctor's office is still positive for strep, a residual of the child's previous true strep infection. Hence, finding strep doesn't necessarily mean your child has strep throat. To bypass this conundrum, doctors will frequently ask kids with "recurrent strep throat" to come in for testing when they feel good. If the strep is still there, confirming colonization, they may give your kids special medicines to try to eradicate the colonization, or carriage, of strep and prevent future overdiagnosis of strep throat.

One clue that a positive strep test reflects carriage and not actual infection is if your child's sore throat comes with runny nose, cough, and other cold symptoms; strep throat usually lacks these more common signs of a viral cold. Another clue is that strep throat is less common in kids prior to school age, although younger kids may catch it in day care.

How contagious is it?

- Moderately contagious, by personal contact and by contact with inanimate objects contaminated by an infected person (see chapter 2) (although it used to be fashionable to blame the pet dog for harboring this germ, Fido has now been vindicated)

Prevention

- Personal and household hygiene (see chapter 9)

Treatment

- Fluids and nutrition
- Acetaminophen or ibuprofen for pain or fever (see chapter 8)
- Oral antibiotics or a one-time intramuscular antibiotic injection (see chapter 5)

Pinkeye (Conjunctivitis)

As the name suggests, the primary symptom and sign of pinkeye is redness of the conjunctiva, the thin mucous-membrane covering of the eye. The redness is due to inflammation and is usually accompanied by pain, itching, and drainage. Infectious conjunctivitis is caused by both viruses and bacteria. Most of the viruses on the long list that cause the common cold can also cause pinkeye, either with or without the symptoms of the common cold; bacterial causes include staphylococcus, streptococcus, and haemophilus, usually without signs of a cold or other problems.

Less common bacterial causes of pinkeye include those germs that cause sexually transmitted diseases (STDs) (see "Parents' Baker's Dozen Most *Feared* Infectious Diseases" below), like gonorrhea and chlamydia, both of which are of concern for eye infections, primarily in babies born to infected mothers; for that reason, all newborns are given antibiotic drops at birth in case their mothers have those germs. Ironically, the eye drops given to babies occasionally cause a chemical irritation of the eyes, a chemical conjunctivitis.

Pinkeye can also result from fungal or bacterial contamination of contact lens solutions; these can become severe and involve deeper layers of the eye, as seen in recent outbreaks that caused infections in dozens of states.

Distinguishing between bacterial and viral causes of pinkeye guides therapy; viral infections resolve on their own, whereas bacterial infections require antibiotics to prevent extension of the infection to other tissues and other complications. Adding to the confusion, noninfectious causes of conjunctivitis, like allergic reactions, can cause pinkeye, as can chemical irritants that get into the eye. Blocked tear ducts also cause conjunctivitis-like signs and symptoms, particularly in young infants.

There are clues that help your child's doctor decide which is the most likely cause of pinkeye. Recent colds in your family or in your kids suggest viral conjunctivitis, as does a clear and watery discharge from the eye. A thicker, more pus-like discharge is more typical of bacteria, usually without cold symptoms. Allergic conjunctivitis is the itchiest of the bunch and often occurs in kids with other allergies, like hay fever; the usual triggers of other allergies also trigger allergic conjunctivitis in kids who are prone to it. Whereas infectious and allergic conjunctivitis often affect both eyes, a chemical irritant in an older child or a clogged tear duct in a baby tends to affect one eye only; infectious conjunctivitis often starts in one eye and is spread by your child to the other by rubbing. Blocked tear ducts are uncommon causes of conjunctivitis after the newborn period. Time is often the best clue; when the infection lasts longer than a few days, bacterial disease is more likely and antibiotic treatment is more likely to be necessary.

How contagious is it?

- Very (if caused by an infection; noninfectious conjunctivitis is not contagious), by personal contact and by contact with inanimate objects contaminated by an infected person (see chapter 2)

Prevention

- Personal and household hygiene (see chapter 9)
- Topical antibiotics (given to all babies at birth) (see chapter 5)

Treatment

- Topical eye washes (sterile saline) for symptom relief
- Acetaminophen or ibuprofen for pain (see chapter 8)
- Topical antibiotics for bacterial infection (see chapter 5)

The Flu

The three types of influenza viruses (A, B, and C) cause most cases of illness diagnosed as the flu, but other respiratory viruses can also cause flu-like symptoms (see chapter 1). Influenza A virus is the most devious of the strains of flu virus because of its ability to mutate quickly and dramatically from year to year and escape both our natural immune defenses and our vaccines; influenza B virus has a less pronounced mutation pattern but also can keep one step ahead of our defenses. The chameleon-like proclivity of influenza A virus for big changes in its surface proteins makes it the only one of the three flu viruses to cause pandemics, the worldwide spread of an entirely new strain of flu virus, resulting in high rates of severe illness and death. This pandemic form of influenza is discussed more completely below under "Bird Flu" (in "Parents' Baker's Dozen Most *Feared* Infectious Diseases").

The attack rate of flu, like many infections, is highest in kids. In fact, kids are such a good target for the virus that one of the best indicators of a community-wide flu epidemic is an overall school absentee rate of 10% or higher. The symptoms of flu are akin to those of a high-octane common cold. The sudden onset of fever, chills, muscle aches, fatigue, and headache is followed by the more common-cold-like signs and symptoms: runny nose, cough, and sore throat. This usually results in several days or more of school or work missed.

Although most cases of the flu resolve uneventfully, complications can be severe and include a sepsis-like syndrome (see "Parents' Baker's Dozen Most *Feared* Infectious Diseases" below) in babies; nervous system involvement, such as meningitis and encephalitis (see "Parents' Baker's Dozen Most *Feared* Infectious Diseases" below); cardiac involvement; and a secondary or confounding infection caused by bacteria. The last is due to the weakening of both surface and whole-body immune barriers (see chapter 4) by the influenza virus.

The chances of getting the flu, and especially of getting a severe case, can be reduced 70 to 90% by receiving the influenza vaccine (see chapter 7). The vaccine, given by injection or by a mist into the nose, prevents many cases of influenza A and B, but it must be repeated each year because of the aforementioned mutations of the viruses. In 2006, the Advisory Committee on Immunization Practices of the Centers for Disease Control and Prevention (CDC) expanded the recommendations for flu vaccine to include all children between the ages of 6 months and 5 years, along with their household contacts and caregivers (see chapter 7).

Several specific drugs are available to treat the flu viruses and prevent household spread (see chapter 6), but they are only moderately effective (reducing the duration of the disease by about a day) and have side effects.

How contagious is it?

- Very contagious, by personal contact and by contact with inanimate objects contaminated by an infected person (see chapter 2)

Prevention

- Personal and household hygiene (see chapter 9)
- Vaccine (see chapter 7)
- Antiviral drugs (see chapter 6)

Treatment

- Rest (see chapter 11)
- Fluids and nutrition (see chapter 10)
- Acetaminophen or ibuprofen for fever and muscle aches (see chapter 8)
- Aspirin and aspirin-containing products should *never* be given (see "Reye's Syndrome" in "Parents' Half-Dozen *Former* Most Dreaded Infectious Diseases That Are Now 'So Last Year'" below and chapter 8)
- Antiviral drugs (see chapter 6)
- Antibiotics for secondary bacterial infections (see chapter 5)

Stomach Flu (Gastroenteritis) and Traveler's Diarrhea

Most cases of stomach flu are caused by viruses in the rotavirus, norovirus, astrovirus, and adenovirus groups; rotavirus is the single leading cause. However, a number of bacteria can also cause gastrointestinal distress, including campylobacter, salmonella, shigella, and certain strains of *Escherichia coli* (as in traveler's diarrhea, discussed below; also see "Parents' Baker's Dozen Most *Feared* Infectious Diseases" below). Parasites, including giardiae and cryptosporidia, can also cause gastroenteritis; these domestic parasites (see chapter 1) are transmitted by contaminated water and are also passed on from child to child. Distinguishing among viral, bacterial, and parasitic causes can be difficult and may be important in what your kids' doctors decide about treatment. Bloody or mucusy diarrhea is more likely to be bacterial, which also lasts longer than the typical viral stomach flu. When symptoms persist for longer than a few days, or if symptoms are severe, laboratory testing of your child's feces may be required to determine the responsible germ and to direct therapy.

Symptoms include vomiting and/or diarrhea and abdominal cramps; fever may also be present, and depending on the severity of the fluid loss, dehydration may complicate the stomach flu. Sporadic cases of gastroenteritis are common, but outbreaks involving many children are frequently reported, as well—associated with food or water contamination or simply because of frequent fecal-oral contact among young children (see chapter 2). Certain bacterial and parasitic causes of gastroenteritis can be contracted from pets (see chapter 2). Day care centers, where kids share toys and so much more, are frequently implicated in outbreaks of all germs that cause gastroenteritis (see chapter 2).

Traveler's diarrhea is probably best described as a somewhat nastier version of gastroenteritis that you and your kids get when you're in another country or shortly after you've returned home. It is typically caused by *E. coli.* The usual strains of *E. coli* that colonize the gut are generally helpful (preventing the overgrowth of more dangerous germs [see chapters 1, 5, and 10]). However, many of the *E. coli* strains that cause traveler's diarrhea and some cases of domestic gastroenteritis have special properties that make them more virulent, including the ability to stick tightly to the intestinal wall and/or to produce toxins that exacerbate the diarrhea and stomach cramps; an even more severe variety of *E. coli* toxin-caused disease (hemolytic uremic syndrome [HUS]) is discussed in "Parents' Baker's Dozen Most *Feared* Infectious Diseases" below.

In most individual cases, the exact cause of traveler's diarrhea is never established (although research studies have identified *E. coli* as the most likely culprit) because there is no need to name the germ in order to treat the patient. Antibiotics taken before departure are effective in reducing the occurrence of traveler's diarrhea, but they are *not* recommended because of the side effects and the potential for overuse of these medicines (see chapter 5).

Most cases of stomach flu and traveler's diarrhea resolve on their own in a few days, and complications are rare—the most common, dehydration, is prevented by attention to fluids and minerals and avoidance of lactose-containing products during the acute infection.

How contagious is it?

- Moderately contagious, by personal contact (requires fecal-oral contamination [see chapter 2]) and by contact with inanimate objects contaminated by an infected person. Bacterial and parasitic infections can also occur by food and water contamination (see chapter 2).

Prevention

- Personal and household hygiene (see chapter 9)
- Traveler's precautions, including consuming only bottled water, avoiding ice and raw fruits and vegetables, and eating foods that have been cooked to a high temperature while they are still hot
- Breast (versus bottle) feeding (see chapter 10)
- Vaccine (rotavirus protection [see chapter 7])

Treatment

- Fluids and nutrition (see chapter 10)
- Rehydration packets for traveler's diarrhea
- Avoiding lactose-containing products during acute illness

- Oral antibiotics or antiparasitics, but only if the illness is due to certain bacteria or parasites, respectively (see chapters 5 and 6). More severe cases of traveler's diarrhea may also require a brief course of oral antibiotics.

Head Lice

The very thought of this creepy, creeping insect infesting your child's head is enough to make most parents squirm and scratch in sympathy—and then blush with embarrassment. Despite the stigma and their reputation, though, these common bugs are not a sign of poor hygiene or of poor parenting. (Incidentally, *body* lice are very different from head lice. Body lice are rare in kids, do result from desperately poor hygiene, and do spread disease. Unless your kids have an exceptional exposure history, they don't have to worry about this bodily cousin of head lice.)

Many kids with head lice are without symptoms; itching is the most common symptom in the others. Head lice aren't dangerous, and they don't spread other diseases (unlike the body lice mentioned above). Occasionally, scratching the scalp can cause secondary bacterial infections, which require antibiotics.

Head lice can be seen with the naked eye as nits (little seedlike eggs that are affixed to hair follicles), but dandruff, normal scalp cells, and other debris on hair shafts can look similar; these other objects comb and brush off more easily than the very clingy nits. Head lice are spread from one head to another or from something in contact with an infested head to another head. Reinfestation of the same child after an initial cure also occurs commonly because of contact with another person who has lice, contact with headgear or linens that are still contaminated, or simply residual bugs that went unnoticed and uneradicated by the initial therapy. This is one of the few infections that children can give to themselves, i.e., they can reinfect themselves by reexposure to contaminated hats, combs, linens, etc. Immunity to head lice does not appear to develop. Pets have no role in harboring or spreading this infection, nor can they catch it from your kids. Head lice survive by taking tiny blood meals from your child's scalp. Nice.

A cultlike following has developed around numerous homemade louse remedies that are designed to smother (mayonnaise, Vaseline, and cooking oil) or dry (hot-air blowers) the lice to death. None of these has been subject to careful, controlled study, but anecdotes of miraculous cures abound. One of the most popular homemade approaches can be viewed at www.headliceinfo.com. A 2006 study of 169 kids with lice compared six different hot-air methods of killing lice and provided useful information, at least for the microbiologists in the crowd. None of the hot air dryers or devices got rid of *all* the little bugs; most of the devices killed most of the eggs, but the already hatched, viable lice were harder to kill. Of course, there are commercial hot-air products that are marketed for treatment of head lice. The one that performed the best in the aforementioned study is also the one that some of the authors have a financial interest in.

Fortunately, except for the social ostracism associated with making your kids' heads smell like salad, most of the recipes for moms' miracles are harmless to try and, if anecdotes can be believed, may be effective. However, using hot air can scald the scalp and can also *ignite* some of the proven antiparasite drugs your kids' doctor may recommend, so hot air should not be used at the same time as antiparasite treatments prescribed by your child's doctor.

How contagious is it?

- Very contagious, by personal contact and by contact with inanimate objects contaminated by an infested person (see chapter 2); reinfestation can also occur in the same child through contaminated inanimate objects.

Prevention

- Personal and household hygiene (see chapter 9)
- Insecticide medicines for household contacts and bedmates (see chapter 6)
- Do not attend school or day care until treatment has been received.

Treatment

- Over-the-counter insecticide medicine (resistance of the germ to this treatment is common [see chapter 6])
- Insecticide medicine prescribed by your child's doctor; these medicines pose risks of adverse effects in kids, especially those under the age of 2 years—doctor's instructions must be carefully followed (see chapter 6).
- Nitpicking (where did you think the word came from?). The removal of nits, by hand or with a nit comb, after treatment with an insecticide medicine improves the chances for successful treatment because some of the little eggs may have resisted therapy.

Skin (Plantar) Warts

Warts are caused by papillomaviruses. Skin warts are benign but a nuisance; those that occur on the bottoms of your kids' feet (plantar warts) can be very uncomfortable. The hands, especially the fingers, are the other most common location for skin warts, and those are usually painless. Flat warts are another variety of skin warts, and they typically occur on kids' faces but can also appear on arms and legs. (Incidentally, papillomaviruses also cause genital warts, a sexually transmitted infection; the strains that cause common skin warts tend to be different from the strains that cause genital disease [see "Parents' Baker's Dozen Most *Feared* Infectious Diseases" below].)

Skin warts are common in kids and are acquired through small breaks in the skin and subsequent close contact with someone else who has warts. Recurrences are common, and kids can spread warts from one part of the body to another by picking at them. Warts usually disappear on their own after several months, but they may last years. Treatment is appropriate for warts that don't resolve on their own and/or are painful because of their location (usually those on the feet). Prevention of spreading from one part of the body to another is also a reason to treat. Treatment is directed at the physical destruction and removal of the warts.

How contagious is it?

- Moderately contagious, by personal contact and by contact with inanimate objects contaminated by an infected person (see chapter 2)

Prevention

- Personal and household hygiene (see chapter 9)
- Wearing sandals in public showers

Treatment

- None necessary for many warts
- Duct tape (see chapter 8)
- Liquid nitrogen freezing in the doctor's office or at home
- Chemical treatments in the doctor's office
- Over-the-counter salicylic acid products (see chapter 8)

Mononucleosis

Most cases of mononucleosis are due to the Epstein-Barr virus (EBV), a member of the herpesvirus family; other cases are due to cytomegalovirus (CMV), also a herpesvirus. Still other cases can't be linked to either of these, and other viruses may be responsible. While infections with EBV and CMV occur in young children, they are usually minimally symptomatic in that age group and are often undiagnosed or thought to be just another viral infection. In contrast, infection with EBV in teens is much more likely to cause the "mono" (mononucleosis) syndrome: fatigue, fever, muscle aches, sore throat with swollen tonsils, swollen glands, and enlargement of the liver and spleen. A typical case may start out very much like the flu. Complications of this infection, while uncommon in otherwise healthy kids, may be serious and may include involvement of the brain and nervous system and rupture of the enlarged spleen. Because of this latter complication, contact sports are forbidden during the acute phase of mono. Fatigue may last for several weeks or even months; this is in contradistinction to chronic fatigue syndrome (CFS), which may last years and is not proven to be associated with EBV or any other single germ (see "Dr. Rotbart's Half-Dozen *Chronic* Diseases for Which Germs May *Someday* Be Found" below).

When mono is diagnosed in teens, it always gets raised eyebrows because of its fame as the "kissing disease." Indeed, mono is transmitted by sharing secretions, a typical route of spread for all members of the herpesvirus family (see chapter 2). Saliva contact by any route, including sharing a straw or a water bottle, is enough to spread the infection. Sharing is bad (see chapter 9).

How contagious is it?

- Moderately contagious, by personal contact (requires sharing respiratory or oral secretions) or by contact with inanimate objects contaminated by an infected person's respiratory secretions (see chapter 2)

Prevention

- Personal and household hygiene (see chapter 9)

Treatment

- Rest (see chapter 11)
- Fluids and nutrition (see chapter 10)
- Avoiding contact sports (until the spleen is normal in size)
- Steroid medicines—only for severe tonsil swelling or other rare complications

Chicken Pox

It was tough for me to decide whether to include chicken pox in this section or in "Parents' Half-Dozen *Former* Most Dreaded Infectious Diseases That Are Now 'So Last Year'" below, with other infections that are on the way out. The introduction of chicken pox vaccine several years ago has dramatically reduced the incidence of the infection, but reluctance by some parents to give their kids the vaccine has slowed the disease's disappearance (see chapter 7). Indeed, parents' nonchalance about this disease has been known to lead them to let their kids participate in "chicken pox parties," where kids are intentionally exposed to friends with the infection to "get it over with." However, besides the common and benign rash illness, chicken pox can pack a punch, with complications that include superimposed bacterial infections, encephalitis (see "Parents' Baker's Dozen Most *Feared* Infectious Diseases" below and chapter 7), and pneumonia.

Chicken pox is caused by the varicella-zoster virus, another member of the herpesvirus family. The symptoms of chicken pox include fever and the often prominent, itchy rash that starts out as small red spots, progressing to small blisters (vesicles). By the time the disease runs its course over about a week, there may be hundreds of these small blisters on the child's skin and mucous membranes, e.g., inside the mouth. The vesicles are filled with live virus; once they all crust over, they are no longer infectious. The disease is more serious and severe in adults, in whom pneumonia is one of the important complications. Like other herpesviruses, once varicella-zoster virus infects your child's body, it remains forever in a dormant state. In adults, the dormant virus may awaken because of stress (see chapter 11), old age, or some other cause of a weakened immune system (see chapter 4). The result is grandmother's shingles, a painful and often debilitating disease that localizes to a single nerve root where the virus set up residence after childhood infection.

How contagious is it?

- Very contagious, by personal contact and by airborne spread (see chapter 2). Exposure to the rash of a patient with shingles can cause chicken pox in patients who have not had the infection previously.

Prevention

- Personal and household hygiene, although this infection is almost impossible to stop from spreading among household members because of its airborne mechanism (see chapters 2 and 9)
- Vaccine (part of routine childhood immunizations; can also be given to kids older than 1 year of age who have not previously had chicken pox following exposure in certain circumstances [see chapter 7])
- Chicken pox gamma globulin (following exposure in at-risk patients, such as those with abnormal immune systems who have not previously had chicken pox, newborns whose mothers have chicken pox immediately before or after delivery, and premature infants [see chapter 7])
- Antiviral medicine (following exposure in at-risk patients, such as those with abnormal immune systems who have not previously had chicken pox, newborns whose mothers have chicken pox immediately before or after delivery, and premature infants [see chapter 6])

Treatment

- Acetaminophen for fever (see chapter 8); *avoid* ibuprofen until further information is available regarding the possible association of ibuprofen with "flesh-eating" streptococcus infections (necrotizing fasciitis) that can complicate chicken pox in children (see "Parents' Baker's Dozen Most *Feared* Infectious Diseases" below).
- Aspirin and aspirin-containing products should *never* be given (see "Reye's Syndrome" in "Parents' Half-Dozen *Former* Most Dreaded Infectious Diseases That Are Now 'So Last Year'" below and chapter 8).
- Colloidal oatmeal baths for itching
- Antihistamines for severe itching (see chapter 8)
- Antiviral medicine (for adults, adolescents, and certain higher-risk children [see chapter 6])

Urinary Tract Infections (UTIs)

The urinary tract consists of the kidneys, the bladder, the ureters (tubes leading from the kidneys to the bladder), and the urethra (the tube leading from the bladder out of the body). Bladder infections are more common and less severe than kidney infections. Symptoms of urinary tract infections (UTIs) are age dependent. The youngest infants may have only fever and fussiness, occasionally with vomiting or diarrhea. Older children and adolescents may also have fever, as well as increased frequency of urination, and may have a burning pain with urination. Kidney infections can cause shaking chills and may even progress to sepsis (see "Parents' Baker's Dozen Most *Feared* Infectious Diseases" below).

Most UTIs are caused by bacteria that enter the urethra from the nearby anus, so most bacterial causes of UTIs are gastrointestinal germs, like *E. coli*. Viruses also occasionally cause bladder infections. Girls are more at risk for UTIs than boys because girls' urethras are shorter and closer to the anal opening; uncircumcised boys have a higher occurrence of UTI than circumcised boys. Other risk factors for UTIs include anatomic abnormalities that cause pooling of urine in the bladder or ureters and kidney. Bubble baths or other harsh soaps may inflame the urethras of young girls and make them more susceptible to UTI.

The diagnosis of UTIs depends on the ability of the child to provide a "clean" specimen, i.e., urine that has not been contaminated by bacteria on the skin. Depending on the age of the child, this can be challenging and may lead to false diagnoses because the urine collection method may be prone to catching skin germs as well as urine germs.

How contagious is it?

- Not contagious

Prevention

- Personal hygiene (see chapter 9); avoiding bubble baths
- Cranberry juice (for patients with frequent recurrent UTIs [see chapter 10])
- Oral antibiotics (for kids with frequent recurrent UTIs and/or an anatomic reason for predisposition to infection [see chapter 5])

Treatment

- Acetaminophen or ibuprofen for pain or fever (see chapter 8)
- Antibiotics, usually oral but intravenous for more severe infections (see chapter 5)
- Urinary anesthetic, a prescription medicine that can be taken by mouth for a short period to relieve severe pain or burning

Parents' Baker's Dozen Most *Feared* Infectious Diseases

AIDS

Although we speak of AIDS as being an incurable viral disease, in fact we have yet to actually "cure" any viral infection. Fortunately, many go away on their own, but the human immunodeficiency virus (HIV), which causes AIDS, is not one of them. Perhaps because of its provocative routes of acquisition by sexual or blood contact, not since the epidemics of plague and polio has an infectious disease caused more panic or paranoia.

Remarkable advances over the past 25 years have allowed us to now view AIDS much as we do other chronic (long-term) diseases—a condition that requires lifelong therapy, but without the immediate death sentence that the diagnosis once held. This change of attitude regarding AIDS has been a double-edged sword; some of the panic and paranoia are gone, but so are some of the precautions and prudence that the early epidemic engendered, particularly among adolescents and young adults.

Throughout the history of the AIDS epidemic, most cases in children in the United States and worldwide have come from exposure during pregnancy and delivery to an infected mother (see chapter 2). The second biggest at-risk pediatric population, adolescents, is at risk for all the same reasons as adults: illicit drug use and sexual contacts (see chapter 2).

The potency of HIV lies in its deviousness—the virus attacks the very immune system that would otherwise contain it. By weakening essential elements of patients' immunity, the virus not only propagates itself more effectively, but also predisposes patients to all of the other complications of weakened immunity. This results in infections by other viruses, bacteria, fungi, and parasites that could not gain a foothold with an intact immune defense system (see chapter 4). HIV infection also results in the production of chemical toxins by the body that cause malnutrition and wasting.

The symptoms and signs of AIDS reflect the mechanisms of HIV damage noted above and include swollen glands, weight loss, enlarged body organs, recurrent diarrhea, neurological impairment, developmental delay, and infections and malignancies of most body organs and systems. No wonder AIDS tops the list of parents' most feared diseases.

The origins of HIV have been hotly debated and widely researched. Genetic fingerprinting indicates that the virus probably originated among African monkeys as a simian immunodeficiency virus that then mutated and "jumped" species to infect humans. Over time, the virus adapted to humans and became more readily spread in its new target species. This mechanism of human acquisition of a "new" germ from a different species occurs uncommonly, yet it has been a consistent explanation for a number of recent "sudden emergences" of previously unknown diseases in humans, such as severe acute respiratory syndrome (SARS), West Nile virus (WNV), and mad cow disease (see "Parents' Baker's Dozen Most *Feared* Infectious Diseases" below for discussions of all three); it is feared that a bird flu outbreak among humans will follow the same path (this is also discussed in "Parents' Baker's Dozen Most *Feared* Infectious Diseases" below).

How contagious is it?

- Not very contagious, by personal contact (requires exposure to blood or other contaminated body secretions, e.g., semen, cervical secretions, and human milk) or exposure to blood-contaminated inanimate objects (e.g., needles) (see chapter 2)

Prevention

- Personal and household hygiene (see chapter 9)
- Condoms (see "STDs" in "Parents' Baker's Dozen Most *Feared* Infectious Diseases" below and chapter 11)

- Antiviral medicines given after exposure to potentially contaminated inanimate objects (e.g., needles) or people (e.g., sexual contact) (see chapter 6)
- Antiviral medicines given to infected pregnant women to prevent transmission to their babies (see chapter 6)
- Caesarian section for infected pregnant women
- Avoidance of breast-feeding by an infected mother (see chapter 2)

Treatment

- Fluids and nutrition (see chapter 10)
- Antiviral medicines (see chapter 6)
- Antibiotics, antiviral medicines, antifungal medicines, and antiparasitic medicines for infections resulting from immune suppression (see chapters 5 and 6)

Bird Flu

If influenza is like a high-octane common cold (see "Parents' Dirty Dozen Most *Hated* Infectious Diseases" above), bird flu appears to have the potential to be a high-octane influenza. In order to understand the bird flu threat, we have to take a step back and understand how influenza becomes pandemic—globally spread and more severe than typical flu.

Each new year brings a new dominant human influenza strain into the community, which is why influenza immunization is required annually (see chapter 7). The typical year-to-year variations in the structure of the virus are minor, leaving some partial immune protection for many people on the basis of their past exposure to flu virus infections or past years' immunizations. These minor variants of flu strains do not become pandemic; they readily spread from person to person, but residual immunity in the population limits the impact.

A pandemic occurs when a *major* change in the structure of the influenza virus surface occurs, usually by mutation, allowing the virus to attack an entirely nonimmune population. That is, because the new strain is almost entirely unlike past strains for which some immunity exists in the population, the potential for spread is virtually unlimited. Unlimited spread, of course, depends not just on the susceptibility of the human population, but also on the genetic ability of the virus to actually be passed from person to person. Most viruses are strictly species specific, meaning they have the genetic capacity to infect only one or a few related species. Hence, a new strain introduced into the community, for example, by exposure to animals or birds, will not become a pandemic if that strain is not readily spread from human to human (see chapter 2).

There were three influenza pandemics during the 20th century (in 1918, 1957, and 1968). The most severe of the three, the pandemic of Spanish flu (named, not because the virus originated in Spain, but because it received very wide press coverage there during a time of wartime news censorship in other countries) in 1918, resulted in worldwide deaths of 20 to 50 million people (675,000 in the United States alone), a catastrophe of unimaginable magnitude. Although the 1918 pandemic influenza strain attacked the usual high-risk groups (the very young, the very old, and those with chronic diseases), it was also remarkably potent in and lethal to young adults. One reason for the predilection for young adults was the close military quarters prevalent during World War I. Very recent research on the original 1918 strain, tested in monkeys, has also shown a more sophisticated scientific reason for the severity in previously healthy young adults: this influenza strain causes a hyperimmune response (called a cytokine storm [see chapter 4]); the patients' own healthy immune systems overreacted to the infection, flooding the lungs and other organs with inflammatory fluid. Flu patients in 1918 drowned internally.

The 1957 Asian flu pandemic, as the name suggests, did first erupt in Southeast Asia, and it caused 1 to 2 million deaths worldwide (70,000 in the United States). Its target victims were more classical; it spread among children in schools and was then brought home to especially affect elderly family members, pregnant women, and other more susceptible individuals. Early detection of the original Asian victims, effective infection control measures (see chapters 2 and 9), and vaccine technology that was unavailable in 1918 (see chapter 7) helped to limit the global impact of this pandemic, as did the likely lower virulence of this strain compared with the Spanish flu strain.

The 1968 pandemic of Hong Kong flu (700,000 worldwide fatalities; 34,000 in the United States) had the most modest impact of the three pandemics, due to improved detection, infection control, and vaccine technology, as well as a stroke of luck—the peak of the outbreak occurred during the winter school holidays, when exposure among children was lowest (see chapter 2).

Many people will remember the fear of a swine flu pandemic in 1976 that never materialized. In that year, 13 soldiers at Fort Dix, NJ, developed influenza of a strain similar to one known to infect pigs (although this outbreak was never proven to have actually originated in pigs). The infection was severe—4 of the 13 patients developed pneumonia, and 1 died—but, remarkably, there was never spread beyond the army base: no cases anywhere else in New Jersey, other states, or internationally. This small outbreak in a very limited locale, however, sparked concerns of a pandemic because of the severity of the illness and the novelty of the influenza strain that was implicated. A national immunization effort was initiated, resulting in nearly 40 million Americans (about one-quarter of the U.S. population at the time) receiving the vaccine. Complications of the vaccine, most notably a rare neurological condition called Guillain-Barré syndrome, forced cancellation of that national vaccine initiative—there were more cases of serious complications from the vaccine than there were cases of swine flu. This miscalculation of 3 decades ago has resulted in ongoing ammunition for the antivaccine movement (see chapter 7), and it illustrates the great caution that must be taken in predicting and attempting to counteract future potential pandemics.

As noted above and in chapter 2, most viruses, and that includes the influenza viruses, are species specific, able to attack only one or a few closely related species. However, with a major mutation, a virus of animal or bird origin could jump species and, if readily spread within the new species, cause widespread disease because of total absence of past exposure. That is thought to have happened with HIV, the cause of AIDS (see above). The potential mechanisms for such a species jump are twofold. First, a human virus and a virus from another species could intermingle in the same host and exchange genes (the scientific word is reassort), creating a new virus with the potency of an animal or bird influenza strain combined with the human spreadability of a human influenza strain. A second strategy for jumping species is simply a mutation within a single animal or bird strain significant enough to allow spread readily in humans.

This brings us finally to bird flu. The fear associated with bird flu has multiple derivations. First, birds were involved in all three pandemics of the past century. The 1957 and 1968 influenza strains are both thought to have been reassortant viruses, i.e., mixtures of genes from human and bird strains. More frightening, however, is the fact that the horribly malevolent 1918 strain was likely an intact bird influenza strain that mutated itself to full human potency without the need for helper genes from human influenza. In other words, the 1918 pandemic proved that a bird strain can mutate on its own and spread like wildfire in people. Furthermore, the current strain circulating prolifically among birds in more than a dozen countries (labeled H5N1) causes a severe disease in birds, with high death rates, an apparent enhancement of the virus' potency since it was first noticed in birds in the mid- to late 1990s. Both this potent strain and the gentler precursor from the 1990s have spread to humans; to date, only a limited number of humans have contracted the bird strain, and those humans have almost all had direct contact with infected birds (usually poultry on large farms in Southeast Asia).

There are two recent and most troubling additional reasons for our bird flu phobia. First, it now appears certain that migrating wild birds can spread the disease well beyond poultry farms, national borders, and even oceanic barriers. Second, there are several well-documented cases of human-to-human spread, although so far they have been limited to very close, usually household, contact with other infected individuals who themselves contracted the virus directly from poultry.

All the ingredients are now there: historical precedent for direct mutation of bird strains resulting in human pandemic spread, widespread bird infections, increased virulence within birds of what was once a gentler H5N1 strain, a new mechanism for global spread (migrating birds), documented human disease from direct contact with birds, and finally, the first inklings of human-to-human transmission.

As of 25 July 2007, there were 319 proven human cases of H5N1 bird flu from 12 countries. This is a bad disease in humans, as it has become in birds;

192 (60%) of the humans with the infection have died. Because children often play with chickens on farms or in their villages as pets, there have been numerous children among the victims. The symptoms include both those expected with influenza (fever, upper and lower respiratory tract involvement, muscle aches, fatigue, and secondary bacterial infections) and those not usually associated with run-of-the-mill seasonal flu (vomiting; diarrhea; bleeding from the mouth and gums; nervous system involvement, including encephalitis; liver dysfunction; and severe pneumonia). High-octane influenza, indeed.

The following "now for the good news" paragraph is going to be a bit shorter. There have been no cases of bird flu in birds or in humans in the United States so far (25 July 2007); the highest concentration of cases remains in Southeast Asia, notably Indonesia and Viet Nam (each of which has reported approximately 100 human cases). At present, spread from human to human appears to be difficult for the virus and limited to intimate household contact with a very sick patient. The swine flu scare shows us that not all predicted pandemics materialize. Our vaccine technology is much more sophisticated than during past pandemics (see chapter 7), and we now have potentially effective (but not yet tested in real-world bird flu cases) antiviral medicines (see chapter 6). Additionally, we have more experience and more success with infection control methods to limit both bird and human spread. An example of the latter is the global response to the SARS epidemic of several years ago, another type of bird virus that originated in Southeast Asia and spread to humans (see "Parents' Half-Dozen *Former* Most Dreaded Infectious Diseases That Are Now 'So Last Year'" below). Chapter 9 summarizes the most current CDC interim guidelines for control of a bird flu pandemic.

How contagious is it?

- Not very contagious (yet!?), by personal contact (requires very close contact with a sick patient—or bird [see chapter 2])

Prevention

- Personal, household, and community hygiene (see chapter 9)
- Vaccines (not yet available but in development)
- Antiviral medicines (only after exposure [see chapter 6])

Treatment

- Acetaminophen or ibuprofen for fever and pain (see chapter 8)
- Aspirin and aspirin-containing products should *never* be given (see "Reye's Syndrome" in "Parents' Half-Dozen *Former* Most Dreaded Infectious Diseases That Are Now 'So Last Year'" below and chapter 8)
- Antiviral medicines (see chapter 6)
- Antibiotics for secondary bacterial infections (see chapter 5)
- Intensive care management until stable

Meningitis and Encephalitis

The brain and spinal cord are covered by three delicate membranes, collectively known as the meninges. When these coverings become infected and/or inflamed, meningitis is diagnosed; when the brain itself becomes infected, the term encephalitis (or, less commonly, cerebritis) is used. Spinal cord infection is termed myelitis. Some infections involve both the brain and its coverings (meningoencephalitis) or the brain and spinal cord (encephalomyelitis).

Meningitis and encephalitis can be fatal infections and may result in permanent brain damage in survivors—hence the understandable fear among parents—yet many patients have benign courses without long-term complications. {With the near eradication of poliovirus in most of the developed world [see "Grandparents' Half-Dozen (Plus One) *Former* Most Dreaded Infectious Diseases That Are Now 'So Last Generation'" below], myelitis is rare; poliomyelitis was the major form of "pure" spinal cord infection.} The outcomes of these infections depend on many factors, including which germ causes the infection, the age of the patient, and the parts of the brain that are involved.

Many different germs can cause meningitis and encephalitis. Viruses are the most common causes of both. Viral meningitis, although unpleasant and debilitating in the short term, is typically self-resolving and without long-term damage. The most common meningitis viruses involved are the enteroviruses, and the most common season is the summer. Herpesvirus is the most often identified cause of encephalitis, although it is responsible for only about 10% of all cases; many other cases are diagnosed clinically but never have a specific virus found. Herpes encephalitis is a bad disease. In babies, among whom the infection is most commonly acquired during delivery from infected mothers (see chapter 2), one of every seven cases is fatal; in older children, the infection is usually a reactivation of the everyday cold sore (see chapter 1 and "A Half-Dozen Leftovers" below), and the death rate for herpes encephalitis in older children and adults is approximately 25%. Among survivors of herpes encephalitis at any age, more than half are left with permanent brain damage. Other well-known viral causes of encephalitis include the arboviruses and the enteroviruses. The arboviruses are mosquito-borne infections (see chapter 2) and include the likes of West Nile virus (see "Parents' Half-Dozen *Former* Most Dreaded Infectious Diseases That Are Now 'So Last Year'" below), St. Louis encephalitis virus, California encephalitis virus, eastern equine encephalitis virus, western equine encephalitis virus, and many others; these viruses tend to be restricted to regions where the primary animal host (e.g., horses or birds) and mosquito vector are found (see chapter 2).

Bacterial meningitis gets a lot more headlines than viral meningitis even though it is far less common. The reasons for the increased awareness are twofold: bacterial meningitis is a much more severe disease, with a high death rate and a high incidence of permanent damage among survivors, and cases of bacterial meningitis can occur in fearsome clusters affecting schools and day care centers. The three most common causes of bacterial meningitis have been targeted by and, with various degrees of success, controlled by vaccines (see chapter 7). *Haemophilus influenzae* type B, once the leading cause of bacterial meningitis in children, has been virtually eradicated from the United States. *Streptococcus pneumoniae* vaccine has proven to be very effective in reducing the occurrence of meningitis due to that common germ, and further reductions can be expected as parents' compliance with vaccine recommendations improves (see chapter 7). *Neisseria meningitidis,* the cause of both meningitis and meninogoccemia, a dread form of sepsis (see below), is also now the target of a vaccine (see chapter 7). Because not all major groups of this germ are included in the vaccine, a more limited success is anticipated, but a success nonetheless. A recent school outbreak of *Mycoplasma pneumoniae* meningitis and encephalitis in Rhode Island called attention to this unusual bacterial germ. *M. pneumoniae* is a common cause of pneumonia but a very uncommon cause of clusters of meningitis cases (although the germ has been associated with sporadic cases of meningitis in the past). Newborns have their own list of bacterial causes of meningitis; group B streptococcus and *E. coli* are the most common, both usually acquired from mothers during or shortly after delivery (see chapter 2).

Bacterial meningitis at any age is associated with both a high fatality rate and a high rate of neurological complications among survivors; deafness, for example, occurs in as many as 20 to 30% of kids who survive this form of meningitis, whereas viral meningitis typically does not leave any long-term complications. The mechanisms of brain damage from meningitis, which, you recall, is infection of the coverings of the brain, include compromise of the blood circulation to the brain and swelling of the brain. Bacteria can also cause infection of the brain itself: encephalitis (or cerebritis). These infections of the brain may accompany meningitis, sometimes complicate sinus infections, and may progress to brain abscess; as with any involvement of the brain itself, the risk of permanent damage is high.

The signs and symptoms of meningitis include fever, stiff neck, difficulty with bright light (photophobia), and, in newborns, a bulging "soft spot" on the skull. Many kids with early meningitis, however, have only fever and vague symptoms, such as decreased appetite, fussiness, lethargy, "not acting like themselves," and vomiting. This can make diagnosis of meningitis very challenging, because it may, in its early stages, look very much like a flu-like illness; the younger the child, the more nonspecific the clinical findings and the more difficult the diagnosis. Because meningitis may occur with sepsis (see below), a more widespread whole-body infection, severe nonneurological signs may also be present, such as very low blood pressure (shock). Although seizures may occur with meningitis, they usually indicate concomitant encephalitis. Other signs of encephalitis include change in behavior, disorientation, coma, localized weakness or loss of sensation, and abnormalities of eye and/or facial muscle function.

How contagious is it?

- Moderately contagious, by personal contact with respiratory secretions (see chapter 2)

Prevention

- Personal, household, and community hygiene (see chapter 9)
- Caesarian section (for pregnant women with active genital herpes at the time of delivery)
- Avoiding fetal scalp monitoring in pregnant women with active genital herpes
- Antibiotics (for pregnant women at risk for group B streptococcus) (see chapter 5)
- DEET or other mosquito repellent during the summer months (see chapter 11)
- Vaccines for prevention of bacterial meningitis (see chapter 7)

Treatment

- Acetaminophen or ibuprofen for fever or pain (see chapter 8); severe pain may require narcotic medicines.
- Antiviral medicines for herpes (see chapter 6)
- Antibiotics (intravenous) for bacterial infections (see chapter 5)
- Steroid medicines are used by some doctors for certain types of bacterial meningitis in the hope of reducing the risk of deafness.
- Intensive care management until stable

Sepsis and Sepsis Syndromes (Including Toxic Shock, Severe Scarlet Fever, and Meningococcemia)

Sepsis is known by many names: blood poisoning is the common lay term; septicemia, systemic inflammatory response syndrome, and septic shock are among the medical terms. Bacteremia, viremia, and fungemia refer to the presence of bacteria, viruses, or fungi in the blood, but the mere presence of these germs in the blood is not itself a life-threatening condition; progression of the disease from circulating germs in the blood to sepsis is life threatening.

The term sepsis refers to a clinical state of whole-body infection in which the germs spread via the blood to infect, or at least affect, many organs. The severe effects of sepsis are multifactorial; they are due to the germs themselves, to toxins that the germs can release, and also to the response of the body to those germs and toxins. Hence, this is another example in which the body's own immune response (see chapter 4) to the infection can be a double-edged sword. The symptoms and signs of sepsis reflect the multiorgan nature of the infection—pneumonia, meningitis, kidney failure, liver failure, and clotting dysfunction are among the severe findings in kids with sepsis. Critically low blood pressure (septic shock) resulting from toxic effects of the germ or the immune response on blood vessel tone contributes substantially to the organ system damage.

While many types of germs can cause sepsis, bacteria are the most common. Staphylococcal and streptococcal sepsis usually originates with localized skin or mucous-membrane infections that spread to the blood, sometimes facilitated by injury or surgery. *E. coli,* klebsiella, and pseudomonas sepsis events typically begin with penetration of these normal gastrointestinal germs through the wall of the intestine into the bloodstream. Initiated by either injury or disease—for example, appendicitis—the germs escape the intestinal tract and are capable of producing very damaging toxins that cause septic shock and further organ system damage. The bacteria that cause meningitis in childhood are also causes of sepsis. With the near eradication of *H. influenzae* type B and the decreasing incidence of *S. pneumoniae,* both the result of successful childhood vaccines (see chapter 7), *N. meningiditis* remains an important cause of sepsis. This germ, like the gut bacteria noted earlier, is also a prolific toxin producer, resulting in a devastating syndrome called meningococcemia, in which shock develops and normal blood-clotting mechanisms are disrupted. Because sepsis is a whole-body infection, with many organ systems involved, meningitis frequently accompanies sepsis, as noted above.

In neonates, normal vaginal bacteria acquired in the birth canal (see chapter 2), such as group B streptococcus and *E. coli,* are the most common causes of sepsis. Viruses can cause sepsis in newborns that is indistinguishable from bacterial sepsis; among those most frequently seen (although still very rare overall) are the enteroviruses and herpesviruses. Fungi are rare causes of sepsis in otherwise healthy kids but are important considerations in patients with chronic diseases, compromised immune systems, and long-term antibiotic use.

Toxic shock syndrome and severe scarlet fever are diseases associated with staphylococcus or streptococcus infections. In these life-threatening illnesses, the germs do their damage by releasing toxins at the site of a local infection or colonization (such as the skin or vagina [see chapter 1]) that then spread to the whole body, even though the bacterium itself may remain local. An example of a local infection resulting in this type of system-wide toxin release is chicken pox, where the virus skin sores become secondarily infected with strep or staph bacteria. Although toxic shock syndrome became a household word associated with tampon use in young women, it was actually first described as a complication of infected surgical wounds in kids. Recently, toxic shock syndrome has been seen in a few kids who developed blisters after wearing a particular type of new soccer shoe. The staphylococcal or streptococcal germs set up shop in a discrete area and, because of favorable conditions for bacterial growth and toxin production in that area (e.g., a tampon that is kept in too long or is made of a material that is too absorbent), massive amounts of toxin are produced and enter the bloodstream. The effects on blood circulation to all organ systems are profound and can result in gangrene of the fingers and toes and damage to major internal organs, as well. Severe scarlet fever may begin with a strep throat (see "Parents' Dirty Dozen Most *Hated* Infectious Diseases" above) or a streptococcus skin infection. The clinical features are similar to those of toxic shock syndrome, but there is a higher rate of bacteria themselves actually getting into the blood (bacteremia) and contributing to the whole-body damage.

How contagious is it?

- Usually not contagious; sepsis develops from a previously localized infection in the same person. Although that localized infection may be contagious as a localized infection, e.g., strep throat, the development of sepsis is dependent on individual factors in the patient.
- The exception is *N. meningitidis,* which has a high likelihood of causing sepsis (meningococcemia) and/or meningitis in anyone who catches the germ from someone else; close personal contact with secretions is required (see chapter 2).

Prevention

- Personal hygiene (see chapter 9), including care of wounds and injuries and prudent use of tampons (see chapter 9)
- Antibiotics for localized infections, e.g., strep throat and skin infections (see chapter 5)
- Antibiotics for anyone with close exposure to a patient with *N. meningitidis* infection (see chapter 5)
- Antibiotics (for pregnant women at risk for group B streptococcus [see chapter 5])
- Vaccines for prevention of bacterial meningitis (which also protect against sepsis caused by the same germs [see chapter 7])
- Caesarian section (for pregnant women with active genital herpes at the time of delivery)
- Avoiding fetal scalp monitoring (for pregnant women with active genital herpes)

Treatment

- Acetaminophen or ibuprofen for fever or pain (see chapter 8); severe pain may require narcotic medicines.
- Antibiotics (intravenous) for bacterial sepsis (see chapter 5)
- Antiviral medicines for neonatal herpes sepsis (see chapter 6)
- Intensive care management until stable

Flesh-Eating Bacteria

Perpetuated by the news media to be attention grabbing, and it certainly is, this tabloid term refers to a medical condition known as necrotizing fasciitis. Necrotizing fasciitis starts out as a localized skin infection but rapidly progresses to involve and destroy the deeper layers of muscle and fascia (connective tissue) of the body.

Usually caused by streptococcal or staphylococcal bacteria, the mechanism is similar to that of toxic shock syndrome and severe scarlet fever (see above) with one important exception: the toxins produced by the bacteria in necrotizing fasciitis act locally to dissolve soft tissue rather than spreading through the blood to affect the whole body. Still, the effects can be devastating, with loss of limbs and other, often massive, tissue damage; in uncontrolled cases, death may result due to the secondary development of sepsis as bacteria gain access to the blood and beyond; the fatality rate in some studies is as high as 25%. Although strep and staph germs frequently initiate the process, other germs may also be involved at the outset or opportunistically join the infection as dead tissue increases and gangrene provides ample growth conditions for more unusual germs, like anaerobic bacteria and pseudomonas. When anaerobic bacteria, like clostridia, are involved, the metabolism of those germs and the toxins they produce may produce pockets of gas in the dying tissues; gas gangrene has a very high death rate due to the wide dissemination of toxins and their effects on blood circulation and other organ systems.

As with toxic shock syndrome and severe scarlet fever, chicken pox sores that become secondarily infected with strep or staph bacteria have been known to progress to necrotizing fasciitis; whether the complication is toxic shock syndrome or necrotizing fasciitis depends on whether the infecting germ produces locally acting or whole-body-acting toxin; fortunately, most staphylococcal and streptococcal germs that infect chicken pox sores or other wounds produce *neither* form of toxin and simply cause an easily treated local skin infection, such as impetigo (see "A Half-Dozen Leftovers" below).

There have been a few reports linking the use of ibuprofen during chicken pox with the development of necrotizing fasciitis. This association is not yet accepted widely as fact, and how such a link might develop is unknown.

How contagious is it?
- Not contagious

Prevention
- Personal hygiene (see chapter 9), including care of wounds and injuries
- Antibiotics for localized infections, e.g., skin infections (see chapter 5)
- Avoiding ibuprofen for treatment of chicken pox-associated fever or pain until further information is available

Treatment
- Acetaminophen for fever or pain (see chapter 8); severe pain may require narcotic medicines.
- Antibiotics (intravenous) (see chapter 5)
- Surgery is *always* required for removal of dead tissues.
- Intensive care management until stable

Bioterrorism

Bioterrorism involves the use of a broad category of potentially devastating germs that generally do not pose a significant public health hazard naturally but have characteristics that lend themselves to being manipulated and used as weapons in a terror campaign.

At the risk of sounding like just another alarmist, there is justification for parental paranoia regarding bioterrorism. In the event of a bioterrorism (or chemical-terrorism) attack, children will likely be disproportionately the most affected victims. In their September 2006 policy statement, the American Academy of Pediatrics delineated some of the important physiologic differences between kids and adults that make kids particularly vulnerable to a bioterrorist attack.

- Kids' immune systems are less well developed than those of adults (see chapter 4); the youngest kids have the least developed immunity.
- On a per-pound basis, kids breathe more air than adults; any airborne germ will have greater access to kids than to adults.
- On a per-pound basis, kids consume more food and water than adults; any contamination of the food or water supply will have greater access to kids than adults.
- Kids literally have thinner skins than adults, with fewer protective layers against agents that do their damage by landing on skin surfaces, and kids have a greater surface-to-body ratio than adults.
- Children are more prone to dehydration from infections of the gastrointestinal tract.
- Kids stand closer to the ground, making any particles that settle more likely to settle on them.
- Kids drink more milk and eat more fruit than adults, both food sources that have been identified as potential targets of terrorist activity.

The CDC has a useful classification scheme for germs that could be usurped from nature and used as weapons; the criteria for grouping these agents are based on their contagiousness, their potential for lethality, and their likelihood of causing societal disruption. The CDC criteria for categorizing potential bioterrorism germs as A, B, or C, from highest to lowest risk, are as follows.

Category A Diseases/Agents. The U.S. public health system and primary health care providers must be prepared to address various biological agents, including pathogens that are rarely seen in the United States. High-priority agents include organisms that pose a risk to national security for the following reasons.

- They can be easily disseminated or transmitted from person to person.
- They result in high mortality rates and have the potential for major public health impact.
- They might cause public panic and social disruption.
- They require special action for public health preparedness.

Category B Diseases/Agents. The second highest priority agents include the following.

- Those that are moderately easy to disseminate
- Those that result in moderate morbidity rates and low mortality rates
- Those that require specific enhancements of the CDC's diagnostic capacity and enhanced disease surveillance

Category C Diseases/Agents. The third highest priority agents include emerging pathogens that could be engineered for mass dissemination in the future due to the following factors.

- Availability
- Ease of production and dissemination
- Potential for high morbidity and mortality rates and major health impact

In the paragraphs below, I describe the CDC category A agents (causing the highest-risk diseases) as a guide for understanding the news headlines that have already appeared and those we hope will never appear.

Anthrax. In 2001, 22 cases of anthrax occurred in the United States, intentionally spread via a powdered form of the germ's spores delivered through the mail; five people died. The societal disruption, as we all too well remember, was profound—closures of government buildings and post offices, emergency rooms flooded with people convinced that their cold and flu symptoms were early inhalational anthrax (see below and chapter 12), and a national run on ciprofloxacin (an effective antibiotic against the germ). Coming as it did on the heels of the September 11 attacks, the anthrax episode contributed to a national sense of vulnerability and victimization.

Historians and scholars believe that anthrax was the biblical fifth plague. In modern times, unless you live on a farm or work with animal hides, the risk of your kids contracting anthrax naturally is near zero. In fact, even among kids on farms, the incidence in the United States is essentially zero—there were no reported cases of naturally occurring disease in this country between 1993 and 1999 and only 236 total cases in the nearly 40 years between 1955 and 1993; there were no cases in 2003 or 2004 (the most recent data available).

Caused by the animal bacterium *Bacillus anthracis,* anthrax can infect us via open sores or wounds (skin anthrax), eating undercooked meat (gastrointestinal anthrax), and breathing spores floating in the air (inhalation anthrax). One of the unique and potentially dangerous features of this bacterium is its ability to encase itself in a protective spore that allows it to survive for years in the environment only to potentially reactivate when conditions are favorable. This explains the extensive and exhaustive cleaning of contaminated buildings before they could be reopened following the 2001 attacks.

The symptoms of anthrax infection depend on the body site infected. Skin anthrax begins by looking like a blister and then progresses to an ulcer that grows and blackens at its base. Gastrointestinal anthrax starts out like the stomach flu (see "Parents' Dirty Dozen Most *Hated* Infectious Diseases" above), becoming more severe with marked stomach pain and bloody diarrhea; this form of the disease is occasionally fatal. Inhalational anthrax is the most severe and potentially lethal form; it begins as flu-like symptoms (hence the emergency room panic of 2001) and advances to severe pneumonia, bleeding, and a sepsis-like picture (see above).

How contagious is it?

- Not contagious from person to person (but potentially very infectious by inhaling spores released into the air [see chapter 2])
- Spores can be "weaponized" to make them more stable, more readily spread in the air, and better able to get deep into victims' lungs, causing the severe inhalational form of the disease (see chapter 2).

Prevention

- Antibiotics (oral) following exposure (see chapter 5)
- Vaccine (experimental; available only to certain military personnel and first-responder health care and emergency personnel)

Treatment

- Acetaminophen or ibuprofen for fever or pain (see chapter 8) (severe pain may require narcotic medicines)
- Antibiotics (intravenous) (see chapter 5)
- Intensive care management until stable (inhalational form of infection)

Smallpox. The "mother of all bioterrorism germs," smallpox is a truly dread disease and a truly perfect bioterrorism weapon. Smallpox is caused by the variola virus and is a uniquely human disease; there are no known "hiding places" for the virus in nature (e.g., animals or the environment) other than humans. Following a highly successful and heroic global vaccination campaign, the last naturally occurring case in the United States was in 1949, and the last naturally occurring case anywhere in the world was diagnosed in Africa in 1977. There have been at least two cases of smallpox acquired among research laboratory workers handling the virus since the worldwide eradication of the natural disease.

Stocks of the virus have been stored in only two research laboratories since the eradication of the natural disease: the CDC in Atlanta, GA, and the Institute of Virus Preparations in Moscow, Russia. There is concern, however, that samples of the virus may have found their way elsewhere. There has long been great philosophical and scientific debate about whether all stocks of this deadly virus should have been (or still should be) destroyed rather than storing them. The debate has greatly intensified since the 11 September 2001 World Trade Center/Pentagon attacks and the October 2001 anthrax attacks, when the fear that smallpox virus could be obtained for terrorist purposes surfaced in a major way. There is uncertainty regarding whether all stocks of the virus can any longer be accounted for, further fueling the paranoia.

Smallpox is characterized clinically by high fever and very severe flu-like symptoms, usually rendering the patient bedridden, followed by development of mouth sores and a rash. The mouth sores and the blister fluid in the rash are both infectious, but the open sores in the mouth and throat herald the most infectious phase of the disease because of the potential for infected droplets getting into the air through coughing and sneezing (see chapter 2). It is not until all of a patient's skin blisters have crusted and fallen off that he or she is no longer contagious. Death occurs in about one-third of patients, especially in very young children, as a result of massive cell damage in the body and intense immune response (see chapter 4).

Because viruses are generally less stable in the environment than are bacteria, it is possible and even likely that a smallpox terrorist attack would involve human "suicide" patients, persons willing to themselves be infected by smallpox to spread the disease by contact with others; further spread would occur from each newly infected person. The CDC considers a single case of smallpox to be proof of a bioterrorist attack.

How contagious is it?

- Moderately contagious from person to person via respiratory secretions and blister fluid; less contagious from inanimate objects contaminated with secretions from an infected person (see chapter 2)

Prevention

- Personal, household, and community hygiene (see chapter 9)
- Vaccine (no longer given routinely since the natural disease was eradicated worldwide; a stockpile is available for emergency situations, including any suspected exposure; even one proven case is considered an emergency).

Treatment

- Acetaminophen or ibuprofen for fever or pain (see chapter 8); severe pain may require narcotic medicines.
- Intensive care management until stable
- Antiviral medicines (experimental; not proven to be effective)

Botulism. You have to appreciate the irony that this era's great cosmetic wrinkle and worry line reducer, Botox, is on the CDC's list of biological agents most likely to cause parents' wrinkles and worries. The bacterium *Clostridium botulinum* is a germ, naturally found in the soil, that produces a nerve toxin, which in turn results in muscle paralysis (the muscles controlled by the affected nerves). The toxin is formed when *C. botulinum* contaminates and grows in improperly preserved or stored food; ingestion of the toxin in the food causes the disease botulism.

It is the botulinum nerve toxin that is used commercially to temporarily paralyze facial nerve endings, reducing the skin wrinkles caused by the underlying muscles served by those nerves. Botox is also used very effectively to treat a variety of noncosmetic, serious nerve- and muscle-related disorders, including Parkinson's disease. However, the potential of botulinum toxin as a natural or bioterrorist food poison remains.

Natural food-borne disease occurs rarely, with only 16 confirmed cases reported in 2004 (the most recent annual data available). In June and July 2007, 4 cases of botulism were reported in association with eating certain commercial canned meat products, prompting a massive recall; other potential cases were being investigated at the time of this writing. Other forms of botulism include an infant variety and a wound infection, neither of which is amenable to terrorist uses, but the potential for large-scale food poisoning with botulinum toxin has raised concerns. The symptoms of the disease are all caused by the progressive paralysis of the body's muscles; the muscle effects, in turn, are related to the nerves directly targeted by the toxin. Death may result from the dysfunction of vital nerve pathways, e.g., those that control breathing; survivors may be affected for many weeks or months.

How contagious is it?

- Not contagious from person to person
- Outbreaks have occurred among many people from a single restaurant or picnic exposure to contaminated food (see chapter 2).

Prevention

- Household and community hygiene (see chapter 9)

Treatment

- Acetaminophen or ibuprofen for fever or pain (see chapter 8); severe pain may require narcotic medicines.
- Antitoxin; a special horse-derived gamma globulin containing botulinum toxin antibody that binds the toxin is available from the CDC (see chapter 7).
- Intensive care management until stable

Plague. The term plague itself conjures images of the Black Death and destruction in long-ago Europe and is impressive enough to serve as the title of this chapter, with the intentional double entendre. The three major forms of the infection, bubonic plague (swollen glands), pneumonic plague (pneumonia), and septicemic plague (sepsis [see above]), are all caused by the *Yersinia pestis* bacterium, a naturally occurring germ in wild rodents. Fleas transmit the bubonic and septicemic varieties of plague from rodents to humans via bites (see chapter 2). Direct exposure to infected rodents and their secretions can also occasionally result in bubonic plague (via contact of infectious material with an open skin sore). From a public health point of view, the far more troubling risk of direct rodent contact is pneumonic plague, which occurs via breathing in infectious particles.

During the Middle Ages in Europe, it was uncontrolled rat populations that resulted in both flea transmission and direct transmission of *Y. pestis* and the resultant deaths of millions. In the modern era in the United States, plague occurs only sporadically (5 to 15 cases each year) and usually in rural areas from exposures to wild rodents (e.g., ground squirrels and prairie dogs) and their fleas; household pets may bring the fleas into the home.

Because inhalation of the germ can cause pneumonic plague and because pneumonic plague can be spread from person to person (whereas the bubonic and septicemic varieties cannot), *Y. pestis* has received attention as a category A potential bioweapon.

Symptoms of all forms of plague begin as rapid development of fever, chills, headache, and other flu-like complaints. Bubonic plague causes swollen and tender glands (lymph nodes). Pneumonic plague progresses to severe pneumonia, respiratory failure, shock, and death.

How contagious is it?

- Moderately contagious from person to person (pneumonic plague only); requires breathing in infectious droplets of secretions from another person's cough or sneeze (see chapter 2)
- Could also be intentionally spread by releasing the bacteria into the air

Prevention

- Personal, household, and community hygiene (see chapter 9)
- Antibiotics (oral) for exposure to a patient with pneumonic plague (see chapter 5)
- Flea control for household pets with potential rodent exposure

Treatment

- Acetaminophen or ibuprofen for fever or pain (see chapter 8); severe pain may require narcotic medicines.
- Antibiotics (intramuscular or intravenous) (see chapter 5)
- Intensive care management until stable

Tularemia. Similar to plague, tularemia finds its natural host in animals and rodents, and transmission to humans is via direct contact with infected animals or via a bite from a tick or other insect that first fed off an animal (see chapter 2). Tularemia can also spread via contaminated water or undercooked meat. The bacterium responsible for this disease is called *Francisella tularensis.*

Tularemia is also similar to plague, and has overlap with anthrax, in some of its clinical features: it has a glandular (lymph node) form, a pneumonia form, a sepsis (also called typhoidal) form, and a gastrointestinal form. Naturally acquired tularemia of *any* form is rare—between 35 and 150 cases in the United States each year. If usurped from nature for terrorist uses, the bacteria would be most dangerous released into the air, where they would cause the pneumonia form of the disease. Symptoms begin (you guessed it) with fever and flu-like features, followed by the signs of severe pneumonia and sepsis (see above).

How contagious is it?
- Not contagious from person to person
- Moderately contagious from infected animals or insects that fed on them (see chapter 2)
- Potentially very infectious if intentionally spread in the air

Prevention
- Personal, household, and community hygiene (see chapter 9)
- DEET or other mosquito repellent during the summer months for prevention of naturally occurring disease (see chapter 11)
- General tick prevention measures for prevention of natural disease (see chapter 11)
- Antibiotics (oral) after exposure to intentionally disseminated tularemia (see chapter 5)

Treatment
- Acetaminophen or ibuprofen for fever or pain (see chapter 8); severe pain may require narcotic medicines
- Antibiotics, intravenous (see chapter 5)
- Intensive care management until stable

Hemorrhagic Fever Viruses. As the name suggests, the viral hemorrhagic fevers are caused by a group of germs that have the capacity to cause severe bleeding. They generally act by damaging the blood vessels, resulting not only in hemorrhage, but in multi-organ system failure (a sepsis-like condition [see above]). For some of the more noteworthy viruses, like Ebola virus and Marburg virus, the origins and hiding places in nature are not known. For other viral hemorrhagic fevers, like yellow fever and Lassa fever, animals or insects are the natural hosts and reservoirs for the germs.

All of these viruses are restricted in their natural spread to the locations of their natural hosts, but for some of the viruses, spread from person to person does occur. With the mobility of individuals in the 21st century, spread from distant locales to nearby neighborhoods may be only a plane trip away. Intentional spread by terrorists is also feasible, because introduction of the virus to just one or a few individuals can result in large-scale outbreaks; naturally occurring Ebola virus clusters in Africa have proven the potential for rapid spread among close contacts.

The high fatality rates of some of these viruses have actually limited their spread—living infected individuals are much more effective at spreading disease than are dead people. Also limiting the potential for a massive outbreak is the absence of evidence for airborne transmission of the germs.

Person-to-person spread occurs by direct contact with an infected individual, his or her secretions, or inanimate objects contaminated by the secretions (see chapter 2).

How contagious is it?

- Moderately contagious; requires direct contact with infected person or secretions or with an inanimate object contaminated with infected secretions (see chapter 2)
- Not very contagious from animal or insect host to humans

Prevention

- Personal, household, and community hygiene (see chapter 9)
- Vaccines (available only for yellow fever and Argentine hemorrhagic fever)
- DEET or other mosquito repellent for prevention of those naturally occurring diseases transmitted by mosquitoes (see chapter 11)

Treatment

- Antiviral medicine (experimental; available only for Lassa and Argentine hemorrhagic fevers)
- Intensive care management until stable

Germs that currently satisfy the CDC criteria for category B diseases and agents (the second-highest priority) are food safety threats (e.g., bacterial causes of gastroenteritis [see above and chapter 2]), water safety threats (e.g., *Vibrio* and *Cryptosporidium* species [see chapter 2]), other animal or bird bacterial diseases (including brucellosis, glanders, Q fever, and psitticosis [see chapter 2]), typhus fever (a louse-borne disease), and viral-encephalitis infections (see above).

Finally, germs that currently satisfy the CDC criteria for category C diseases and agents (the third-highest priority) are new infections that could be manipulated by terrorists, e.g., hantavirus (a rodent disease that can cause severe pneumonia in humans [see chapter 2]), and multidrug-resistant tuberculosis (TB) (see below).

Tuberculosis (TB)

Had this book been written a few years ago, TB would have likely been found in "Grandparents' Half-Dozen (Plus One) *Former* Most Dreaded Infectious Diseases That Are Now 'So Last Generation'" below as a disease our kids' grandparents once feared but that was now either completely controlled or well on its way out. However, as those grandparents would say, what goes around comes around.

Although it was never really gone, TB is back in the United States (and many other developed nations) as a result of the prominence of AIDS (which puts patients at risk for more contagious and more resistant cases of TB) and the influx of new immigrants into our urban areas. More than one-quarter of cases in kids in the United States in recent years have occurred among foreign-born immigrants.

TB is caused by the *Mycobacterium tuberculosis* bacterium. The disease has several forms but is primarily an infection of the lungs. When kids first contract the infection, unless they are immune compromised (see chapter 4) or otherwise uniquely susceptible, there are usually no symptoms. Occasionally, the first infection can progress to cause severe pneumonia with total-body signs of illness: fever, chills, loss of appetite and weight, and night sweats. In other cases, the asymptomatic first infection may lie dormant in the lungs and glands (lymph nodes) of the chest and reactivate later in life (e.g., during times of weakened immunity), causing pneumonia and the other signs of lung and whole-body illness noted above. Very young infants and adolescents after puberty have a relatively higher, but still low overall, risk of reactivating dormant TB even without a specific immune abnormality.

Recently, facilitated by the concomitant AIDS epidemic of the past 25 years, TB strains have emerged that are resistant to many formerly effective antibiotics. Coupled with the socioeconomic hardships (crowded living quarters, poor hygiene, and drug abuse) of many new immigrants into urban areas, the incidence of TB has steadily risen among both children and adults.

How contagious is it?

- Moderately contagious from patients coughing up sputum; kids are generally less contagious because their coughs are not as deep, but they are very susceptible to infectious cough secretions from adults (see chapter 2).

Prevention

- Personal, household, and community hygiene (see chapter 9)
- Vaccine (not recommended in the United States except in very unusual circumstances where there is unavoidable risk of exposure and other methods of prevention, e.g., personal, household, and community hygiene, are not possible)
- Antibiotics (oral) (see chapter 5) following exposure

Treatment

- Antibiotics (oral) (see chapter 5); multiple different antibiotics are given simultaneously for a prolonged period of time, often under direct observation by a health care professional to ensure compliance.

Lyme Disease

Named for the town in Connecticut where a puzzling cluster of cases of arthritis in children was first observed in 1975, Lyme disease is caused by the bacterium *Borrelia burgdorferi,* which is transmitted by deer ticks. The distribution of the disease in the United States (the highest number of cases is in the Northeast, followed by fewer in the upper Midwest and fewer still in the upper Northwest) tracks the distribution of the ticks capable of spreading the germ. Kids, perhaps because of their proclivity for the outdoors, their low-to-the-ground stature, and their intimacy with pets who may transport the ticks, have a disproportionately high infection rate.

Lyme disease has become one of the most exhaustively and heatedly debated and controversial infectious diseases in recent memory (see below). The source of the controversy lies in the clinical and laboratory diagnostic methods and their interpretation. Some, particularly parent advocacy groups and doctors treating large numbers of patients with suspected Lyme disease, think the infection is *underdiagnosed;* they suggest that Lyme disease may explain many cases of nonspecific and chronic conditions, such as chronic fatigue, neurological and psychiatric abnormalities, sleep disorders, chronic joint problems, and immune abnormalities. Others, including the major medical bodies, like the Infectious Diseases Society of America (IDSA), the National Institutes of Health, the American Academy of Pediatrics, and the CDC, believe the disease is *overdiagnosed* as a result of misinterpretation of laboratory test results. Conspiracy theories abound on both sides of the argument.

The controversy regarding chronic Lyme disease was recently fueled (ignited may be more accurate) by the release in fall 2006 by the IDSA of guidelines for the diagnosis and treatment of the infection. The IDSA, representing many infectious-disease doctors around the country, specified precise treatment protocols for each recognized stage and form of Lyme disease; these were based on extensive reviews of the published literature. Most disturbing about the guidelines to many Lyme disease advocacy groups, however, was the discounting (dismissal) of the existence of any chronic (long-term) infection due to Lyme disease. This paragraph is from the IDSA guidelines published on-line on 2 October 2006 in the IDSA journal *Clinical Infectious Diseases:*

> There is no convincing biologic evidence for the existence of symptomatic chronic *B. burgdorferi* infection among patients after receipt of recommended treatment regimens for Lyme disease. Antibiotic therapy has not proven to be useful and is not recommended for patients with chronic (≥6 months) subjective symptoms after recommended treatment regimens for Lyme disease.

A representative response statement of outrage from a prominent Lyme disease advocacy group, the Lyme Disease Association (http://www.lymediseaseassociation.org), issued on 16 November 2006 gives you an idea of the ferocity of this controversy and illustrates the increasingly political nature of the battle:

> The national non-profit Lyme Disease Association (LDA), representing more Lyme disease patients than any organization in the United States, applauds Connecticut State Attorney General Richard Blumenthal for beginning an investigation into the IDSA Lyme disease guidelines development process. In an unprecedented move, the Attorney General's office filed a Civil Investigative Demand (CID) to look into possible anti-trust violations by the IDSA in connection with exclusionary conduct and monopolization in the development of the Lyme guidelines.

The proven and well-established (dare I even say undisputed) early clinical signs and symptoms of Lyme disease include a fever and a characteristic rash (a large, red, circular eruption, often looking like a target, at the site of the instigating tick bite). Treatment of kids at this early stage is almost always curative and prevents further complications of the infection. Some patients may progress to a more widespread illness, usually because their early disease doesn't produce the characteristic rash or, for other reasons, isn't recognized. If the germ gets into the blood and spreads to other parts of the body, patients may develop a more extensive rash, as well as involvement of the nervous system (including meningitis and/or facial-nerve paralysis), joints, and heart (although the last is rare in children).

How contagious is it?
- Not spread from person to person
- Most tick bites are not infectious.

Prevention
- DEET or other mosquito repellent during the summer months (see chapter 11)
- General tick prevention measures (see chapter 11)
- Antibiotics (oral) (see chapter 5), recommended with strict limitations: a child 12 years old or older (the only age group studied) and only after a confirmed bite by a deer tick that is still attached and engorged in a geographic area where there is a very high rate of Lyme disease

Treatment
- Antibiotics (oral) (see chapter 5) for early disease and some cases of late disease
- Antibiotics (intramuscular or intravenous) (see chapter 5) for patients with persistent or recurrent arthritis, meningitis, encephalitis, or heart involvement

Hepatitis

Hepatitis occurs as an alphabet soup (types A through E and more) variety of infections, and the recent availability of vaccines for hepatitis A and hepatitis B has brought the disease to the forefront of parents' attention. Hepatitis means an inflammation (often an infection) of the liver. Symptoms range from benign and self-resolving (e.g, fever, fatigue, and yes, flu-like symptoms) to florid and fatal, depending on the virus responsible and the ability of the child to contain the infection. Many patients with hepatitis develop jaundice, a yellow discoloration of the skin and eyes due to buildup of a pigment that is normally removed by a healthy liver. Some forms of hepatitis may persist and establish chronic liver disease and even liver cancer.

A brief overview of the hepatitis viruses and their unique features follows.

Hepatitis A Virus. Spread by contact with fecal contamination from an infected person, hepatitis A is the form of hepatitis that causes restaurant and other food-borne and waterborne outbreaks. The infection is also frequent in day care centers (see chapter 2) and can be spread sexually. Hepatitis A usually spontaneously resolves without long-term adverse effects. Many children who acquire the infection never develop any symptoms, and their infections are discovered only when they are passed on to adult contacts; infected adults are almost always symptomatic and may have prolonged (weeks) disability. Severe disease is rare, and there is no chronic state of infection.

How contagious is it?

- Very contagious; requires oral ingestion of fecally contaminated material from another individual or from a contaminated inanimate object (see chapter 2)

Prevention

- Personal, household, and community hygiene (see chapter 9)
- Gamma globulin (antibody [see chapter 7]) given after exposure
- Vaccine; routine childhood immunization (see chapter 7)

Treatment

- None

Hepatitis B Virus. As with hepatitis A, hepatitis B is frequently asymptomatic in children while it usually causes symptoms in adults. Unlike hepatitis A, however, the outcome for patients with hepatitis B is highly variable and potentially bad. Infection is transmitted by blood and other body secretions, including semen, cervical secretions, saliva, and sores. The routes for acquiring the disease (see chapter 2) reflect the secretions containing the virus: sexual transmission, drug (particularly intravenous) abuse, and through the birth canal, where infection is passed from mother to baby with an efficiency of 75% or higher. Hepatitis B is very common in Southeast Asia, China, and Africa; kids emigrating to the United States from those regions have high rates of infection, an important issue for many adoptive families. Whereas blood transfusion was a risk many years ago, effective screening in the United States has made this route of infection virtually nonexistent (see chapter 2). Breast-feeding by an infected mother poses no additional risk to the baby.

Children are particularly prone to developing chronic, lifelong infection—as many as 90% of babies infected during delivery, and one-quarter to one-half of kids infected under age 5, become chronic hepatitis patients. Of all kids with chronic hepatitis, 25% develop liver cancer or cirrhosis (permanent scarring and often ultimate failure of the liver).

How contagious is it?

- Very contagious from mother to baby during delivery (see chapter 2)
- Very contagious with blood exposure from shared needles or other sharp objects (see chapter 2)
- Moderately contagious with sexual exposure (see chapter 2)

Prevention

- Personal, household, and community hygiene (see chapter 9)
- Hepatitis gamma globulin (antibody [see chapter 7]) given to newborns of infected mothers and after certain exposures
- Vaccine; routine childhood immunizations (see chapter 7) and after exposures; also *before* anticipated exposures, e.g., occupational or travel to a high-prevalence region of the world

Treatment

- Antiviral medicines (see chapter 6) for chronic infection

Hepatitis C Virus. Like hepatitis B, hepatitis C is more feared for its chronic infection than for the illness it causes during the initial infection; again, many patients, in this case both kids and adults, are initially asymptomatic. More than half of infected children, however, develop a persistent infection state. Over the long term, cirrhosis and liver failure may ensue, resulting in hepatitis C being one of the leading causes of liver transplantation in this country.

The major route of transmission of hepatitis C is via blood exposure or exposure to blood-contaminated secretions; intravenous-drug use and sexual contact are both risk factors (see chapter 2). Birth to an infected mother is much less of a risk factor for hepatitis C transmission than for hepatitis B, but it can still occur. Rarely, blood transfusions in the United States still result in hepatitis C infection (less than 1 per 1 million) (see chapter 2). Breastfeeding by an infected mother poses no additional risk to the baby.

How contagious is it?

- Not very contagious from mother to baby during delivery (see chapter 2)
- Very contagious with blood exposure from shared needles or other sharp objects (see chapter 2)
- Moderately contagious with sexual exposure (see chapter 2)
- Almost no risk with blood transfusions (see chapter 2)

Prevention

- Personal, household, and community hygiene (see chapter 9)

Treatment

- Antiviral medicines (see chapter 6) for chronic infection

Hepatitis D Virus. Hepatitis D virus acts as a Trojan horse with hepatitis B virus and can infect only patients already infected with hepatitis B. In such patients, hepatitis D can convert a stable and quiescent infection into a severe and rapidly fatal form of liver failure. The risk factors, transmission, and strategies for prevention are similar to those for hepatitis B (see chapters 2 and 9). Hepatitis D is uncommon in kids and very rarely transmitted from mother to baby during delivery.

Hepatitis E Virus. Rare in the United States (only three cases have been reported to date), hepatitis E is transmitted in other countries, similarly to hepatitis A, by the fecal-contamination route (see chapter 2). Hepatitis E virus is also found in animals, unlike the other hepatitis viruses discussed, which may explain some human cases (i.e., contamination of human food or water with animal waste [see chapter 2]). Clean water is the best preventive measure.

Other Viruses That Cause Liver Infection. Many other viruses can affect kids' livers, including EBV, which causes infectious mononucleosis (see "Parents' Dirty Dozen Most *Hated* Infectious Diseases" above); CMV; and the chicken pox virus (see "Parents' Dirty Dozen Most *Hated* Infectious Diseases" above). The hepatitis from these viruses is usually mild and resolves with the resolution of the infection from the rest of the body. The enteroviruses can cause a severe and fatal hepatitis that develops in the newborn infected while still in the womb.

Rabies

Although mail carriers are less concerned than in the old days, rabies has endured as a source of societal and parental anxiety despite the reduction of disease occurrence in family pets (and even unrecognized neighborhood pets) to near zero. That endurance in our collective consciousness is probably a good thing, because rabies hasn't gone away, it has just found new ways to infect us. Actually, the more accurate statement is that we've found new ways to become infected as our incursions into the wilderness have become more bold and brazen.

With immunization of dogs and the availability of vaccines and gamma globulin antiserum (antibodies [see chapter 7]) for humans after potential exposure, there have been fewer than 50 cases of human rabies in the 15-year period from 1990 to 2004, and the vast majority have been contracted from bats. Of the 15 human cases in the United States since 2000, 14 have come from bats. Animal control officers find the rabies virus in more than 7,000 wild animals throughout the United States each year (see chapter 2), notably raccoons, bats, foxes, coyotes, skunks, etc., and occasionally in small rodents (squirrels, rats, mice, and hamsters) and rabbits.

The worrisome characteristic of bat transmission to humans is that in most cases there is no history of a bite; handling the bat, dead or alive, or exposure to bat droppings and even to air in a bat-infested area (e.g., a cave or cabin) have all been implicated as potential mechanisms for humans to contract rabies. If you find a bat, dead or alive, in your child's room or in the family cabin where your kids have spent time, contact your doctor or state health department immediately for advice.

Rabies virus is transmitted from the saliva of animals, enters the skin or mucous membranes of exposed individuals, and infects their nervous systems, causing a rapidly fatal disease if protective vaccine and gamma globulin antiserum (antibodies [see chapter 7]) are not given in time. The symptoms can be as nonspecific as aggressiveness, irritability, abnormal behavior, and disordered thought processes and as unique as excessive saliva production and difficulty swallowing (foaming at the mouth). Rabies meningitis and encephalitis (see above) may cause seizures, weakness or paralysis, and abnormal movements of facial and other muscles.

How contagious is it?

- Not contagious from person to person
- Minimally contagious from infected animals (see chapter 2); many more people have exposures to bats and raccoons each year than the number who actually develop the disease.

Prevention

- Community hygiene (see chapter 9) and animal control
- Immunization of domestic animals (pets)
- Avoiding contact with wild (or stray) animals, dead or alive
- Immediate and vigorous cleansing (soap and water) of any bite wound (to prevent virus from entering the nervous system)
- Rabies immune gamma globulin (antibodies [see chapter 7]) immediately following exposure
- Rabies vaccine as quickly as possible following exposure

Treatment

- None; there has been no proven treatment once symptoms have begun.
- In July 2005, a single patient (a 15-year-old girl) was reported to have survived rabies (contracted from a bat bite) after treatment with antiviral medicines (see chapter 6) and intentional induction of coma with anesthesia medicines; this first ever success in treatment of a patient after rabies symptoms had already begun has not yet been repeated.

Severe *E. coli* Food Poisoning and Hemolytic Uremic Syndrome (HUS)

It's hard not to be impressed with the number of restaurant, fast food, and supermarket outbreaks of severe *E. coli* disease that have been in the news in recent years. Just when you thought your kids should be eating more fruits and vegetables, apple juice, spinach, lettuce, and green onions are implicated as spreading this dangerous, and potentially lethal, bacterium.

If you've been following along, you'll remember from chapter 1 that *E. coli* is the most numerous and ubiquitous of our *normal* intestinal bacteria. However, the devil is in the details. Certain strains of *E. coli* contain genes for producing toxins and other products that the benign, and even helpful, normal strains of *E. coli* lack. These toxins can cause severe intestinal disease and, when they get into the bloodstream, severe disease beyond the intestines. The result is everything from typical gastroenteritis (see "Parents' Dirty Dozen Most *Hated* Infectious Diseases" above) to prolonged (even chronic) diarrhea, traveler's diarrhea (see "Parents' Dirty Dozen Most *Hated* Infectious Diseases" above), and, the most worrisome of all, hemorrhagic colitis (severe, bloody gastroenteritis) and HUS.

The strain of *E. coli* most commonly associated with hemorrhagic colitis and HUS is O157:H7. The source of most outbreaks of O157:H7 diarrhea is animals shedding the bacteria in their feces (see chapter 2); cattle are the most common, but the germ has also been found in sheep and deer. Undercooked meat, unpasteurized dairy products, and apples and vegetables harvested from fields where infected cattle graze have all been identified as sources of *E. coli* O157:H7 outbreaks, as have petting zoos and day care centers (see chapter 2).

The *E. coli* O157:H7 germ produces a toxin similar to that produced by shigella, another cause of severe gastroenteritis (see "Parents' Dirty Dozen Most *Hated* Infectious Diseases" above). When this shiga toxin is confined to the intestines, severe and bloody diarrhea, abdominal cramps, and occasionally fever are the results. However, when the toxin gets into the blood circulation, the distant organ most often targeted is the kidney. Blood vessels of the kidney are damaged by the toxin, resulting in anemia (from the destruction of red blood cells passing through the damaged vessels), clotting problems (from the destruction of blood platelets, the cells responsible for clotting, as they pass through the damaged vessels), and kidney failure. As many as 10% of kids with *E. coli* O157:H7 infections develop HUS, half of whom require dialysis and 5% of whom die. Kids seem to be the most vulnerable, although a similar severe syndrome in adults can also occur. Diabetes has been reported to develop at a higher incidence among kids who survive *E. coli*-associated HUS.

Because certain antidiarrhea over-the-counter medicines (see chapter 8) work by decreasing the normal contractions and propulsive activities of the intestines (gut motility), toxins produced by *E. coli* may remain in the intestines longer, and do more damage, in patients treated with these medicines.

How contagious is it?

- Moderately contagious by food and/or water contaminated with fecal material from infected animals or humans (see chapter 2)
- Moderately contagious by personal contact with fecal material from infected animals or humans or with inanimate objects contaminated by infected animals or humans (see chapter 2)

Prevention

- Personal and household hygiene (see chapter 9)
- Community hygiene (water and food sanitation [see chapter 9])
- Avoiding undercooked meat, unpasteurized dairy products, and unpasteurized juices (see chapter 9)
- Avoiding antimotility over-the-counter diarrhea treatments (see chapter 8)

Treatment

- Fluids and nutrition (see chapter 10)
- Intensive care management until stable
- Antibiotics have not been shown to be helpful.

Sexually Transmitted Diseases (STDs)

Some things never change. STDs (also known as venereal diseases) have been among the most durable and determined scourges of humanity, recorded in history since, well, the beginning of recorded history. Even though we have identified new STDs over time, the old ones remain, reemerge, and resist our control.

Our kids are at risk for STDs at every stage of their childhood, from exposure to infected mothers in the womb and during delivery to sexual abuse and stupid adolescent behavior. STDs are all transmitted by the same route, as the name suggests (see chapter 2). However, within the sexual-transmission modus, there are two subcategories: STDs that are caught by contact with sores (or lesions) and those that are caught by contact with secretions (or discharge). This is an important distinction to make for a number of reasons, not the least of which is predicting the effectiveness of condom prevention (see chapter 11).

The major STDs of both subcategories are reviewed below; because of significant overlap in the contagiousness, prevention, and treatment aspects of these diseases, the bulleted summaries are presented for all the STDs at the very end of the discussion.

STDs Primarily Transmitted by Contact with Sores or Lesions. *Syphilis.* One of the "granddaddies" of the STDs, syphilis is caused by a bacterium called *Treponema pallidum*. Infection in the womb may cause stillbirth, premature birth, and/or birth deformities. Other babies show no signs until years later, when they may develop abnormalities of the brain, bones, teeth, and eyes. In older children and adolescents, the infection occurs in three stages if not treated. Beginning with a genital sore (chancre), the infection may progress to a second stage involving glands (lymph nodes) and causing a flu-like illness that can be prolonged and recurring, along with a rash over the whole body. In the third stage, major organ systems are involved, including the heart. The brain may become infected at any of the three stages.

Genital herpes. Herpes simplex viruses come in two types, cleverly called type 1 (HSV-1) and HSV-2. HSV-1 most commonly causes oral cold sores (fever blisters [see "A Half-Dozen Leftovers" below]), and HSV-2 is the usual cause of genital herpes, but some crossover exists. (I won't go into great detail as to how and why the crossover occurs.) As noted earlier regarding herpes encephalitis (see "Parents' Dirty Dozen Most *Hated* Infectious Diseases" above), viruses in the herpesvirus family are lifelong infections—the immediate symptoms resolve quickly, but the viruses remain dormant ("latent") in a child's nerve roots forever. Recurrent oral cold sores and recurrent genital lesions are the result of reactivation of the dormant virus from the nerve roots during times of body stress (e.g., fever and sun exposure for HSV-1 and menstruation and pregnancy for HSV-2).

Genital herpes is also associated with blisters (vesicles) that contain millions of viruses and typically are inflamed, painful, and itchy. This is an insidious and very contagious infection, spread from the infected blisters (see chapter 2). However, blisters may not be evident to the patient or even objectively present on a doctor's examination, and there may be no symptoms, yet a person may still be contagious because the virus can be shed from infected mucous-membrane cells. This subclinical form of genital herpes is particularly troubling for newborns, who may be born through infected birth canals to women with no signs of infection. Neonatal herpes is a severe infection with high death rates and with many permanently brain-damaged babies among the survivors (see "Meningitis and Encephalitis" in "Parents' Dirty Dozen Most *Hated* Infectious Diseases" above). When genital herpes is acquired later in childhood, its manifestations, like those in adults, are usually limited to recurrent genital sores.

Genital warts. Genital warts are the most common sexually transmitted disease in this country and are found in nearly half of sexually active adolescent girls. The papillomaviruses that cause genital warts are cousins of those that cause common skin and plantar warts (see "Parents' Dirty Dozen Most *Hated* Infectious Diseases" above). Transmission is via direct sexual contact with infected genital warts, but as with herpesvirus infections, warts may be very subtle or even invisible infections of the mucous membranes. The hazard of genital warts lies far beyond the immediate infection and even beyond the warty growths that can be quite prominent; papillomaviruses also become dormant (although they are not in the herpesvirus family) and are strongly associated with abnormal Pap smears (cervical dysplasia) and cervical cancer. Additionally, babies born to women with genital warts are at risk for developing warts on the vocal cords (respiratory papillomatosis) because of aspiration of infected material in the birth canal. A vaccine for prevention of genital warts and, thereby, cervical dysplasia and cervical cancer has recently been approved and recommended for use in preteen and adolescent girls (see chapter 7).

Chancroid. Chancroid is a relatively uncommon infection in the United States that is strongly associated with urban prostitution and drug use. The ulcer that forms is very painful and tender and very contagious with direct contact. This is a bacterial infection (*Haemophilus ducreyi*) and tends to travel with HIV, facilitating infection with that virus, i.e., the presence of the bacterial chancroid ulcer helps the AIDS virus get in.

STDs Primarily Transmitted by Contact with Secretions (Discharge). *Gonorrhea.* Gonorrhea, the other granddaddy of the STDs, is caused by the bacterium *Neisseria gonorrhoeae* and transmitted through contact with secretions from mucous membranes of the genital tract or rectum. Person-to-person sexual contact is the most common (see chapter 2), but occasionally contact with contaminated inanimate objects, such as bed sheets, has resulted in childhood infection; the presumption must be that sexual contact has occurred until another explanation can be proven. Babies born through infected birth canals can develop gonococcal eye infections and occasionally more widespread infection (see chapter 2). Symptoms in older children and adolescents include the contagious discharge (pus-like fluid from the vagina or penis), pain on urination, itching, and inflammation, but the disease may also cause no symptoms yet remain contagious. More serious complications may occur with spread of the bacteria to major organs, but that complication is rare. Not rare is the complication of pelvic inflammatory disease in girls, which can result in fallopian tube scarring and infertility.

Chlamydia. Chlamydia is clinically similar to gonorrhea; they are spread the same way (see chapter 2) and frequently coexist in the same patient. Caused by the bacterium *Chlamydia trachomatis,* this is the most commonly *reported* STD. (The evidence is that more people actually have been infected with genital warts, as noted above, but many of those people don't know they're infected and don't seek treatment. Genital warts is not listed as a reportable disease by health departments and the CDC, whereas chlamydia is required to be reported to health departments.)

Like gonorrhea, chlamydia causes a discharge, pain on urination, itching, and inflammation, but more patients with chlamydia are without symptoms than is the case with gonorrhea, explaining the rapid spread and high occurrence rates. Again, pelvic infection and resultant fallopian tube scarring and infertility are the biggest risks of this infection in women; men occasionally also develop infertility because of scarring of the tubes carrying sperm. Infection of babies in the birth canals of infected women (see chapter 2) can cause eye infections and pneumonia.

AIDS. See above.

Trichomoniasis. Affectionately known as trich, trichomoniasis is an extremely common parasitic infection of the genital tract (said to be the second most common STD in the United States). The infection causes vaginal discharge, burning, and itching. Men are more frequently without symptoms but may also develop pain with urination and itching. Complications of trichomoniasis are rare, but as with other STDs, the infection may facilitate the spread and infectiousness of the virus that causes AIDS.

How contagious are STDs?

- Very contagious, by personal contact (see chapter 2)

STD Prevention

- Personal and household hygiene (see chapter 9)
- Sexual abstinence
- Condoms (more effective for secretion-transmitted infections than for sore-transmitted infections [see chapter 11])
- Laboratory screening tests for pregnant women (syphilis, gonorrhea, chlamydia, and HIV)

- Clinical screening for genital herpes
- Caesarian section (for women with active herpes lesions at the time of delivery [see "Meningitis and encephalitis" above])
- Vaccine (papillomavirus [see chapter 7])

STD Treatment

- Antibiotics (oral) (syphilis, chancroid, chlamydia, and gonorrhea [see chapter 5])
- Antiviral medicines (herpes and AIDS [see chapter 6])
- Antiparasitic medicines (trichomoniasis [see chapter 6])
- Topical therapy (genital warts [see chapter 8])

Parvoviruses (Fifth Disease)

Fifth disease got its name because it was the fifth of the childhood rash illnesses described in the old days. Measles was the first, scarlet fever the second, German measles was sometimes known as third disease, etc. By fifth disease, they must have been running out of new names, and the specific germ that caused the infection hadn't yet been identified. It has since been discovered that fifth disease is caused by parvovirus B19. For pet lovers, rest assured that although it is a cousin of the parvoviruses that cause diarrhea in cats and dogs, parvovirus B19 is not acquired by humans from animals and is not passed on to animals from humans.

Placing fifth disease in this category (among parents' most *feared* infectious diseases) was tricky because, truth be told, most parents don't spend much if any time thinking about this infection, and in most kids, the infection is fleeting and benign. So what's to fear? The justification comes when Mom (or a school teacher or other woman in close contact with young kids) is pregnant.

Like German measles [rubella; see "Grandparents' Half-Dozen (Plus One) *Former* Most Dreaded Infectious Diseases That Are Now 'So Last Generation'" below], parvovirus B19 is spread by respiratory secretions (see chapter 2) and causes a characteristic rash in children (the parvovirus rash includes a "slapped cheeks" red facial blush and a lacy, itchy eruption on the rest of the body) and usually little else but a mild fever. A small percentage of kids may get joint pain and swelling, and occasionally fifth disease is mistaken for juvenile arthritis.

However, also like rubella, parvovirus B19 can cause significant damage to the fetus when a mother is infected during the first half of her pregnancy. The infected fetus may develop total-body swelling (termed hydrops fetalis) due to anemia and heart failure, failure to adequately grow in the womb, and fluid accumulations around the lungs and heart. Stillbirth due to these conditions in the fetus occurs in about 5% of cases.

About half of all women are immune to parvovirus infection by the time they reach child-bearing age, the result of their own childhood infections with the germ. The infection is most contagious *before* the rash begins, meaning that a pregnant school teacher, for example, may be exposed to the virus before a case of fifth disease is even recognized in the classroom. By the time the rash appears, a child may have been contagious for as long as a week.

Although serious damage to the fetus may occur, pregnancy termination is not considered a viable option for this infection. When all the numbers are crunched, the chances of fetal damage and loss are very low: first, remember that preexisting immunity protects 50% of pregnant women exposed to the virus; secondly, contagiousness in the workplace is only moderate, with 20% communicability from an infected child or adult to a susceptible pregnant woman (household exposures are more risky, with a 50% chance of spread to a woman who is not immune); finally, there is a very low (5%) chance of fetal damage and loss even if a pregnant woman is susceptible *and* becomes infected. When all those odds are multiplied by each other, the risk to a pregnant woman for fetal damage and loss (when it's unknown if she is immune from past exposure) following a workplace (e.g., school) exposure to parvovirus is about 1 in 200, and about 1 in 80 for a household exposure. (Of course, if it's known that a woman is immune from past exposure, her risk is zero; she's immune!) Again, since kids are most contagious *before* they develop a rash, it is very difficult for pregnant women to avoid exposure during a community-wide outbreak of parvovirus B19.

How contagious is it?

- Moderately contagious, by direct personal contact or contact with respiratory secretions from an infected individual (see chapter 2)

Prevention

- Counseling of pregnant women who are exposed to children at home or at work
- It is usually not effective or practical for pregnant women to avoid occupational or home exposures because community-wide outbreaks extend the exposure risk beyond the obvious places, 50% of pregnant women are immune, and the most contagious period is before the infection is apparent.

Treatment

- No treatment is necessary for children who are otherwise healthy; special treatment considerations occur for certain high-risk kids (those with sickle-cell anemia, other chronic hemolytic anemias, or immune system abnormalities).
- If a fetus is suspected of infection because of abnormal findings on ultrasound, intrauterine blood transfusions (given to the fetus in the womb using a special needle) or treatment of the fetus with medicines for heart failure have occasionally been used successfully.

Parents' Half-Dozen *Former* Most Dreaded Infectious Diseases That Are Now "So Last Year"

It seems like only last year that we were in a collective national frenzy about one or another impending disease of doom only to wake up this year and find no mention of it on the news or in the carpool. This section is devoted to yesterday's panics, because infections have a way of cycling back and catching us unaware if we let down our guard—and also because it's always fun to ask, "Whatever happened to. . .?" Additionally, in some, but not all, cases, impressive infectious-disease detective work resulted in identification of the germs responsible for these once-dreaded diseases and the ways in which those germs are spread. Those discoveries led to putting some of these conditions on the back burner of our consciousness. You'll recognize most of these afflictions from the cover stories of your favorite national newsmagazines and newspapers.

Mad Cow Disease

The story of mad cow disease is fascinating, itself worthy of a book. The medical name for this neurodegenerative illness in humans is Creutzfeldt-Jakob disease (CJD), and it is caused by a heretofore-unknown type of infectious agent: a prion. Unlike every other germ known to humankind, prions contain no genetic material, that is, they have no DNA (deoxyribonucleic acid) or RNA (ribonucleic acid), the vital ingredients of life required for reproduction and all other functions of an organism. Prions, instead, are simple protein molecules that exist in all of us but can become distorted by contact with other distorted proteins. A chain reaction results in many distorted prion proteins that then clump in the brains of infected animals or humans, causing brain deterioration that clinically resembles Alzheimer's disease but advances much more severely and rapidly. Under the microscope, affected brains show not only the clumps of abnormal proteins, but holes that look like a sponge; hence the formal scientific description of mad cow disease as a "spongioform encephalopathy."

Now comes the exciting part of the story; enter medical epidemiologists (disease trackers), their good detective work, and a dramatic societal intervention to halt a disease. It turns out that CJD has been known for many years as a rare (about 1 case per 1 million people), sporadic (occurring randomly and without clustering of cases) type of dementia of the elderly. However, in Britain, several young people developed a CJD-like disease that was distinctive, not only for its attack on a younger age group, but also for several unique neurological abnormalities that it caused.

Here's the reason we pushed the panic button in 1996: it was determined that these rogue prion proteins can originate in animals' brains, and when those brains are consumed by humans, the proteins can be passed on and cause distortion of human brain prion proteins. Hence, "mad" cows (behaving strangely, slowly wasting away, and dying because of their own rogue prion accumulation) could cause mad humans when the latter ingested the former. A link was made to the consumption of beef, specifically portions of meat that could have been contaminated with brains or spinal cords of the cows. Ground beef, with its source from many cows and who knows what parts of those cows, was implicated. An epidemic of hundreds of thousands, perhaps millions, of human CJD dementia cases was predicted based on exposure to mad cows.

More tracking and more detective work identified specific steps in the beef rendering, processing, and treatment methods that had changed in the years leading up to the clustering of cases of CJD. Corrections of those possible shortcomings, as well as the sacrifice of hundreds of thousands of cattle that may have been exposed to affected animals, appear to have stopped the feared epidemic. Fewer than 200 total cases of the human variety of mad cow disease, CJD, occurred. Ongoing surveillance of cattle herds around the world has detected occasional animals with the abnormal prion protein; those animals and any cows exposed to them are culled from the food supply and/or killed.

Like every great infectious-disease tale, this one may not yet have reached its conclusion. There are other animal prion diseases; scrapie in sheep and chronic wasting disease in deer and elk are the best known. So far, there have been no human cases of CJD or other diseases associated with consumption of those animals, but as with ongoing surveillance of cattle, vigilance is warranted.

There is no treatment for CJD; prevention depends on government and societal regulation of the foods we eat and simple common sense; sick animals of any species (e.g., those that hunters can too easily track, shoot, and kill) should not be eaten.

West Nile Virus (WNV)

DEET became a household word (see "Meningitis and Encephalitis" in "Parents' Baker's Dozen Most *Feared* Infectious Diseases" above and chapter 11) in 1999 as WNV began spreading from the East Coast of the United States to the west. WNV is an arbovirus, defined as a virus spread by insects, in this case mosquitoes. Mosquitoes bite infected birds, the natural hosts of WNV, and then bite animals (e.g., horses) or people. Although most WNV infections in humans are without symptoms, about 1 in 5 infected individuals will develop a fever and flu-like illness and about 1 in 150 or fewer infected people will develop meningitis or encephalitis (see "Parents' Baker's Dozen Most *Feared* Infectious Diseases" above). The panic button in this case was pushed by the severe neurological illness in elderly adults, often resulting in death. Although these severe cases were rare, they were dramatic.

WNV has now assumed the pattern of other arboviruses, establishing itself first as a seasonal cause of epidemic infections and then migrating across the country. The East Coast states with the highest number of cases early in the U.S. WNV experience now have the fewest cases, and the westward movement of the infection to new, nonimmune populations has resulted in the highest concentration of WNV in the most recent years occurring on the West Coast. The usual next step in arbovirus distribution is the settling of the infection in an area where conditions are most favorable and the establishment of endemicity, or a low annual level of disease cases. Such has been the pattern for St. Louis encephalitis virus, California encephalitis virus, western equine encephalitis virus, eastern equine encephalitis virus, and others. Although it remains to be seen if and where WNV will settle, the panic that this new and dangerous arbovirus would take over the country and forever ruin picnics appears to have subsided with the recognition that we've seen this kind of thing before.

There is no specific treatment for WNV other than intensive care management of severely affected patients; prevention is by mosquito control and bite prevention (see chapter 11).

SARS

On 5 May 2003, *Time* and *Newsweek* magazines both led with cover stories on SARS; each cover featured a picture of a wide-eyed, frightened woman wearing a protective surgical mask. This dual prime-time recognition of SARS was remarkable, not only for its synchrony, but because the outbreak had begun in China only 3 months before, in February 2003; this was one of the quickest media panic reactions in memory. By the end of July 2003, SARS was gone. *Gone!* There have been no cases reported since then (except for research laboratory accidents). Whew! That works out to one national newsmagazine cover for every 3 months of the epidemic.

SARS stands for severe acute respiratory syndrome, and it is caused by a coronavirus, usually a common-cold germ (see "Parents' Dirty Dozen Most *Hated* Infectious Diseases" above). The disease is characterized by a fever and flu-like illness (like so many other infections!), followed by severe pneumonia. During its few months on the scene, SARS infected 8,000 people worldwide, killing about 10% of them. Eight people developed the infection in the United States, all of whom had traveled to Southeast Asia; none of the U.S. patients died. That works out to one national newsmagazine cover for every four patients with the disease in the United States.

SARS spreads by close personal contact with the respiratory secretions of infected patients, like the more benign coronaviruses that cause the common cold (see chapter 2). The origin of this virus is thought to have been an unidentified animal or bird source in China. Other than intensive care management of patients with severe lung disease, there is no treatment for SARS. Worldwide travel restrictions and careful screening and quarantining of people with suggestive symptoms may have had a major role in limiting the epidemic, or those measures may have had a minimal role and the epidemic may have disappeared on its own. Sometimes, things work out okay and we just don't know why, which I find reassuring and humbling.

Ebola Virus

Here's another "we just don't know" story that is less reassuring because there is good reason to believe that we haven't heard the last of this one. The *Newsweek* cover story of 22 May 1995 was distinctive, not only because it profiled an exotic disease that had never occurred in humans in the United States (and 12 years later still hasn't), but also because the mandatory token test tube shown in the cover picture spelled the virus incorrectly ("E. bola"). Named for the Ebola River that runs through the Democratic Republic of the

Congo (formerly Zaire), where the virus was first recognized in 1976, Ebola virus causes a severe, often fatal, form of hemorrhagic fever (see "Bioterrorism" in "Parents' Baker's Dozen Most *Feared* Infectious Diseases" above).

At the time of the 1995 *Newsweek* story, Africa was going through its fifth outbreak of Ebola hemorrhagic fever since its first discovery in Zaire nearly 20 years earlier. The 1995 outbreak again centered in the Democratic Republic of the Congo and caused 315 infections in humans, more than 80% of whom died. Since then, there have been eight additional outbreaks, all in Africa, cumulatively involving nearly 900 people; the death rate remains between 50 and 90%. Symptoms begin generically, with fever, red eyes, and the ubiquitous flu-like illness. Those early complaints are followed quickly by gastrointestinal symptoms and then the signature bleeding, both internal and external hemorrhaging.

The virus is thought to be spread to humans from animals, perhaps monkeys, by direct contact. Where the monkeys get the virus, i.e., where the virus originates, remains a mystery. Humans spread the infection to other humans by direct contact with blood and other body secretions (see chapter 2); it's usually the caregivers of infected and sick victims who themselves become the next victims of the infection. There is one major factor that has probably limited the spread of the disease to several dozen or several hundred in each outbreak, sparing the multitudes in Africa and beyond who could become infected: the rapid lethality of the virus. Although dead and dying patients can spread the virus to close contacts, the potential for mass epidemics is far greater when the human vectors of infection are living, breathing, moving from place to place, and interacting person to person (see chapter 2). The great fear with all such viruses is that a mutation could make it more readily transmitted (e.g., by respiratory secretions, which are easier to spread than blood [see chapter 2]) or just enough less lethal to extend the life expectancy of contagious individuals and therefore the number of contacts those contagious persons might have.

There is no known treatment other than intensive care management of those infected (unfortunately, intensive care units are not available to many of the victims). Prevention is via avoidance of contact with those infected or their secretions.

Legionnaires' Disease

Just as Africa was experiencing its first Ebola virus outbreak in 1976, an American Legion convention was held in Philadelphia to celebrate our nation's independence. More than 200 delegates to that convention, or their family members, developed a severe infection within a few days to a couple of weeks after the meeting. The severest of the cases developed pneumonia, with other organ systems (liver and kidney) also involved in some; most patients were older, and men predominated, perhaps reflecting the demographics of the conclave. Nearly 3 dozen of the infected died. On 16 August

1976, *Time* magazine ran this cover story: "Tracing the Philly Killer." The mandatory test tube was in the picture.

Disease detectives from the CDC and experts from around the country descended on Philadelphia seeking the cause of this severe pneumonia; it was elusive and initially seemed destined to be another medical mystery that would never be solved. Terrorism was entertained as a possibility. Finally, a bacterium was identified in some of the lung specimens and then confirmed to be present in all of the cases. This newly recognized germ was named for the conventioneers, *Legionella pneumophila*. Further investigations found the likely source: contaminated water in one of the convention hotels. We now know that this bacterium likes warm water and grows avidly unless the water is sufficiently heated and the tanks are sufficiently cleaned (see chapter 2).

Outbreaks have since been identified in hospitals, government facilities, and other large buildings; infections are also associated with hot tubs and spas. In all cases, the germ gets into the patients' lungs when they breathe the mist or vapor from contaminated water sources. Another previously mysterious disease called Pontiac fever, named for the town in Michigan where people working in a health department building developed a mild respiratory and flu-like illness, has now been proven to be caused by the same bacterium, *L. pneumophila*.

Although several thousand cases of Legionnaires' disease and Pontiac fever are now recognized to occur in the United States each year, they undoubtedly occurred long before the 1976 outbreak in Philadelphia—we just didn't recognize them as distinct from other cases of community-acquired respiratory infections or pneumonia until the dramatic appearance at the convention. The panic associated with the uncertainty and the unsolved mystery is now gone. We know that prevention can be accomplished, and the occurrence rate can be reduced, by proper care and treatment of water sources. Now, when a patient presents with symptoms consistent with legionellosis (the new combined disease name for Legionnaires' disease and Pontiac fever), diagnostic tests are available and the patients can be successfully treated with antibiotics. Score another one for the good guys.

Reye's Syndrome

It is now more than 20 years since the FDA mandated that Reye's syndrome warnings be placed on all aspirin-containing products (and all products containing salicylates, the active ingredient of aspirin). That marked the culmination of another true public health investigative triumph, no less dramatic than the association of smoking with lung cancer and heart disease.

In the late 1960s and through the 1970s and early 1980s, hundreds and even thousands of cases of Reye's syndrome occurred in children in this country each year. Reye's syndrome is a devastating brain and liver disease in which brain swelling causes severe neurological complications and often death. The disease typically follows an otherwise unremarkable case of influenza, chicken pox, or stomach flu. Annual epidemics of those infections triggered coincident bursts of cases of Reye's syndrome.

Heralded by vomiting beginning within a few days of flu, chicken pox, or stomach flu, Reye's syndrome rapidly progresses to cause fatty accumulations in the liver (and often other organs) and severe fluid accumulation in the brain, resulting in seizures, abnormal breathing patterns, and coma. Delay in diagnosis was common because the vomiting was often mistaken for a symptom of the triggering virus infection itself. Intensive care units in children's hospitals across the country housed cases of Reye's syndrome all winter, the peak season for both flu and chicken pox.

Health departments doing survey research on cases of Reye's syndrome in the late 1970s began to suspect an association between kids taking aspirin for their flu or chicken pox fevers and the development of this severe disease. Soon, the evidence became overwhelming for an association, prompting a formal Surgeon General's advisory in 1982 and the FDA label warning requirement in 1986. Since then, Reye's syndrome has become rare, although cases still occur. Some of the residual cases may be due to the failure to recognize which over-the-counter products contain salicylates. The National Reye's Syndrome Foundation provides a list of many over-the-counter products that contain aspirin or aspirin-like chemicals (salicylates) and should therefore be avoided in kids (http://www.reyessyndrome.org/aspirin.htm).

The mechanism by which aspirin interacts with common viral infection to cause this very severe disturbance of liver and brain function has never been determined, but the avoidance of aspirin has proven to be protective. Unless instructed by a doctor, children should not ingest aspirin or products containing salicylates. It has not been established whether topical salicylate products, like many acne medicines and wart removal products (see chapter 8), pose a risk for Reye's syndrome.

Grandparents' Half-Dozen (Plus One) *Former* Most Dreaded Infectious Diseases That Are Now "So Last Generation"

Ask your kids' grandparents what scared them most when you were growing up. If you, like me, grew up in the 1950s or early 1960s, the answer is swimming pools. The fear of contracting paralytic poliomyelitis from other kids in public swimming pools, or any other public place where lots of kids hung out, changed the way my generation was raised. There were other germ worries, too, that were almost as great as polio. This section reviews seven diseases that no longer keep our kids' grandparents awake at night; they probably never even crossed the minds of most of your generation because they have been controlled for so long.

This section will not discuss vaccines; those are covered in chapter 7. Rather, I will briefly profile seven vaccine-preventable diseases that we shouldn't have to deal with anymore (after all, that's what vaccine-preventable means!). I do this for two reasons: first, because it's important to appreciate the tragedies we are avoiding in case your kids' grandparents forgot to tell you, and secondly, because some of these diseases are reemerging as parental reluctance regarding vaccinating their kids grows. Parents should know what they may be choosing when they choose not to immunize.

Poliomyelitis

That I am about to describe poliomyelitis to you, informed readers of the 21st century, is itself remarkable. Polio was a household word (whispered, the way AIDS is today) for more than half of the 20th century in the United States. From the images of President Franklin Delano Roosevelt (who is widely believed to have suffered from childhood poliomyelitis, although some argue he may have had another paralytic disease known as Guillain-Barré syndrome) in his wheelchair to the images of hospital wards filled with iron lung machines breathing for paralyzed children, polio was the "AIDS of the 1900s through the 1960s."

Polio has been known for thousands of years; Egyptian artwork depicting kings and priests with withered limbs, crippled and walking with canes, dates back to antiquity. Epidemic polio was reported in Europe and the United States in the 19th century, but the outbreaks were limited and received little attention. Twenty-six cases were reported in Boston during the summer of 1893 and 132 cases in Vermont in 1894, the largest outbreaks recognized up to that time.

Poliomyelitis is caused by the polioviruses, members of the enterovirus group. Enteroviruses are spread via the fecal-oral route (see chapter 2), meaning that feces from an infected individual carry the virus, which is then spread to a susceptible person who ingests it. Enteroviruses, including the polioviruses, are summer-fall infections, their spread facilitated, perhaps, by the fewer layers of clothing and closer contacts with others in public places during that season. Hence, the swimming pool paranoia (subsequent debate questioned the role of pools, arguing that chlorination probably inactivated most polioviruses shed into the water).

The vast majority of poliovirus infections cause no symptoms or only minor common-cold-type complaints that resolve on their own, resulting in lifelong immunity. A very small minority, 1% or less, of poliovirus-infected people develop paralytic disease. The rest become immune without ever having significant illness. The strong predilection of polio for young infants and children is explained by their sharing each others' fecal germs much more prolifically than adults (see chapter 2) and the greater likelihood that a child has not yet developed immunity because of many fewer summers exposed to the virus than an adult.

In one of the ultimate ironies of public health medicine, improvements in community sanitation and water purity at the turn of the 20th century probably resulted in the dramatic *increase* in the magnitude of polio outbreaks seen during the next 60 years. As the opportunities for fecal contamination of public places and public water declined with our cleaner 20th century society, natural immunity from exposure to circulating polioviruses also declined. Fewer cases of infection without symptoms were occurring, and fewer people (especially children) were becoming naturally immune. Nursing mothers were also less likely to have been exposed to poliovirus and therefore less likely to passively protect their babies by breast-feeding (or by passing antibodies to their fetuses during pregnancy [see chapter 4]) than were mothers in the less sanitary 19th century.

In 1916, an unprecedented epidemic of 27,000 cases of paralytic poliomyelitis occurred in the United States (remember, the previous numbers were in the several dozens of cases in 1893 and 1894), with more than 9,000 in New York City alone; almost all the victims were under 5 years old. Panic was everywhere. Paralyzed kids were quarantined, all children under the age of 16 were barred from public places in New York City, and public health officials dumped millions of gallons of water on city streets each day in the hope of washing away the contamination. Immigrants were blamed for importing polio from Europe; hate crimes increased. Six thousand children died from polio during the 1916 epidemic, and most of the remaining 20,000 were maimed for life, in many cases relegated to living in places aptly but cruelly named the New York Society for the Ruptured and Crippled, the Home for the Destitute Crippled Children, and the Home for the Incurables.

Franklin Roosevelt is said to have contracted his poliomyelitis in 1921 during his visit, as an adult, to a Boy Scout camp. It was through the impetus and inspiration of Roosevelt that the public embraced the effort to find a cure or prevention for polio. As epidemics of polio continued through the decades of the 1930s and 1940s, the Mothers' March of Dimes raised a fortune in door-to-door contributions to fund vaccine research; government money was negligible or absent entirely. The early 1950s again saw outbreaks of 15,000 to more than 20,000 cases of paralytic disease each summer and more than double that number of nonparalytic, but still terrifying, cases.

By then, however, there was hope fueled by March of Dimes-funded research discoveries. Three poliovirus strains were identified and successfully grown in the laboratory by John Enders and his colleagues at Harvard. Growth of the virus allowed production of large quantities of experimental vaccines. The *Time* magazine cover story of 29 March 1954 pictured vaccine pioneer Jonas Salk and asked, "Is this the year?" Indeed, 1954 was the year that the first volunteers, nearly 2 million kids, received a shot of the still-experimental Salk killed polio vaccine (see chapter 7). It was licensed for general use in 1955, and the effects of the Salk vaccine were dramatic, with the number of reported cases plummeting over the ensuing few years. A setback, known as the Cutter incident (named for a California vaccine manufacturer), occurred shortly after the Salk vaccine was introduced. A bad batch of Cutter polio vaccine that was supposed to contain only killed vaccine actually contained some residual live virus. The bad batch of vaccine caused more than 200 cases of poliomyelitis; the mistake was quickly identified, and no further cases of killed-vaccine-related infection developed. Sabin's attenuated live virus vaccine (see chapter 7) was introduced in 1961.

By 17 November 1961, *Time* magazine and the nation were closer to celebrating an end to summer's panic when John Enders, the Nobel Prize-winning virologist whose work paved the way for development of the Salk and Sabin polio vaccines, was profiled on the cover beneath the heading, "Medicine gains on viruses." There were 2,525 cases of paralytic poliomyelitis in the United States in 1960 and only 61 in 1965. The last case of domestically acquired natural polio in the United States was in 1979. Worldwide, the efforts to eradicate polio are progressing, but not as quickly as had been hoped. Approximately 2,000 cases of paralytic poliomyelitis were reported for 2006, a slight increase in the total over 2005 and four times the number reported in 2001 (the year in which the world came closest to polio eradication with fewer than 500 cases). Hampered largely by political and access obstacles in countries with residual persistent low levels of cases—India, Nigeria, Pakistan, and Afghanistan—the World Health Organization's ongoing Global Polio Eradication Program is the largest public health initiative in history.

Measles

Prior to the introduction of measles vaccine in 1963 (see chapter 7), this viral infection was a rite of passage for all children; most kids had been infected by 6 years of age, and 90% of kids had had the infection by the time they were 15 years old; there were 3 to 4 million cases every year in the United States alone (corresponding approximately to the birth rate). Historically, references to measles can be found as early as the 7th century, and measles is referred to in 10th century writings as "more severe than smallpox." In the modern era, the disease has been judged by many to be the greatest killer of children worldwide.

Measles is caused by a virus of the same name and is spread by respiratory secretions (see chapter 2). The symptoms of early disease are fever, cough, runny nose, pinkeye (conjunctivitis), a total-body rash, and spots in the mouth. The danger of measles rests with its complications, which may be severe and include ear infections, pneumonia, diarrhea, encephalitis (see "Parents' Baker's Dozen Most *Feared* Infectious Diseases" above), and a degenerative neurological condition called subacute sclerosing panencephalitis. Although the severe complications occur in only a small percentage of the total number of measles cases, the huge number of total cases in the prevaccine era in the United States resulted in a large number of permanently brain-damaged children (from the encephalitis that occurs in 1 of every 1,000 cases of measles). Between 1 and 3 of every 1,000 measles patients died of pneumonia or encephalitis, an average of 450 childhood deaths in the United States each year in the prevaccine era.

With the introduction of vaccine in 1963 (see chapter 7), the incidence of measles in the United States fell by more than 99%. Success was a double-edged sword, however, as parents forgot what natural measles wrought and began questioning the need for (and safety of) the vaccine (see chapter 7).

With falling immunization rates, cases of measles and measles-related deaths resurged. From 1989 to 1991, there were almost 56,000 cases in the United States and more than 120 deaths, almost all in young, unvaccinated children; U.S. measles deaths in 1990 were at the highest level in more than 20 years. Outbreaks in the United States since that time have been among unvaccinated groups, e.g., religious sects and college students.

Worldwide, the story was much bleaker until 5 years ago, with as many as 50 million cases reported each year and 1 million deaths. The death rates from measles in developing countries, where malnutrition, dehydration, and vitamin A deficiency (see chapter 10) exacerbate the infection, are higher than the mortality in developed countries. In 1999, before an international consortium of health organizations targeted measles in the developing world, there were nearly 900,000 deaths; by 2005, the death toll had fallen to under 350,000, with the lives of 2.3 million children saved in that 5-year period since the beginning of the worldwide immunization initiative.

German Measles (Rubella)

German measles in children and adolescents is typically a benign infection that causes a rash, swollen glands (lymph nodes), and fever; complications in these kids are rare. As with parvovirus B19 infection (see "Parents' Half-Dozen *Former* Most Dreaded Infectious Diseases That Are Now 'So Last Year'" above), it is the *unborn* child of an infected pregnant woman that is at greatest risk of severe disease.

The name German measles comes from the fact that German investigators first distinguished this infection as something other than a variant of measles, which had been the prevailing theory prior to the early 19th century. Caused by the rubella virus, German measles is spread by respiratory secretions (see chapter 2) from person to person.

The congenital rubella syndrome was first described in 1941, when a cluster of cases of neonatal cataracts was observed during an Australian epidemic of rubella infections. Subsequent studies showed that infection of the fetus results in a wide variety of birth defects, including of the eyes, ears, heart, and brain. The resulting impairments of the newborn include blindness, deafness, abnormal heart structure, mental retardation, and perhaps psychiatric disorders (see "Schizophrenia" in "Dr. Rotbart's Half-Dozen *Chronic* Diseases for Which Germs May Someday Be Found" below). The virus is spread to the fetus across the placenta by the mother's blood (see chapter 2). Birth defects occur in the babies of 85% of women who contract German measles during the first trimester of pregnancy, half of the babies whose mothers become infected during the first half of the second trimester, and one-fourth of the babies of those infected during the last half of the second trimester.

Prior to the availability of vaccine, rubella occurred in epidemics every 5 to 10 years as new susceptible kids entered the population, i.e., when enough children who had never been exposed to rubella accumulated in the

U.S. population, rubella spread rapidly. In 1964 and 1965, there was an epidemic of more than 12 million infections in the United States, resulting in 20,000 newborns with congenital rubella syndrome; 11,600 of the babies were born deaf, 3,600 blind, and 1,800 mentally retarded. The CDC estimates that the cost of lifelong care for a single infant with congenital rubella exceeds $200,000 in today's dollars.

Today, because of the vaccine introduced in 1969 (see chapter 7), there are fewer than 25 cases of German measles reported each year in the United States. Small clusters of cases continue to occur periodically among unvaccinated populations.

Pertussis

First recognized in the 1500s, pertussis is due to an insidious bacterium, *Bordetella pertussis,* that infects the upper respiratory tract and produces a toxin that paralyzes the body's respiratory cells' ability to clear germs from the airways. The resulting disease starts out like the common cold (see "Parents' Dirty Dozen Most *Hated* Infectious Diseases" above), distinguishing itself by the severe "whooping" cough that develops several days to a week later. Whooping cough, once heard, is impossible to forget. Kids, especially young infants, develop short bursts of uncontrollable coughing, followed by a desperate sounding attempt to catch their breath; the whoop is the noise that is emitted by the gasping child during this deep breath in. Many kids with whooping cough turn blue because they cough out more air than they are able to breathe back in; kids often vomit after these bursts of coughing and whooping. Air hunger is not a pretty thing. These whooping episodes occur on average 15 times a day for several weeks.

In addition to the parental horror during the severe coughing episodes, pertussis also has an impressive list of severe complications, including pneumonia, seizures, brain damage (from lack of oxygen), and physical aftereffects of the severe coughing (including rib fractures, nose bleeds, hernias, and loss of bowel and bladder control). Death occurs in 1% of all cases of pertussis in babies younger than 2 months and in 1 of every 200 to 300 infants between 2 months and 1 year of age. Before the introduction of the vaccine in the 1940s (see chapter 7), between 175,000 and 200,000 cases of whooping cough occurred in the United States each year; the vaccine reduced that number to an average of less than 3,000 cases per year between 1980 and 1990. In 2004, in an era of increasing parental skepticism about vaccine safety (see chapter 7) and a shift in the target population of the disease (more older children and adolescents became infected, likely the result of waning immunity from earlier childhood vaccination), almost 26,000 cases of pertussis were reported in the United States, the highest number in 45 years.

Worldwide, pertussis incidence has also fallen in the vaccine era, but high numbers of cases continue to occur because of difficulties in adequately immunizing children in developing countries. In the 3 years 2000 to 2002, there were an estimated 40 million global cases of pertussis and 300,000 deaths due to the infection each year.

Mumps

If you read the newspapers in 2006, you realized that a significant mumps epidemic occurred, and you might wonder if this disease really belongs in the "so last generation" category. You're right, and I may have to relocate this discussion for future editions of this book. Infectious diseases are a moving target.

Beginning in Iowa in December 2005, an outbreak of mumps fueled largely by college students spread to 45 states and involved nearly 6,000 people, the majority of whom were 18 to 24 years old. Until this outbreak, fewer than 300 cases of mumps had occurred in the United States each year since 2000.

Mumps virus spreads by respiratory secretions. Symptoms start out like a common cold or flu but progress to cause swollen parotid (saliva-producing) glands, giving infected patients a "chipmunk" appearance. In its heyday, prior to vaccines, mumps was the leading cause of this parotitis. Since the vaccine was licensed in 1967 and recommended for routine use 10 years later, other viruses were statistically more likely to cause these swollen glands—that is, until this recent outbreak began. The swollen glands are uncomfortable, but the real hazard of mumps is the occurrence of complications, such as meningitis or encephalitis (see "Parents' Baker's Dozen Most *Feared* Infectious Diseases" above), testicular inflammation in boys, ovary and breast inflammation in girls, and deafness. About half of all boys who contract mumps after the onset of puberty develop testicular inflammation—pain and swelling, accompanied by nausea and fever, with sterility as a rare complication; 5% of girls develop ovarian inflammation not associated with sterility. During World War I, mumps was the third most common reason for hospitalization of soldiers.

The incidence of mumps dropped from more than 200,000 cases per year before vaccines to fewer than 3,000 cases annually between 1983 and 1985. In 2004, there were only 258 cases in the United States. The 2005–2006 outbreak is thought to have been due to a combination of waning immunity among college age students, the close dormitory living conditions of the students, and a subset of students who were underimmunized due to varying immunization requirements among different colleges.

Tetanus

"Tetanus shot" (see chapter 7) is as commonly recognized a term among patients visiting doctors' offices and emergency rooms as is "open wide." The shot, a modified form of the bacterial toxin that causes tetanus, is part of routine childhood immunizations and is also given to susceptible people (who have gone 5 to 10 years without a booster shot) after dirty wounds. Since its introduction in the 1940s, the vaccine has reduced the number of U.S. cases of tetanus to an all-time nadir of 20 in 2003.

Neonatal tetanus, a devastating complication of unsterile births, has been essentially eliminated from this country (two cases in the past 15 years, both in babies born to unvaccinated mothers) but remains a worldwide problem, with an estimated 200,000 cases each year.

Tetanus, also known as lockjaw, is the result of contamination of a wound by the soil bacterium *Clostridium tetani,* which, once established in the wound, produces a nerve toxin that circulates throughout the body and paralyzes muscles, usually beginning with the facial and neck muscles. The jaw muscles typically are locked, stiff neck follows, and total body rigidity often develops; 10% of cases result in death.

Diphtheria

When the *Corynebacterium diphtheriae* bacterium infects the respiratory tract, it sets up shop and begins producing a toxin. The toxin acts locally, producing a severe sore throat with a characteristic membrane that is tightly adherent to the throat and can obstruct breathing. Swollen neck glands (lymph nodes) trying to drain the infection may become so large that, along with swelling in the rest of the neck, they can give a "bull neck" appearance. The toxin can also get into the blood and act on distant organs, causing a sepsis-like picture (see "Parents' Baker's Dozen Most *Feared* Infectious Diseases" above). The nervous system and the heart are particularly susceptible to toxin damage. Death occurs in one-fifth of children under the age of 5.

The infection is transmitted from person to person (see chapter 2). In the 1920s in the United States, 100,000 to 200,000 cases of diphtheria and 15,000 deaths occurred each year. Since routine vaccine use began in the United States in the late 1940s, the incidence of diphtheria has fallen dramatically. Only two or three cases were reported each year between 1980 and 2004 and only five cases since 2000. The cases since 1980 have been in unvaccinated or undervaccinated patients.

Worldwide, diphtheria continues to be a problem. After the Soviet Union disintegrated, all 15 new states experienced outbreaks of diphtheria, mostly among unimmunized adults. The reasons for this epidemic are not entirely clear, but between 1990 and 1995, nearly 160,000 cases and 5,000 deaths were reported from that part of the world.

Dr. Rotbart's Half-Dozen *Chronic* Diseases for Which Germs May *Someday* Be Found

The Inspiration of Peptic Ulcer Disease

My late father was one of millions of Americans, 25 million each year to be more precise, who suffered from peptic ulcer disease (he died of something else). His generation was under the universal impression that peptic ulcer disease was the result of some bad combination of a stressful life, spicy foods, stomach acid, and bad genetics. Treatments were oriented toward addressing those presumed causative factors; milk became the preferred remedy, although it may actually have stimulated acid production in patients hoping for acid neutralization.

Then, two Australian physicians identified a bacterium, *Helicobacter pylori*, in the stomachs of patients with peptic ulcer disease. This finding was dismissed by most of the medical community until one of the Australian doctors, Barry Marshall, embarked on the definitive proof: he fed pure cultures of the bacteria to himself and documented the development of his own peptic ulcer disease. Peptic ulcer disease is now recognized as an *H. pylori* infection and treated with antibiotics—and cured. Marshall, our Australian hero, and his mentor, Robin Warren, were awarded the Nobel Prize in medicine in 2005. Marshall's words, from his autobiography on the Nobel Prize website (http://nobelprize.org/nobel_prizes/medicine/laureates/2005/index.html), provide insight into the difficulties associated with finding new causes for diseases when those findings dispute conventional wisdom. I think this excerpt nicely sets the stage for my subsequent discussion of other possible infectious diseases waiting to be discovered.

> But 1984 was a difficult year. I was unsuccessfully attempting to infect an animal model. There was interest and support from a few but most of my work was rejected for publication and even accepted papers were significantly delayed. I was met with constant criticism that my conclusions were premature and not well supported. When the work was presented, my results were disputed and disbelieved, not on the basis of science but because they simply could not be true. It was often said that no one was able to replicate my results. This was untrue but became part of the folklore of the period. I was told that the bacteria were either contaminants or harmless commensals.
>
> At the same time I was successfully experimentally treating patients who had suffered with life threatening ulcer disease for years. Some of my patients had postponed surgery which became unnecessary after a simple 2 week course of antibiotics and bismuth. I had developed my hypothesis that these bacteria were the cause of peptic ulcers and a significant risk for stomach cancer. If I was right, then treatment for ulcer disease would be revolutionized. It would be simple, cheap and it would be a cure. It seemed to me that for the sake of patients this research had to be fast tracked. The sense of urgency and frustration with the medical community was partly due to my disposition and age. However, the primary reason was a practical one. I was driven to get this theory proven quickly to provide curative treatment for the millions of people suffering with ulcers around the world.
>
> Becoming increasingly frustrated with the negative response to my work I realized I had to have an animal model and decided to use myself. Much has been written about the episode and I certainly had no idea it would become as important as it has. I didn't actually expect to become as ill as I did. I didn't discuss it with the ethics committee at the hospital. More significantly, I didn't discuss it in detail with Adrienne [Dr. Marshall's wife]. She was already convinced about the risk of these bacteria and I knew I would never get her approval. This was one of those occasions when it would be easier to get forgiveness than permission. I was taken by surprise by the severity of the infection. When I came home with my biopsy results showing colonization and classic histological damage to my stomach, Adrienne suggested it was time to treat myself. I had a successful infection, I had proved my point.

For those of us in infectious diseases, this act of courage (some might say lunacy) provided further reason to deepen our belief that someday all diseases will be found to be caused by germs—hopefully the type that can be treated. Here are six current candidates to someday become infections like the one that causes peptic ulcer disease.

Coronary Artery Disease (CAD) (Hardening of the Arteries)

Although risk factors for coronary artery disease (CAD) explain many cases, they do not explain all of them. There are patients lacking family predisposition, abnormal cholesterol levels, exposure to tobacco smoke, diabetes, and high blood pressure who nevertheless develop CAD. In other parts of the world, CAD seems to stratify according to wealth; the poorest have higher rates of heart disease, suggesting potential environmental factors yet to be determined.

The enthusiasm for the infection hypothesis to explain CAD and hardening of the arteries has ebbed and flowed for many years. The most recent crescendo came several years ago, when several seemingly convergent pieces of evidence pointed to germs as a cause of heart disease.

Inflammation (the body's immune response to infection and other irritants [see chapter 4]) has been implicated in the formation of arterial plaque, or hardening. Laboratory studies and animal models have shown that the presence of inflammation correlates with accelerated plaque formation and artery damage. Measures of body inflammation, such as the high-sensitivity C-reactive protein, predict heart attack risk.

Additionally, a specific germ has now been identified as a potential cause of or contributor to CAD. Many investigators have confirmed the footprints (protein or DNA from the germ) of *Chlamydia pneumoniae* bacteria in the plaque of patients with CAD; living chlamydia organisms have even been recovered from a few patients. Still other patients have high levels of antibody in their blood that is specifically targeted against *C. pneumoniae,* suggesting at least past exposure to the germ.

This would seem to be conclusive evidence of a causal role for this germ in CAD. But much as was the case with *H. pylori* and peptic ulcer disease, the debate now centers on whether the finding of bacteria in CAD patients is a coincidence or true evidence for causality (see chapter 12). *C. pneumoniae* is a recognized and extremely common cause of community-acquired pneumonia and a "cousin" of the germ that causes the STD (see "Parents' Baker's Dozen Most *Feared* Infectious Diseases" above). Many argue that chlamydia simply have an attraction for plaque, i.e., they get into the body, perhaps causing mild respiratory infection in the past; they circulate widely through the blood; and then they accumulate in sticky arterial plaque, where they find a cozy place to rest or even grow, but they don't necessarily *cause* plaque or CAD.

Others say both hypotheses may be true. Chlamydia may be attracted to already existent plaque and accumulate there but also perpetuate and even accelerate the formation of more plaque by stimulating an inflammatory (immune) response (see chapter 4) that targets the chlamydiae and in so doing damages the vessels in which the germs are nestled.

The question could be answered with antibiotic trials, as it was in part for peptic ulcer disease. To date, several antibiotic trials have been conducted and have led to mixed results and mixed interpretations, but a clear benefit of chlamydia antibiotic therapy in plaque formation or heart attack risk reduction has *not* been seen. Now the debate centers around the timing of antibiotics in relation to the infection, the dosage and duration of therapy, and other limitations of study design that might have prevented seeing a positive effect versus the equally strong possibility that chlamydia has nothing at all to do with CAD.

C. pneumoniae is not the only germ to be implicated as a possible cause of CAD. Recent interest has also focused on, of all things, gingivitis, inflammation of the gums (see chapter 9), and its more severe form, periodontitis (in which the inflammation damages teeth and bone). A laundry list of mouth bacteria are found in inflamed gums and periodontal tissues. Again, the hypothesis is based on finding a higher incidence of CAD in certain populations of people with bad gum disease. Either direct invasion by the germs into heart vessel walls or damage to those walls triggered by the inflammatory response to oral disease could be implicated. There are no stronger data, or clinical trials, to support this theory so far. CMV and other herpesviruses have been studied for years as potential causes of CAD, largely stimulated by the observation that a bird herpesvirus causes CAD in chickens. Again, there is some circumstantial evidence in humans for the presence of CMV or its footprints in patients with CAD, but 2 decades of research has failed to establish a causal role. Perhaps it's just another theory for the birds.

Finally, what's good for the goose . . . *Helicobacter pylori*, our now familiar nemesis that causes peptic ulcer disease, has been sought in CAD as well, with no luck so far. Any discovery of a germ as the cause of CAD will be important for kids' health because the inciting infection for adult CAD, if one exists, may well occur in childhood.

Schizophrenia

For many years, infection in the womb or in early infancy has been hypothesized as a potential cause of childhood or adult onset psychiatric disorders, including schizophrenia.

Most attention has focused on rubella, because congenital rubella syndrome [see "Grandparents' Half-Dozen (Plus One) *Former* Most Dreaded Infectious Diseases That Are Now 'So Last Generation'" above] is a proven cause of numerous defects of the nervous system, including deafness, blindness, and mental retardation. While schizophrenia has received more study than other psychiatric or behavioral conditions, autism, impaired social skills, and even separation anxiety have been hypothesized to result from rubella infection in the womb.

The strongest evidence for a role of rubella in schizophrenia comes from follow-up study of adults who were born with congenital rubella infection during a rubella pandemic in 1964 [see "Grandparents' Half-Dozen (Plus One) *Former* Most Dreaded Infectious Diseases That Are Now 'So Last Generation'" above]. There were 11 adults with schizophrenia among the 53 available for study, a 10- to 20-fold increase in the occurrence of schizophrenia compared with that expected in the general population, although no control patients were included in this evaluation (e.g., adults born in 1964 without congenital rubella). Unfortunately, although there were 20,000 cases of congenital rubella in the United States that year, few were available for long-term follow-up in this study.

Influenza virus infection of pregnant women (see "Parents' Dirty Dozen Most *Hated* Infectious Diseases" above) has also been implicated as a potential cause of schizophrenia in their offspring. Although many studies attempted to confirm a 1988 publication linking the virus to the disease without producing consistent results, a 2004 investigation used better methods to verify that influenza infection had actually occurred in the pregnant women being studied. Sixty-four adult patients with schizophrenia were included in the study and about double that number of controls without schizophrenia. Titers of antibody to influenza virus were measured in stored blood obtained from their mothers during their pregnancies 30 to 38 years earlier. Of those who experienced infection in the first trimester of pregnancy, there was a sevenfold increase in the likelihood of schizophrenia developing in their offspring later in life compared with controls; exposure to influenza later in pregnancy was not associated with an increased risk of schizophrenia.

The same investigators, using the same blood specimens as in the influenza study, found a 2.5-fold increase in the risk of schizophrenia if pregnant women were exposed to *Toxoplasma gondii,* a common parasite also known to cause congenital neurological abnormalities (see chapter 2).

HSV-2 (see "Parents' Baker's Dozen Most *Feared* Infectious Diseases" above), the cause of genital herpes, has been linked to schizophrenia in a small study of 27 schizophrenic adults and double that number of nonschizophrenic controls for whom blood samples had been stored from the time of their mothers' pregnancies. There was a higher incidence of HSV-2 antibodies in the schizophrenic patients' mothers' sera than in those of the controls. A similarly designed study done subsequently by another group failed to confirm that association.

Finally, a potentially unifying hypothesis, with no supportive data to date, suggests that any, or many, infections during pregnancy might reactivate latent genes, perhaps from a primordial infection with a retrovirus (retroviruses are germs that go backwards in their reproduction scheme, from RNA to DNA instead of the usual DNA-to-RNA sequence; HIV is the prototypic retrovirus) that has become part of our inherited chromosomal DNA, and that those reactivated genes may predispose to schizophrenia. This one has a long way to go.

Tourette's Syndrome

This perplexing and disturbing neurological condition is notable for causing uncontrollable movements and tics, as well as uncontrollable utterances of obscenities or other inappropriate sounds. Long thought to be genetic, because the disease runs in families, there is some recent evidence that the disease might be triggered by *S. pyogenes* (group A streptococcus, the same germ that causes strep throat [see "Parents' Dirty Dozen Most *Hated* Infectious Diseases" above]). Anecdotal reports of kids developing Tourette's syndrome immediately after strep throat or of patients with preexisting Tourette's getting worse with strep throat have fueled the interest. The theory is that the antibody immune response (see chapter 4) to the strep germ may cross-react with proteins in the brains of patients, causing damage.

Similar anecdotal reports link individual cases of Tourette's with recent *M. pneumoniae* and Lyme disease (see "Parents' Baker's Dozen Most *Feared* Infectious Diseases" above) infections. Anecdotal reports are severely limited by the risk of coincidence (see chapter 12), particularly when very common infections are studied, i.e., common infections are common, in patients with neurological disorders as well as in the general population (see chapter 12).

Recently, a large population-based study, the type of research much less likely to be confounded by coincidence, examined kids between 4 and 13 years old with Tourette's and found a strong association with strep infection occurring within 3 months of the onset of the syndrome (a 3-fold increase in occurrence of recent strep in Tourette's patients compared with controls), as well as within 12 months of onset (a 2.5-fold increase). Most impressive, perhaps, is the finding that among children with Tourette's, the chances of having had two or more strep infections within the previous year were more than 13 times greater than in control subjects. If some strep is bad, perhaps more strep is worse.

OCD

The story linking infection with obsessive-compulsive disorder (OCD) is very similar to that with Tourette's, and there is even some overlap between the patients, i.e., there are patients with both problems.

OCD is a disturbing anxiety condition, often of childhood onset, that is notable for obsessive and unwanted thoughts and compulsive and repetitive behavior patterns. The compulsive rituals, such as hand washing, cleaning, counting, and double-checking, are undertaken in an attempt to make the obsessive thoughts go away.

The evidence for infection, and particularly for infection with the strep throat germ *S. pyogenes* (see "Parents' Dirty Dozen Most *Hated* Infectious Diseases" above), began to accumulate with the observation that patients with Sydenham's chorea, a known movement disorder complication of strep throat infection, had a higher than usual occurrence rate of OCD. As with Tourette's, a number of studies have reported individual patients or groups of patients in whom strep infection seemed to either trigger or worsen OCD.

The same population-based study that looked at Tourette's and strep (see above) also examined OCD and found very similar results. Patients with OCD (a small percentage of whom also had Tourette's) had a nearly fourfold-increased occurrence of group A strep infection in the 3 months prior to the onset of OCD. The association was not, however, present for infection within a year prior to the onset of OCD, nor was the occurrence of multiple strep infections preceding OCD as strong an association as that seen in the same study with Tourette's syndrome patients.

The mounting evidence for a causal or contributory role for strep infections in OCD, Tourette's syndrome, and other tic disorders has resulted in the coining of a new medical term to describe these patients: PANDAS (for pediatric autoimmune neuropsychiatric disorders associated with streptococcal infection). How the PANDAS saga unfolds "bears" watching. (Sorry.)

Chronic Fatigue Syndrome (CFS)

All attempts at defining CFS and finding a cause for it have been complicated by the nonspecificity of its clinical findings and the overlap with a variety of other medical and psychiatric conditions. Patients who suffer from CFS, and suffer is the right word, know exactly how to describe their disease, but it is difficult to find unifying descriptions for all patients.

Dominated by unremitting or recurring severe fatigue, patients with CFS also have a variety of other chronic complaints, including headaches, muscle and joint aches, mood and sleep disorders, memory lapses, tender glands (lymph nodes), and difficulty with normal thought processes (e.g., focusing on work).

Many (many!) infectious agents have been associated with cases of CFS; if there is one common piece of history that a majority of patients give, it is that their CFS symptoms seemed to follow an infection. In fact, postviral fatigue syndrome is another moniker by which this disease has become known. The leading candidate for the germ that causes CFS was for many years EBV. The severe fatigue, tender glands (lymph nodes), and low-grade fevers seen in patients with infectious mononucleosis (see "Parents' Dirty Dozen Most *Hated* Infectious Diseases" above), due to EBV, made it a logical suspect. However, despite many years of intense investigations, a consistent link between CFS and EBV has yet to be established. Studies that purported to find an association were refuted by more studies that did not. Other candidate germs have met similar fates, including the enteroviruses, human herpesvirus 6 (HHV-6), *C. pneumoniae* [see "Coronary Artery Disease (Hardening of the Arteries)" above], and hepatitis viruses. None has been proven to be the cause of CFS, and large studies have disproven most of them.

A single-germ cause of CFS may never be identified, but the intriguing complaint that many CFS patients bring of an often trivial virus-like illness that immediately precedes the abrupt onset of CFS suggests that there may be a common pathway that different germs may take in triggering CFS. Either an immunological response gone awry (see chapter 4) or a genetic predisposition activated by germs may result in CFS. As of now, the search goes on.

Diabetes Mellitus

There are many parallels between the evidence for a germ cause of diabetes mellitus and the evidence for infection in CFS, most notably, the pattern of great excitement when one or another new research study or laboratory method seemingly identifies "the germ," followed by the sobering failure to consistently confirm the results (see chapter 12).

In the case of diabetes, it is the enteroviruses, and specifically the coxsackievirus group of enteroviruses, that have received the most attention. Animal studies and some human studies have suggested a link and a hypothesis as follows. Coxsackieviruses attack the pancreas cells that produce insulin. In genetically prone individuals, an overexuberant immune reaction (see chapter 4) ensues, and the body's own cells or antibodies attack the pancreas, along with the virus. To date, as with the other diseases discussed in this section, studies refuting the association have also been published, and definitive proof for an infectious cause of diabetes is still lacking.

A Half-Dozen Leftovers

There are a few common childhood infections that are neither among the most hated nor among the most feared. They don't fit into the category of former most dreaded either, because they are still very much with us. Here, then, are the leftovers.

Viral Infections of the Mouth: Cold Sores, Fever Blisters, Stomatitis, and Hand-Foot-Mouth Syndrome

Herpes simplex viruses (HSVs) cause genital herpes (see above) and brain infections (see above) but most commonly cause cold sores and fever blisters on the lips and mouth. The first infection a child gets with HSV may be quite dramatic: a mouthful of painful blisters that can make drinking and eating very difficult, accompanied by fever; this condition is called stomatitis, meaning an inflammation of the mouth. Dehydration can complicate stomatitis, the result of painful swallowing and reduced fluid intake. Like all viruses in the herpesvirus family (see chapter 1), HSV becomes dormant in the body; in the case of HSV, the hiding place for the virus is the nerve roots of the mouth and lips. During times of stress, e.g., fever, sun exposure, menstruation, or final exams, the dormant virus reactivates, making its way down the nerve root and back to the mouth or lips. Rather than the widespread mouth infection (stomatitis) sometimes associated with the first herpesvirus eruption, reactivation episodes usually are limited to one or a few painful blisters.

Another group of viruses, the enteroviruses (see chapter 1), also cause stomatitis, sometimes accompanied by blisters on the hands and feet (hand-foot-mouth syndrome). The enteroviruses also cause summer colds and meningitis (see above); unlike the herpesviruses, enteroviruses do not become dormant and do not reactivate; when the mouth infection is gone, it's gone, but while it's there, it can also cause dehydration due to pain on drinking.

How contagious is it?

- Moderately contagious by direct personal contact with sores or contact with respiratory secretions from an infected individual; also contact with eating utensils, toothbrushes, etc., from an infected individual (see chapter 2)

Prevention

- Personal and household hygiene (see chapter 9)

Treatment

- Careful attention to maintaining fluid intake
- In most cases, no antiviral treatment is necessary, although both topical and oral antiviral medications are available for severe cases of herpes stomatitis or frequent and painful recurrent cold sores/fever blisters (see chapter 6).

Viral Skin Rashes: Roseola and Roseola Impersonators

Most cases of roseola are caused by another member of the herpesvirus family (see chapter 1), HHV-6; a related virus called HHV-7 may cause some cases of roseola. A benign infection, roseola achieved fame early in the description of childhood illnesses because of its propensity to cause high fevers in young children, who, if they read the textbook correctly, would then develop a whole-body "rosey" rash just as the fever broke. A flu-like illness accompanies the fever in some kids, although most feel pretty good despite the fever. There are many variations on the theme, with the timing of fever and rash not always as predictable as in the textbook. Part of that variability has to do with the fact that viral rash illnesses in kids are extremely common and are caused by many more viruses than just the two responsible for roseola. As a result, your children's doctor may tell you that this new rash is roseola—again! "But doctor, my daughter's already had roseola!" Yes, that's true, but maybe the last one wasn't roseola and maybe this one isn't roseola either. We don't routinely test for specific viruses in kids' rash-and-fever illnesses, but as long as they get better, it really doesn't matter what the name of the virus was that caused this particular rash. In as many as 10% of kids with roseola, the fever spikes high enough that they may have what's known as a febrile seizure, a usually benign generalized seizure that does not recur once the fever resolves.

How contagious is it?

- Moderately contagious by direct personal contact or contact with respiratory secretions from an infected individual (see chapter 2)

Prevention

- Personal and household hygiene, although these may not work because the most contagious period may be before the development of the fever or the rash (when the child does not appear ill) (see chapter 9)

Treatment

- Acetaminophen or ibuprofen for fever (see chapter 8)

Bacterial Skin Infections (Impetigo and Others)

Skin infections can occur at cuts and scrapes, at bug bites, in hair follicles, and through any other "opening" or skin defect into which germs can creep, including eczema and allergic skin rashes that kids are prone to scratch. Impetigo refers to a specific kind of skin infection, typically due to *Staphylococcus aureus* or *S. pyogenes* (group A strep), that causes blisters and weepy-looking sores. The germs that cause impetigo are very readily spread from one part of the skin to another and from one child to another. The area around the nose and mouth seems to be most commonly involved, but any and all areas of the skin are susceptible.

How contagious is it?

- Very contagious by direct personal contact or contact with clothes, towels, or linens from an infected individual (see chapter 2)

Prevention

- Personal and household hygiene (see chapter 9)
- Covering infected areas will limit the spread from one area of the skin to another and from one child to the next.
- Trimming nails; avoiding scratching

Treatment

- Topical antibiotics (see chapter 5)
- Oral antibiotics (see chapter 5) if more extensive or unresponsive to topical antibiotic therapy

Fungal Infections of the Skin: Ringworm, Athlete's Foot, and Jock Itch

Dermatophytes are fungi that like (phyte) the skin (dermato). These germs creep into hair follicles or settle on moist areas of the skin, like the feet and groin. They get their resilience from their ability to grow on the tough outer layer, called keratin, of skin, nails, and hair. A number of different species are involved (see chapters 1 and 6), but all respond well to therapy. Scalp infections are the most difficult to treat and require oral medicines; all other skin fungal infections can be treated topically (see chapter 6). Ringworm gets the ring part of its name from the circular form that most cases assume; it may be a hollow circle or a filled-in circle, defined by raised and moist-looking bumps that are typically quite itchy. The worm part of the name is a misnomer; there is no worm involved. Athlete's foot and jock itch are red, scaly, itchy, and sometimes painful rashes that thrive in the moist crevices between the toes and in the groin.

How contagious is it?

- Moderately contagious by direct personal contact or contact with clothes, shoes, or hats from infected individuals; also from shower floors in the case of athlete's foot (see chapter 2)

Prevention

- Personal, household, and community hygiene (see chapter 9)

Treatment

- Topical antifungal creams and ointments (see chapter 6)
- Oral antifungal medicines (see chapter 6) for scalp ringworm and for other skin fungal infections that do not respond to topical therapy

Fungal Infections of Mucous Membranes: Thrush and Vaginal Yeast Infections

Candida albicans is the most common yeast form of fungus (see chapter 1); this germ has a propensity to grow on the mucous membranes lining the mouth and the vagina. Low levels of the yeast are present in most people all the time, but the yeast may emerge to grow in higher concentrations and cause symptoms. The most common conditions that facilitate the emergence of yeast infections in otherwise healthy kids (i.e., with no immune abnormalities) are antibiotic use and hormonal variations. In babies, yeast in the mouth (termed thrush) is often accompanied by a yeast diaper rash as the swallowed germ exits at the other end. Antibiotics for ear infections and other infant illnesses predispose to thrush, but babies can develop the yeast infection with no previous antibiotic use; maternal hormonal variations may influence a baby's risk for thrush via breast-feeding. In adolescent girls, vaginal yeast infections may be precipitated by antibiotics, menstruation, and pregnancy. Thrush may be painful in older kids and adults (we also worry that thrush in an older child or adult may herald an abnormal immune system; see chapter 4), but babies usually seem unaffected except for the white plaques that parents see in their babies' mouths. Vaginal yeast infections are itchy and painful.

How contagious is it?

- Probably not contagious; yeasts are present in most peoples' mouths and vaginas but require favorable growth conditions before emerging with symptoms.

Prevention
- Personal hygiene (see chapter 9)
- Avoiding unnecessary antibiotic use

Treatment
- Topical antifungal creams and ointments (see chapter 6)
- Oral antifungal medicines (see chapter 6) for infections that do not respond to topical therapy (rarely needed)

Pinworms

The parasite *Enterobius vermicularis* is the most common worm infestation of kids in the United States. The little worms are ingested as eggs that are transmitted from one child to another (frequently in day care centers [see chapters 1 and 2]). The eggs make their way through kids' intestines, maturing along the way, and then emerge as small (the size of a headless thumb tack) adult creepy crawlers in and around the anus. The female adult worms lay eggs on the skin around the anus. Pinworms are very itchy, which is usually the way that parents first notice them and the way that the worms are spread: hand to mouth from child to child.

How contagious is it?
- Moderately contagious by direct personal contact or contact with clothes, linens, or towels from an infected individual (see chapter 2)

Prevention
- Personal, household, and community hygiene (see chapter 9)

Treatment
- Oral antiparasite medicines (see chapter 6)

Conclusions

That concludes our tour through the plagues that plague parents and grandparents most. Some of the infections that didn't make the lists in this chapter are discussed elsewhere in this book in different contexts. For example, it is beyond the scope of this chapter to discuss all the possible infectious maladies of travel abroad (although a number of the important ones made it to the lists). Travelers' infections are almost as varied as the places that people can travel. However, many of the exotic worms and germs from abroad are noted in chapters 1 and 6. For a country-by-country description of infection risks and prevention strategies, including vaccines for travelers, I refer you to the CDC's *Health Information for International Travelers*, also known as the Yellow Book, available free of charge online at: http://wwwn.cdc.gov/travel/contentYellowBook.aspx.

Most importantly, perhaps, in bringing this chapter to an end is the recognition that infectious diseases are constantly moving targets. Some come, while others go. Diseases that seem to go away may come back. New diseases emerge, and old diseases are discovered for the first time using new technologies. The arrangements of infections in the arbitrary sections above may well change by the next edition of this book. We are in a perpetual and precarious equilibrium with the germs around us; we prevail more often than do the germs, but it's an ongoing battle.

II. WEAPONS IN THE WAR

4

Homeland Security: Our Natural Defenses against Germ Invasions

By now you're probably terrified, and I take full responsibility. The plethora of pathways for germ invasion discussed in chapter 2, coupled with the litany of loathsome and lethal diseases available to your kids in chapter 3, have unsettled you. Well, now it's time to resettle and reassure you. Your kids' health is in the good care of their own innate immunity.

Imagine an immune system that is so highly evolved that billions, no *trillions,* of germs that coevolved with it are left stranded at your kids' portals, frustrated by their inability to break through. Imagine a Star Wars-like defense shield around your kids' bodies, constantly on guard against nearly infinite microbial terrorist threats. And, finally, imagine a biological system so sophisticated and intricate as to . . . well, defy imagination; a highly complex and interwoven network of chemical and physical resources that continues to baffle and astound scientists who have pondered human immunity for centuries.

Our bodies are fortresses, built to withstand ferocious onslaughts of microscopic enemy invaders. Our skin serves as the outer walls and moat, resisting the swarms of germs in the environment around us and on us. Our mucosal surfaces, those more delicate coatings of our respiratory, genital, and gastrointestinal tracts, are akin to the walls of the inner courtyards of the fortress, vigilantly guarding against deeper penetration by germs already living peaceably inside us.

In the event that a breach in the castle walls occurs, our remarkable immune system is immediately summoned into action. (If I use the word remarkable a lot in this chapter, it's because there are few synonyms that convey the truly extraordinary nature of our immune system.) White blood cells respond to the alarm within minutes, rushing to the site of the infringement to digest the invaders while simultaneously signaling other cells to join in the fight. Antibodies are produced that help eliminate the current infection and protect against future incursions by the same germs. These elaborate, coordinated defense mechanisms have allowed humankind, and our evolutionary forbears, to survive and thrive for epochs before the recent conveniences of antibiotics and vaccines.

This chapter lays the groundwork for understanding human immunity—what your kids bring to the table in their "negotiations" with the germ world. Of course, as you read on, you'll be asking yourself, "What can we do to boost, or at least avoid depressing, our kids' immune systems?" For the answers to that question, you can patiently wait to reach Part III of this book ("Wear Your Boots in the Rain"), on the role of nutrition, nutritional supplements, vitamins, minerals, herbal remedies, stress, sleep, exercise—and even wardrobe and the weather—in kids' abilities to fight off infections. Then again, you can cheat and skip ahead to Part III and read chapters 10 and 11, where the hope and hype of immune-boosting strategies are presented.

This chapter also probes the "my child is constantly sick" syndrome. If your child has 10 or more episodes of infection a year, is that abnormal? (Hint: no!) Finally, the signs of truly excessive susceptibility to infection that parents need to be concerned about are reviewed. Certain patterns of infection *do* suggest a possible immune system deficiency that could require immediate intervention. Distinguishing these "immunocompromised" kids from normal kids who just happen to get repeated infections can make the difference between life and death.

The "Outer" Immune System: Physical Barriers to Infection

Our kids' bodies are covered, and their inner cavities are lined, with protective coatings that form the first line of defense against potential invaders.

The Skin

In a recent study from New York University that was trumpeted around the world, microbiologists molecularly analyzed the bacteria found on the forearms of humans and discovered that "the skin is home to a virtual zoo of bacteria" (Martin Blaser, lead investigator, as quoted by the BBC on 6 February 2007)—and the forearms are "clean" compared with other mentionable (like the underarms) and unmentionable areas of skin. We have long known that the quantitative load of bacteria on our kids' skin is astronomical, numbering in the trillion range (see chapter 1). What Blaser and his colleagues learned was that the diversity of bacterial species on the skin is much greater than previously recognized; they discovered germs on the skin of normal adults, living right out there in the open, that had never previously been known to exist, yet the skin on these forearms was not infected and the humans attached to the forearms were not ill. The important difference between colonization and infection (see chapters 1 and 2) is most evident on the skin and mucous membranes (see below), where the vast majority of germs live with your kids in harmony. However, when the skin barrier is breached, e.g., by a penetrating wound or a scraped knee, the heretofore harmless germs on the skin are allowed deeper entry, and that's when the trouble begins.

The skin is the largest organ of the body, comprising about one-sixth of total body weight. When intact, the two main layers of the skin, the epidermis and the dermis, combine to form a virtually impenetrable barrier to germs—the outer walls and moat of your child's fortress. The epidermis ranges in thickness from 0.05 mm on the eyelids (about the thickness of a pencil line) to as thick as 1 mm on the soles of the feet, and the epidermis itself has sublayers. The most superficial sublayer is composed of dead skin cells, which are being constantly shed and entirely replaced every couple of weeks by new cells that make their way up from the lowest sublayer of the epidermis. The keratin component of the epidermis gives the skin its toughness and makes it waterproof—and virtually germ proof! Beneath the epidermis lies the dermis, the thicker (0.3- to 3.0-mm) layer of skin that contains sweat glands, oil glands, hair follicles, collagen, and cells that make the skin elastic. The dermis also contains blood vessels, small capillaries that can spread germs from a penetrating wound to the rest of the body via the bloodstream. Lying just beneath the dermis are the subcutaneous (under the skin) tissues, where larger blood vessels course; germs that reach the subcutaneous tissues and those larger blood vessels have ready access to the rest of the body. That's what makes the barrier function of the skin so vital to your kids' health.

Mucous Membranes

Lining all body cavities, mucous membranes are akin to the walls of the inner courtyards of your kids' immune fortress. These thin, flimsy-looking skin-like structures compensate for their delicate constitution by secreting a thick mucus coat that supplements the protective function of the membranes. The result is a highly effective barrier to the penetration of respiratory, gastrointestinal, and genital germs—those that live there natively (colonizing) and those foreign fighters that seek entry. The mucus physically traps germs and also carries special types of antibodies (see below) that can actually bind to germs and inactivate them. If invaders are able to get through the mucus, they find a layer of cells that are tightly connected, one to the next, by junctions that make germ passage between them very difficult.

The mucous membranes of the respiratory tract have an additional protective mechanism: they are lined with tiny hairlike structures called cilia that beat in a wavelike motion to move germs and other foreign matter up and out. The gastrointestinal tract membranes also have a unique weapon: the highly acidic nature of stomach juices kills many would-be germ invaders that travel in food and water; one risk factor for severe salmonella infections (see below and chapter 3), in fact, is the recent consumption of acid-neutralizing foods or medicines.

The "Inner" Immune System

Although highly effective, the skin and mucous membranes are not always able to prevent germs from breaking through and entering deeper tissues of our kids' bodies. All human infections result from a victory of germs over the protective barriers that are designed to keep them out. Once the germ gets in, the job of containing it and/or killing it falls to the inner immune system, a sophisticated network of blood cells, tissue cells, antibodies, and specialized body chemicals designed for defense.

Cells

The blood is made up of three different classes of cells, as well as the liquid (plasma) component that keeps the cells flowing through the arteries, veins, and capillaries of the body. Red blood cells carry oxygen to all organs of the body. Platelets are fragments of cells that circulate in the blood and repair damaged blood vessels by triggering a clot at the site of the injury. White blood cells are the body's major soldiers in the war against germs. Whereas red blood cells and platelets are each of a uniform type within an individual, with variations in their appearance only a function of their maturity, white blood cells are quite heterogeneous, having evolved into several different types, each with its own distinctive appearance and unique immune mission. Since red blood cells and platelets do not have immune functions, the rest of this discussion focuses on white blood cells.

White blood cells can be grouped into the "eaters," cells that ingest the bad guys, and the "facilitators," cells that secrete chemicals that help the eaters eat. The eaters, scientifically known as phagocytes, include the neutrophils and the monocytes/macrophages. Neutrophils are the body's main inflammatory cells, those cells that first rush to the scene of an infection or an injury. Neutrophils surround and consume invading germs and then release potent chemicals from tiny packets stored within the cells; the chemicals finish the job on the germ. Neutrophils are most effective at fighting bacterial and fungal infections. Monocytes circulate in the blood to tissues that have been invaded by germs; once the monocytes have reached the affected areas, they leave the blood vessels and migrate into the infected tissue, where they are now called macrophages. Once in the tissues, macrophages consume germs, clean up infectious debris from the battle that is taking place, and release chemicals that signal other cells to join the fight.

Among the white blood cells that serve as facilitators in the immune system, the lymphocytes are by far the most important. Like every component of human immunity, lymphocytes are highly specialized and diversified in their functions. B-cell lymphocytes produce antibodies, the molecules that specifically target each new germ species that gets into your kids (see below). B cells are remarkable (there's that word again). Each B cell is programmed to produce antibodies against only one germ. When that germ is detected by a B cell, the B cell matures and clones itself to produce many daughter cells (called plasma cells) that also produce the antibody against that one germ. Millions of antibody molecules are generated against each invading germ, the result of the amazing repertoire of B cells in the body. In contrast to B cells, which recognize germs, T-cell lymphocytes recognize infected (or malignant) human cells in the body and destroy them before the infection (or malignancy) can spread. T-cell lymphocytes are further specialized into those that actually kill the abnormal cells and those that help in the killing. The important thing to know about T cells is that they function by recognizing molecular discrepancies on abnormal cells that distinguish those cells from normal cells; this allows T-cell lymphocytes to be precise in their killing without harming the healthy cells of the body. Healthy cells are seen as "self" by the T-cell lymphocytes and left alone.

Occasionally this remarkable T-cell system for distinguishing self from nonself breaks down. There may rarely be overlap between the fingerprint that an invading germ leaves on an infected cell and the self fingerprint on normal cells, and T cells may mistakenly attack healthy cells of the body "thinking" that those cells are infected. Often, a prompting infection can't be identified in these self-attacks—the body just seems to turn on itself. These autoimmune reactions are responsible for some forms of arthritis and heart disease, lupus, and diabetes, among other disorders. Additionally, T cells are responsible for recognizing nonself markers on transplanted organs and transfused blood cells: it is the T-cell lymphocyte that causes organ rejection and transfusion reactions.

Antibodies

Antibodies are like laser-guided molecular missiles, able to seek out and latch onto target molecules (called antigens) that exactly fit the structure of the antibody, like two pieces of a puzzle coming together. Antibodies are produced by B lymphocytes in response to those cells detecting a germ invader that has penetrated the first-line defenses of the skin or mucous membranes (see above). The antibody is released into the blood circulation and homes in on the target antigen; once the antibody is bound to the target, immune cells "know" this target must be eliminated, and it is. As the now antibody-coated invader passes by phagocytic (eater) cells, those cells recognize the germ as foreign, and they ingest and destroy it. A special type of antibody, immunoglobulin A (IgA) (see below), is found in mucous-membrane secretions, patrolling the respiratory, genital, and gastrointestinal tracts for germs which those antibodies are preprogrammed to attack, adding an additional level of protection to the first-line defenses.

Antibodies are part of a family of molecules known as immunoglobulins. When blood is processed, the fraction of the plasma (the liquid component of blood) that contains most of the antibodies is called "gamma globulin"; this antibody-rich blood component can be pooled from many donors and is used to prevent and treat a number of infectious diseases (see chapters 3 and 7). Each of the classes of immunoglobulins (IgG, IgM, IgA, IgE, and IgD) has its own role in the immune response. Deficiencies of IgG are those most commonly associated with an increased risk of severe infections (see below).

Allergies and asthma are examples of overexuberant immune responses, in which certain immune cells overproduce antibodies, an overreaction to provocations by specific triggers that the body sees as more threatening than they really are. The result is that what should have been a completely benign encounter with pollen, grass, or dander results in symptoms that are entirely the result of the body's overreaction rather than anything the triggering antigens themselves cause. The body sees pollen as a threat, but the real threat is the body's own overenthusiastic response to the pollen.

Chemical Messengers of the Immune System

As noted above, antibodies are molecules produced and secreted by B lymphocytes. Lymphocytes and other immune cells also secrete a variety of messenger molecules that serve as the communication network for the immune system. Lymphocytes (facilitator cells) signal phagocytes (eater cells), phagocytes signal lymphocytes, lymphocytes signal other lymphocytes, and the resulting concerted response protects your kids from germs that they have seen before and new ones that are trying to gain a foothold. The nomenclature of these messenger molecules is confusing and not especially important for this discussion, except that some of them have been used in treatment of viral infections (see chapter 6). The term cytokine refers to a diverse group of these substances that act on or stimulate (kine; e.g., kinetic) cells (cyto). Subgroups of cytokines include lymphokines, monokines, and chemokines, each of which is derived from or acts on a different cell type. The names of specific cytokines that you may have heard of include interferon, interleukin, and tumor necrosis factor. Other, similar molecules, also termed cytokines by some, include cell growth factors that act on immune cells, as well as cells outside of the immune system; examples of growth factors that are used for the treatment of certain immune deficiencies are given below.

The complement system is a collection of about two dozen protein molecules that comprise yet another branch of the immune system; it is so named because it complements the actions of antibodies and immune cells. When one of those proteins is called into action by the immune system, a cascade of changes occurs in the entire set of complement proteins, resulting in a battering-ram-type structure that punches holes in the wall of invading germs, facilitating their destruction. Remarkable, isn't it?

Immune Organs and Tissues

While many of the components of the immune system are circulating through the bloodstream at any one time, others are in place in specific organs and tissues of the body. Specifically, lymphocytes divide their time between the blood and the so-called lymphoid tissues.

The lymph nodes are bean-shaped structures that are distributed throughout the body. They are commonly referred to as glands, e.g., "my daughter has swollen glands in her neck"—an imprecise term. The term glands should be reserved for body structures that have as their primary purpose the production and secretion of specialized chemicals (e.g., glands that secrete saliva, hormones, sweat, or digestive juices); these typically have nothing to do with the immune system. In this book, when I use the term glands to refer to lymph nodes, I'll parenthetically indicate that the more appropriate term is lymph nodes.

Lymph nodes are the waste-processing centers for germ disposal. As germs or the remnants of germs enter a lymph node, lymphocytes in the node process the germ for disposal and trigger the production of antibodies and more lymphocytes, which are released from the lymph nodes into the blood circulation to hunt for more of the same type of germ. Lymph nodes are connected by tiny vessels called lymphatics to form lengthy chains of a germ sewage system that parallels the blood vessels, coursing throughout the body. Lymph nodes are the hubs of this disposal system, concentrated in areas where lots of germs and their byproducts drain: the armpits (where all hand and arm germs are processed), the groin (where foot and leg germs are processed), the neck (where head and face germs are processed), and the chest and abdominal cavities. Within the intestines, for example, small patches of lymphocyte-rich tissues serve as yet additional barriers beneath the mucous membrane and intestinal cells lining the gut.

Other lymphocyte-rich organs within the body also serve as monitoring stations for circulating germs; these include the spleen, the tonsils and adenoids, and the appendix. While the last appears to be a remnant of some primordial function and is often more trouble than it's worth (the appendix traps fecal matter and becomes inflamed [appendicitis], often requiring surgical removal), the importance of the other lymphocyte-rich organs is increasingly recognized. Patients with dysfunctional or absent spleens are prone to very severe bacterial infections, and the trend several decades ago to liberally remove the tonsils and adenoids from patients with frequent sore throats and colds has been tempered by the recognition that those structures are part of important defense mechanisms against more serious respiratory infections.

The bone marrow is the production factory for all blood cells, including all circulating cells of the immune system. A primordial cell type in the bone marrow, the stem cell, has the ability to become any blood cell in the body and does just that. Another organ, the thymus, lies just under the sternum (chest bone) and serves as a preparatory center for T lymphocytes; these cells enter the organ as immature foot soldiers and leave as mature warriors ready for the immune battles ahead. I know what you're thinking—this is really remarkable!

The Immune System in Action: How It All Fits Together

While each of the components of human immunity is impressive in its own right, the most amazing aspect of the immune system is its coordinated response to invading germs. Because different classes of germs use different strategies of attack (see chapters 1 and 2), I'll illustrate the coordinated immune response with four representative scenarios.

Salmonella Gastroenteritis

Fresh farm eggs seemed like the perfect choice for your homemade ice cream birthday party; all the neighborhood kids were invited. Unbeknownst to you, however, the salmonella bacteria that are so ubiquitous in poultry (see chapter 2) silently and secretly contaminated the unpasteurized farm eggs and joined the happy mix in the blender. The germs, now tastefully coated with chocolate and pralines, are swallowed and enter your child's stomach (and the stomachs of all the friends at the party), where the usual acid barrier is partially neutralized by the rich creamy mixture. Some of the germs are killed on site, but others pass the stomach and enter the intestines. Like most bacteria, salmonellae are able to survive outside of cells, and they rapidly establish themselves on the lining of the intestine. Thick intestinal mucus partially immobilizes the germs, but they continue to reproduce and reach a critical mass despite the yeoman efforts of the outer immune system. They begin to damage the cells of the intestine and creep between the junctions of those impaired intestinal lining cells.

The inner immune system is called into action. Neutrophils, lymphocytes, and other immune cells rush to the scene. Some of those cells indiscriminately attack any invader, regardless of type; other cells initiate a specific salmonella strategy, triggering the production of antibodies locally in the intestine, as well as antibodies that circulate in the bloodstream to preemptively search for any salmonellae that may escape the confines of the gut and get into deeper tissues or the blood. As a result of the germs damaging intestinal cells, and as a result of the immune battle being waged, diarrhea develops—the physical sign of loss of absorptive capacity of the intestines. Cramps testify to the ongoing struggle between germs and immunity in the gut. A low-grade fever reflects the release from immune cells of cytokines that are sending signals to, and communicating with, other elements of the immune system. Inflammation advances; more and more immune cells are summoned, germs are killed, chemicals are released, and the cycle repeats. Most of the germs are killed on site, on the intestinal surfaces, from which they are shed in the watery bowel movements, along with the rest of the debris.

However, a few very hearty and tenacious salmonellae are able to penetrate through the now damaged junctions between the intestinal cells and get into the deeper layers of the intestinal wall. There, once again, reinforcements quickly arrive. The would-be vanguard of penetrating bacteria is challenged by phagocytic cells, by complement proteins, and by antibodies; other salmonella germs are trapped in the local lymph nodes of the deeper intestinal layers, where they are digested and disposed of via the elaborate lymphatic system draining the abdomen.

Within a few days, almost all of the salmonellae are dead. The bloodstream is flooded with antibodies guarding against any rogue salmonellae that might have managed to find their way into a damaged capillary or vein in the inflamed intestinal wall; those antibodies also confer lifelong immunity against infection with the same strain of salmonella. The immune cells retreat from the intestinal tissues; signaling between immune elements ceases, and the fever, diarrhea, and cramping resolve. The parents of your kids' friends begin to forgive you. Another successful immune campaign has been waged.

Croup

The barking from the other side of the house wouldn't have bothered you as much if you owned a dog, but unless your kids' goldfish has learned a new trick, something very alarming is happening in your toddler's throat. Indeed, the parainfluenza virus outbreak at the day care center has come home. The virtually limitless variety of respiratory viruses have a certain comforting predictability in the way they are transmitted (see chapter 2) and in the ways they cause illness. Your child's hands touched the hands of his playmate; the playmate had parainfluenza infection, and his hands were full of the virus, having just wiped his nose and mouth clean of the snotty goo coming from them. Hands went to your darling's mouth and nose, and the virus was implanted on the cells of the mucous membranes in your child's respiratory tract. Once on the inner lining of the nose or mouth, the virus has only a few hours to penetrate the outer immune system. In that short time, the virus must try to creep through the mucus on the lining of the membranes, defy the beating "cilia" (those hairlike structures that line the airways [see above]), and avoid being coughed or sneezed out of the body into a tissue (or, as is more often the case, onto your kids' own hands or shirt sleeves or the floor).

If the viruses do manage to overcome those hurdles, they must land on a cell that is receptive to them (see chapter 2) and have the cell take them in. This is a devious strategy that viruses use; they alight on tiny cell surface structures that are intended for other purposes, and the unsuspecting cell envelopes the virus (see chapter 1). Once inside, the virus begins reproducing itself, making hundreds, thousands, and even hundreds of thousands of new viruses, using the cell machinery that it has usurped. Cells begin to die, and the viruses amassed inside are released and spread to neighboring cells of the respiratory tract, and more cells die. The croupy cough reflects the preferred location of parainfluenza viruses—the voice box and trachea—and the spasm of some of the smooth muscles in the airway due to the inflammatory response. Other respiratory viruses (see chapters 1 and 3) prefer other sites: rhinoviruses in the nose, respiratory syncytial viruses in the small airways of the lung, and influenza viruses in the lung itself.

However, ultimately the virus outsmarts itself: so much of the virus is produced within the cell that viral footprints begin to peek out from the cell's surface. That is the fatal flaw in the virus' strategy. Once a cell begins to express a viral characteristic, your child's inner immune system is immediately alerted and responds. T lymphocytes speed to the infected cells and begin destroying them, along with the viruses they contain; the T cells use cytokines to mobilize B cells, which in turn produce antibodies against the parainfluenza virus; cytokines stimulate B cells to replicate themselves, as well. The antibodies specifically made against this strain of virus circulate in the bloodstream and are released into the mucus, killing free-floating viruses that have not yet infected cells. Special types of cytokines, called interferons, are produced that protect neighboring cells from being infected by the free-floating viruses released from infected cells. Still more cytokines are released, more signals are sent, and more immune cells arrive: T cells, B cells, and phagocytic cells.

Finally, after several days, the viruses die off, destroyed along with the cells they had infected. The barking subsides as the immune cells retreat from the battle site and rejoin the bloodstream or the lymph nodes. There's even better news: a permanent "memory" of the infection has been created. For most virus infections, there will never be a second attack. The antibodies and programmed B cells that were generated in response to the initial infection become a lifelong protection against infection by the same strain of respiratory virus. Two or three years from now, when this same parainfluenza virus infects the new kids at day care, the virus will again get implanted into your child's nose, but now her body is ready. No longer naïve to this strain, the circulating antibodies in her bloodstream and in the mucus of her respiratory tract immediately bind to the virus, never letting it perch on your daughter's respiratory cells. Her memory B lymphocytes, circulating in the bloodstream and residing permanently in the lymph tissues of her tonsils and adenoids, instantly reactivate their antibody production. The levels of antibody are boosted by this new parainfluenza challenge, ensuring that no matter how many parainfluenza viruses get in, none will get past the antibody blockade and into her voice box or trachea cells.

Tempering this victory, however, is the reality that there are so many strains of respiratory viruses (see chapters 1 and 3) that the next infection, against which your child has no memory immunity, may be lurking just around the corner. And, among the viruses-in-waiting are numerous other strains that also cause croup!

Chicken Pox

It turns out those little spots and blisters on your preschooler's best friend's face weren't bug bites after all. Chicken pox has long been one of childhood's rites of passage (see chapter 3). Caused by the varicella-zoster virus, chicken pox enters your child's body in the same way that respiratory viruses do—via the nose and mouth. Unlike respiratory viruses, though, chicken pox is able to efficiently spread beyond the respiratory tract cells it initially infects and get into the bloodstream, where it rapidly flows to all corners of the body. The widely distributed itchy skin sores are only the outward indicators; similar lesions occur throughout the inside of the body, everywhere that the blood takes the virus.

With little effective resistance by the outer immune system, your kids depend on their inner immune systems to contain and kill the millions of virus particles that are coursing through their bloodstreams. Rapidly disseminated to the near and far corners of the skin and inner organs, each deposit of virus results in a pock, and each of the pocks (pox) becomes an immune battleground. Lymphocytes are mobilized to each pock; the B cells produce antibody, and the T cells kill infected tissue cells and the viruses within. B cells and T cells signal each other and other inflammatory cells via cytokines. In a matter of a few days, the viruses within the pox begin to die, and the sores scab over; when all the lesions are scabbed, your child is no longer contagious.

Once your child has had chicken pox, the antibodies she has formed circulate forever in her bloodstream and usually protect her from a second episode; the rare second cases are typically very mild, reflecting effective but partial immunity. B-cell lymphocytes retain memory of the infection as well; if your daughter is exposed to a classmate with chicken pox, her B cells will be stimulated into action and produce a boosting of the circulating antibodies that target the chicken pox virus. This lifelong protection against second episodes of chicken pox is identical to the protection afforded by immune memory against most viral and other infections.

However, chicken pox infections have several unique twists that illustrate the diversity and enormity of the challenges facing the immune system. Even though the scabs form and the viruses in the pox are dead, the virus isn't gone. Varicella-zoster virus is a member of the herpesvirus family (see chapters 1 and 3). Like other members of this ignoble clan, the chicken pox virus establishes permanent residence in your kids' bodies—once infected, always infected. Under assault by the immune system, the virus retreats into protected sites—the nerve roots—where it remains dormant for years. With any weakening of the immune system, either by age or by another concomitant disease, the chicken pox virus reactivates along the nerve root, causing the painful condition called shingles. Chicken pox may also reactivate as shingles during periods of extreme and long-term stress, evidence that chronic stress may weaken the immune system (see chapter 11).

The immune system's surveillance capabilities are substantial; any chicken pox virus that tries to escape the confines of the nerve root, either by entering the bloodstream or by creeping down the length of the nerve, is immediately stopped in its tracks by a healthy immune system; this immune system function works much the same as the aforementioned immune memory protection against second chicken pox infections. However, as humans age, the immune surveillance capacity declines, and elderly patients with shingles attest to the miserable results of this decline.

Complicating the chicken pox picture further is the fact that each pock on the skin during the initial infection creates a breach in the outer immune system, facilitating the entry of skin bacteria and the development of secondary bacterial skin infections and even system-wide bacterial infections, like pneumonia and sepsis (see chapter 3). Five to 10% of all chicken pox infections are complicated by the development of a secondary bacterial infection that requires antibiotics and, occasionally, hospitalization. Severe bacterial complications, including toxic shock syndrome, scarlet fever, and necrotizing fasciitis (see chapter 3), may even be fatal. Hence, not only must the immune system respond to the viral challenge of chicken pox, it must also often mobilize inflammatory and phagocytic cells to defend against opportunistic bacteria at each new skin sore. The availability of an effective vaccine that protects against chicken pox (see chapter 7) is helping to make this childhood rite of passage a thing of the past.

Pregnancy

Okay, so maybe you haven't thought of pregnancy as an infectious disease—and it's not, of course. But it is among the most remarkable of immune challenges that the body has to face. Recall from the discussion of T-cell lymphocytes above that the immune system has an elaborate mechanism for distinguishing between self, i.e., those tissues that are a normal part of the individual, and nonself, those cells that don't belong and must be eliminated. If a kidney or heart is donated by a husband to his immunologically healthy wife, for example, the organ would be rejected unless the match was nearly perfect on a genetic basis—highly unlikely (and probably incestuous if it did occur). As a result, recipients of donor organs or bone marrow (see below) usually must be given medicines to severely suppress their immune response and prevent organ rejection.

And yet, neither the donor sperm from the same husband whose kidney would be rejected by his wife nor the resultant fetus, which is immunologically different from both mother and father, is rejected by the recipient woman. Pregnancy occurs and is tolerated by the woman's immune system in a way that continues to defy our knowledge of immunity. It seems the uterine cavity (the womb) is an immunologically privileged site in the woman's body where foreign antigens (nonself markers) are ignored by the woman's immune system—no attack by T cells, no antibody formation by B cells, no phagocytosis, no cytokine signaling. The pregnancy is allowed by the immune system to continue unrejected. The resulting baby has a full set of organs that would each be rejected by the same woman if they were given as transplants but, when carried for 9 months in the womb, are protected and nurtured. Remarkable, eh?

In addition to the extraordinary immune tolerance that occurs during pregnancy, that blessed state of incubation actively protects the baby once born. Beginning at about 32 weeks of gestation (a full-term pregnancy is ≥37 weeks), Mom passes antibodies circulating in her bloodstream across the placenta to her fetus. Those antibodies, known as passive immunity, protect the newborn from all infections that the mother has ever had. That is, not only do the pregnant woman's antibodies from past infections protect her, they now protect her newborn, as well. Passive immunity lasts for about 6 months before those antibodies disappear from the baby's circulation. The result is that the first 6 months of a baby's life are typically characterized by many fewer infections than the next 6 months and beyond, when the baby is on his own.

However, breast-fed babies aren't really on their own as long as they're nursing. Mom also actively secretes antibodies in her breast milk, making each nursing event after birth a protective infusion of immunity. There is much more on breast milk in chapter 10.

Breakdown of the Immune System—How It All Falls Apart

The vital importance of an intact and healthy immune system should be obvious to all by now—the germs lurking at your kids' portals are limitless and formidable. Unfortunately, each major and minor component of the elaborate defense mechanisms that comprise our immunity is potentially subject to malfunction. Some children are born with deficiencies of one or more immune attributes; these are termed primary immune deficiencies. Other children can acquire deficiencies as a result of malignancy, infection, or toxicities from medicines and other exposures (collectively, these conditions are called secondary immune deficiencies).

Table 4.1 presents the categories of immune deficiency based on the component that has broken down, examples of each category, characterization of each example as a primary (inherited) or secondary (acquired) deficiency, and the resultant susceptibilities to infections.

There are between 80 and 100 different primary immune deficiency disorders, and thus, Table 4.1 is necessarily simplistic. However, important generalizations can be made. Inherited disorders of skin or mucous membranes, while not technically deficiencies of the immune system, functionally behave as if they were. Kids with these conditions, as well as those with damaged barriers that occur after birth (e.g., thermal and chemical burns, traumatic injury, chicken pox sores, and bug bites), have impaired ability to resist infections that enter through the skin and mucous membranes. Cystic fibrosis, for example, is a complex disorder of secretory functions that results in a thick, choking respiratory mucus that impairs bacterial clearance from the airways and lungs.

Bacteria are usually contained by phagocytes and antibodies, and hence, deficiencies of phagocytic cells or of B-cell lymphocytes (those cells that produce antibodies) predispose to bacterial infections. In contrast, most viruses are contained by T-cell lymphocytes; deficiencies of those cells predispose to severe infections with members of the herpesvirus family and many others. When both B-cell lymphocytes and T-cell lymphocytes are impaired, the resulting combined susceptibilities to infections call to mind the "bubble boy" images of a child forced to live in a germ-free protective chamber to prevent even the most minor infections from becoming life threatening. Indeed, two such bubble boys, Ted DeVita and David Vetter, did live in plastic encasements for years before finally succumbing; their stories generated extensive media coverage in the 1970s and 1980s, as well as a popular film starring John Travolta in a fictionalized version of their composite stories. We have come a long way in treating severe combined immune deficiencies since the bubble boys (see below).

Secondary immune deficiencies (those that occur after birth, i.e., that are not congenital) can be as devastating as those that are inherited. AIDS (see

Table 4.1 Examples of immune deficiencies and the infections they predispose to

Missing or malfunctioning immune component	Example of deficiency	Type of deficiency		Type(s) of infection
		Primary (inherited)	Secondary (acquired after birth)	
Skin or mucous-membrane barriers				
Generalized	Eczema	X		Bacterial infections of the skin
	Bechet's syndrome	X		Bacterial infections of mucous membranes
	Thermal and chemical burns		X	Bacterial and fungal infections of skin and/or mucous membranes
Localized	Cystic fibrosis	X		Chronic, recurrent lung infections
B-cell lymphocyte deficiency				
Severe	Bruton's agammaglobulinemia	X		Bacterial infections of ears, sinuses, lungs, brain, and bloodstream; viruses that are normally cleared by antibodies, e.g., enteroviruses
	Common variable immune deficiency	X	X[b]	
Mild	Selective IgA deficiency	X	X[b]	Mild, but recurrent and difficult-to-treat, bacterial infections of ears, sinuses, lungs
	IgG subclass deficiency	X		
T-cell lymphocytes	DiGeorge anomaly	X		Severe viral infections, especially herpesviruses
	AIDS		X	
	Steroid medicines		X	

Both B-cell and T-cell lymphocytes	SCIDS[a]	X		Severe and life-threatening bacterial, viral, fungal, and parasitic infections of all types
	Chemotherapy		X	
Phagocytic cells				
Inadequate amount	Cyclic neutropenia	X		Severe bacterial infections
	Chemotherapy and other medicines		X	
Defective function	Chronic granulomatous disease	X		Severe bacterial and fungal infections
	Leukocyte adhesion defect	X		
	Absent or defective spleen (asplenia)	X	X[b]	
Complement	Terminal cascade deficiencies			Severe bacterial infections

[a]SCIDS, severe combined immunodeficiency syndrome.
[b]Both inherited and acquired forms may occur.

chapter 3) is due to the human immunodeficiency virus, and as the name implies, this virus is particularly nasty because it attacks the very immune components designed to defend against it—the T-cell lymphocytes (see chapter 3). Whereas the human immunodeficiency virus destroys T cells, steroid medications, for example, those used to treat asthma and some forms of juvenile arthritis, suppress T-cell function. Patients who lose their spleens as a result of injury or surgery are susceptible to severe infections with bacteria that otherwise are targeted for removal by phagocytic immune cells in the spleen.

One of the most perplexing areas of practice for pediatricians occurs as a result of the deficiency category of "B-cell lymphocyte deficiency, mild" noted in Table 4.1. That's because many, many children have recurrent minor infections of the ears, sinuses, and even lungs. Are these the result of an immunodeficiency disorder or part of the natural history of being a child? There is more about that below in the discussion of the "my child is always sick" syndrome.

Treatment of Immune Disorders

Now that we have a very global view of immune disorders, it's time for a similarly global view of their therapy. What can be done for children with true deficiencies of immunity? First and foremost, infections must be prevented from occurring if at all possible, and they must be treated when they do occur. Remember that what are typically trivial infections in healthy hosts can become life threatening in children with abnormal immunity. The staples of our preventive and therapeutic strategies are antibiotics (see chapter 5) and antiviral, antifungal, and antiparasitic medicines (see chapter 6). Patients with known susceptibilities are frequently prescribed prophylactic medicines directed at the most likely and most worrisome germs associated with particular disorders (see chapters 5 and 6). While some patients with known immune deficiencies have an inadequate response to vaccines, many immunologically impaired patients do have partial or even complete vaccine responses, and guidelines for such patients are available. Care must be taken in vaccinating immune-compromised children, however, because live attenuated vaccines (see chapter 7) contain viable viruses that can cause severe disease in patients with certain immune deficiencies and therefore must be avoided in those individuals.

The immune system is difficult to replenish when deficiencies occur, but great advances have been made in recent years, obviating the need for bubble boys. Antibody supplementation of patients with B-cell lymphocyte disorders can be accomplished by regular administration of gamma globulin, a blood product derived from healthy donors (see above). One of every six children born with severe combined immunodeficiency syndrome (SCIDS) has an absent enzyme at the root of the disorder—that enzyme can now be given as replacement therapy.

For the remainder of children with SCIDS, and for several other important immune deficiencies, bone marrow transplantation (BMT) may be curative. Recall from the earlier discussion that all immune cells originate from the bone marrow. Patients with inherited defects in their immune systems, particularly those with B-cell and T-cell abnormalities, have bone marrow that fails to produce the required cells or cell competencies. Receiving bone marrow from a donor with a healthy immune system can often reconstitute the bone marrow of the immune-deficient patient. However, the BMT procedure is fraught with difficulty. Critical to its success is finding a donor who is genetically closely matched with the recipient. If the match is not close enough, two potentially severe complications may occur. Some immune-deficient patients have enough residual immunity that their cells may attack the transplanted cells, causing rejection. In other immune-deficient patients, the bone marrow from a poorly matched donor may actually attack the patient (termed graft-versus-host disease). Either complication can result in further immune compromise and require treatments that themselves can be dangerous and even life threatening. Nevertheless, BMT is increasingly being used successfully for treatment of immune deficiency disorders, as bone marrow banks have been established to help find closely matched donors.

A variation on the theme of BMT is the use of umbilical stem cells for reconstituting the bone marrow of immune-deficient patients. Stem cells are those parent cells from which all bone marrow cells derive. It is these stem cells that live in normal bone marrow and by their programmed maturation become all of the blood cells that circulate in the body. Stem cells are present in normal bone marrow, in fetuses, and in umbilical cord blood, where they are particularly numerous and easy to harvest. There is also evidence that umbilical cord stem cells may not have to be as closely matched to a potential recipient as, for example, bone marrow cells must be prior to BMT. Umbilical cord stem cells are now also being banked and typed and are being made available to patients with immune deficiencies, and successful cures of inherited disorders have been increasingly reported using this method.

There is one more BMT variation. A highly experimental intervention involves the removal of bone marrow cells from a patient with an immune deficiency that is linked to a specific gene deficit. The missing gene is then inserted into the patient's own bone marrow cells, and the marrow is reinfused into the patient. This so-called gene therapy has been tried in only a few patients with mixed results to date. It will become more widely applicable only when additional specific genes responsible for immune deficiencies are identified and when the safety of the procedure can be improved.

Neutrophil deficiencies, either inherited or acquired during chemotherapy, for example, predispose patients to severe bacterial and fungal infections. Two potent stimulators of neutrophils have been purified and made into medicines for patients with low neutrophil counts. These are naturally occurring (that is, all humans have these mediators normally in their systems) growth factors called granulocyte colony-stimulating factor and granulocyte-macrophage colony-stimulating factor. Another chemical mediator, the cytokine interferon, has been used to stimulate the faulty phagocytes in patients with chronic granulomatous disease (see Table 4.1).

My Child Is Constantly Sick—Is There Something Wrong with Him?

How many infections are too many? Fortunately, primary immune deficiencies are quite rare. Other than IgA deficiency, which may occur in as many as 1 in 500 people (most of whom are asymptomatic and never know they have a deficiency), the individual disorders occur on the order of 1/10,000 to 1/100,000 children. SCIDS (the bubble boy disease) occurs in only 1/500,000 children. Hence, it is extremely unlikely that your child has an immune problem. What about all those colds, ear infections, and sinus symptoms, then?

Data regarding the expected number of infections in immunologically healthy young children have been known for many years thanks to community watch programs that have tracked entire towns or cities. One such study, known as the Tecumseh Study (named for the town in Michigan where it was conducted), followed 1,000 people from 1965 to 1971 and then resumed in 1976 and continued until 1981. The results were newsmakers back in the day. Kids younger than 2 years of age developed on average six respiratory infections each year. The incidence dropped off slightly between ages 3 and 4 years (just under 5 respiratory infections each year) and still further between the ages of 5 and 9, when kids had an average of only 3.5 respiratory infections each year. The vast majority of all of these infections were viral (i.e., they shouldn't need antibiotics [see chapter 5]). For the first 2 years of life, boys had slightly higher rates of infections than girls, a trend that reversed itself beginning at age 3; the reasons for this gender flip have never been understood, but they have been reproduced in other studies. After age 9, the average incidence of infections in kids dropped to as low as 2.4 respiratory infections each year through the teens, climbing again among young adults, particularly females, in their early to mid-20s, presumably reflecting the demographics of parents of young children. Further confirming the role of young kids in the infections of young parents, women who worked outside the home had fewer annual respiratory infections than those who worked at home.

These data from Tecumseh confirmed even earlier studies among families in Cleveland between 1948 and 1957. Approximately 100 families were followed weekly, and again, kids under 2 years of age had the highest frequency of respiratory infections—an average of 7 to 8 such events. The incidence dropped to just over six annual infections on average between ages 5 and 9 years and to five infections each year between ages 10 and 14 years. A continued fall between ages 15 and 24 years was observed (about 4.5 infections each year) with, again, a blip upwards between the ages of 25 and 29.

Group child care/day care centers (see chapter 2) amplify the incidence of infections in young children. Kids in day care centers consistently are found to have more infections each year than those in home care. Studies done in North Carolina and Pittsburgh demonstrated that day care kids have between 7 and 10 infections on average each year regardless of age group and that the number of infections in immunologically healthy kids in day care settings can reach as high as a dozen or more each year. As seen in the community and family studies noted above, most illnesses were respiratory virus infections. Two-thirds of all kids studied in the Pittsburgh day care studies had more than 60 days of respiratory illness per year and at least four severe (defined as a rectal temperature of at least 102°F for at least 3 days, an infection that lasted more than 10 days, or an illness that required a visit to a physician) infections each year. The good news is that when kids are in day care for more than 1 year, their numbers of infections go down with each successive year and may, in fact, remain lower than those of home care kids once they all reach school age; that is, day care kids get their infections early and are protected to some extent from exposures to respiratory viruses when they enter school.

So, how does that compare with your kids? Hopefully, you're reassured by the fact that even though it seems like your kids are constantly sick, in fact they probably don't have more than the 6 to 12 annual respiratory infections that average, healthy young kids are expected to have. Having now reassured you, I need to add a sobering closing thought. Some kids do have too many or too severe infections. Appreciating the warning signs of immune deficiency is important for both you and your child's doctor. As a general rule, if your child is growing and developing normally, he or she probably doesn't have a severe immune deficiency. As another general rule, it's not the number of infections so much as it is the type and severity of infections that raise concern regarding immune deficiency.

That said, the other rules for identifying kids whose immune systems need to be evaluated are summarized in Table 4.2. These criteria, developed by the Jeffrey Modell Foundation (http://www.info4pi.org/), a partner of the National Institutes of Health and Centers for Disease Control and Prevention in advancing knowledge about immune deficiencies, are widely accepted by specialists in infectious diseases and immunology.

Table 4.2 Modell Foundation warning signs for immune deficiency[a]

1. Two or more *serious* sinus infections within 1 year
2. Two or more months on antibiotics *with little effect*
3. Two or more pneumonias within 1 year
4. Failure of an infant to gain weight or grow normally
5. Recurrent deep skin or organ abscesses
6. Persistent thrush[b] in mouth or elsewhere on skin, *after age 1*
7. Need for *intravenous* antibiotics to clear infections
8. Two or more *deep-seated* infections
9. A family history of primary immunodeficiency

[a]Italics added for emphasis.
[b]A yeast infection common in young infants; see chapters 1 and 3.

Remarkable

Now, it's time to return to the big picture. The facts that you have reached the age of parenthood and that your kids have reached the age of having many colds and ear infections every time they go back to school after summer vacation are both testimonies to the wonders of human homeland defenses. Despite the seas of germ ooze that live in and on you and your kids, the nonsterile food and water you consume, and the snotty and sniffly people you encounter on the stairwells, subways, and schoolyards of your lives, you and your kids are here, alive and kicking. Your kids spend most of their childhood uninfected, growing, developing, and thriving.

Yes, it's true that, thanks to antibiotics (see chapter 5) and vaccines (see chapter 7), the life expectancy for your kids' generation will continue the upward trend of the past decades and infant mortality will continue downward (see the introduction). Still, even before the advent of antibiotics and vaccines, even when infectious diseases were the leading killers on the battlefields of the world and in the cities of this country, our homeland defenses did a darn good job of getting most people well into older adulthood.

Yes, it's also true that sometimes things go wrong. Sometimes the immune system overreacts and causes allergic reactions. Rarely, the immune system goes haywire and attacks the body's tissues, causing autoimmune diseases. And even more rarely, immune deficiencies can be inherited or acquired and result in a truly abnormal number and severity of infectious disease. However, considering the denominator (the number of times everything goes well), those small numerators are forgivable.

The next time your kids recover from a cold, or scab over a healed cut, or resolve their diarrhea, give a nod to the remarkable immune system that again did its job with little fanfare and even less appreciation—and remember, you'll hear and read lots of advice and advertisements about ways to boost your kids' immunity, from nutritional supplements to dietary miracles to what they should wear on cold days and hot. Read chapters 10 and 11 before changing anything in the delicate balance that is your child's immune health.

5

Germinators: Antibiotics as Miracle and Menace

Recall from the introduction to this book that the germ theory is only 150 years old—that is, we have only known the real cause of infections for less than 2 centuries. Despite our relatively recent awakening, ancient healers were, as usual, ahead of the curve. Records from China, Egypt, and Greece dating back more than 2 millennia demonstrate the empiric use of natural substances to treat skin, respiratory, and gastrointestinal infections (although of course no one knew what an infection was or that one was present!). However, it wasn't until germs were identified in the mid-19th century that scientists had a true target for drug development. Even so, efforts at antibiotic discovery and development prior to 1928 were rudimentary and mostly doomed to failure.

Dr. Fleming's Wondrous Discovery

The story of the first important antibiotic, penicillin, illustrates that no matter how disciplined the approach, and regardless of how systematic the study, serendipity has a huge role in the advance of science. The details are variably told. What is certain is that microbiologist Alexander Fleming was studying *Staphylococcus aureus* (staph) bacteria in the hope of finding a drug to fight the scourge of World War I battlefield infections due to this germ. Staph would be a model for all germs—whatever molecular weapon could be developed to kill staph would teach the scientific world about killing germs in general. Earlier magic bullets against bacteria had proved to be too toxic for human use.

Here's where the story gets a little muddy (or moldy, as the case may be—read on), but the specifics are less important than the punch line. The legend has it that Fleming returned to his laboratory after either a long vacation or a weekend away to find bacterial culture plates that he had left sloppily unwashed because of his notoriously bad laboratory hygiene. He noticed that one of the plates was not submerged in the disinfectant in the sink. Another version has it that he needed to show a colleague the work he was doing and retrieved a culture plate from the discard pile. Fleming himself describes it this way, perhaps somewhat defensively.

> The contamination in 1928 of a culture plate by spores of a species of *Penicillium* was the beginning of the study of penicillin. Such contamination is not uncommon in a bacteriological laboratory and is usually regarded as a reflection on the technique of the bacteriologist. Sometimes, however, it is unavoidable, as in this particular instance when the culture plate had to be opened for examination under a dissecting microscope and then left for future examination. (Alexander Fleming, *Penicillin: Its Practical Application*, Blakiston Company, Philadelphia, PA, 1946.)

In other words, Fleming left the plate exposed to air intentionally in order to return to it and study it further. Whatever.

Although the true version of events may have been lost in the myth and lore of this wondrous discovery, the outcome is undisputed. A common fungus mold, *Penicillium notatum,* took hold on the neglected culture plate and began growing, much as it might in your own sink if left unattended for days. Surrounding the mold was an area where the staph bacteria had once grown but were now dead. Something secreted by the mold killed *S. aureus*! It was then left to chemists Howard Florey and Ernst Chain to purify the active mold substance responsible for killing the bacteria and begin the efforts to manufacture the miracle drug—in time to save hundreds of thousands, some say millions, of lives in World War II. For their efforts, Fleming, Florey, and Chain were awarded the Nobel Prize in Medicine in 1945; Alexander Fleming was knighted in 1944.

With the first evidence of the extraordinary lifesaving potential of penicillin, the race to find new antibiotics was on, as was the golden era of drug discovery. The vast and confusing armamentarium of antibiotics that has resulted will be summarized and, hopefully, simplified for you below. Before that, however, we need to answer important questions about this most important class of medicines and the germs it treats.

- What are bacteria?
- What are antibiotics?
- When does my child need antibiotics, and when doesn't she?
- What are the risks of antibiotics to my child?
- What are the risks of antibiotics to society?
- Should my child take antibiotics to *prevent* infections?
- Is my child exposed to antibiotics in the environment?

What Are Bacteria?

For purposes of this chapter's discussion and scope, the most important definition of bacteria is that they are the only germs killed by antibiotics. More generally, as described in chapter 1, bacteria are self-contained, single-celled organisms that live everywhere around us and even in us—usually harmlessly. When bacteria progress from symbiotic cotravelers with our kids to invasive causes of infection and disease, they become eligible for antibiotic treatment. The diseases caused by bacteria are extensively reviewed in chapter 3, including all of the infections used as examples below.

Whereas the overview of bacteria in chapter 1 focused on the body sites of bacterial infection, antibiotic therapy is better understood by thinking of bacteria in terms of the germs' biology. Indeed, it is the exploitation of the biological properties of bacteria that has directed antibiotic development and that determines which antibiotics are best used against which infections. The three most important biological properties of bacteria from an antibiotic perspective (and from your child's doctor's perspective when he or she decides what to write on the prescription pad) are color, shape, and oxygen growth requirements.

Color

Although a number of bacteria are distinctively pigmented naturally, the more important color characteristic from a treatment point of view is that which results from a special laboratory staining technique. Called the Gram stain for the scientist who discovered it, the sequential exposure of bacteria to several different chemicals results in either a gram-positive (purple) or gram-negative (pink) appearance under the microscope. How could this seemingly contrived laboratory phenomenon predict a bacterium's susceptibility to antibiotics? Because the staining pattern is a reflection of the biological components of the bacterial cell wall, and those components, in turn, influence the penetrability of antibiotics into the germ. Examples of gram-positive (purple) bacteria are the *Staphylococcus, Streptococcus, Listeria, Clostridium,* and *Bacillus* species. Gram-negative (pink) bacteria include *Neisseria, Escherichia coli, Pseudomonas, Campylobacter, Shigella, Salmonella, Haemophilus,* and *Legionella* species.

Shape

Bacteria come in several different shapes. The two most common are round (coccus) and rod shaped (bacillus). The name of the bacterium sometimes gives a clue to the shape—*Staphylococcus aureus* and *Streptococcus pneumoniae,* for example, are round. *Bacillus anthracis* (the cause of anthrax) and *Bacillus cereus* (a cause of food poisoning) are rod shaped. Most bacilli, however, don't have telltale names; they include *E. coli, Pseudomonas, Cornybacterium diphtheriae* (the cause of diphtheria), *Shigella, Salmonella,* and many others. A few important disease-causing bacteria are spiral shaped and are known as spirochetes; they include the bacteria that cause syphilis, leptospirosis, and Lyme disease.

Oxygen Requirements

If a bacterium grows well in room air, it is considered to be aerobic. In contrast, an anaerobic bacterium does not grow well, or perhaps at all, in room air, preferring an environment in which oxygen is lower or absent. Examples of aerobic bacteria are most of the bacteria that cause human disease, e.g., the staphylococci and streptococci, *E. coli*, *Pseudomonas*, *Shigella*, and *Salmonella*. Infections by anaerobic bacteria, while not as common as those by aerobic ones, are still very important clinically. Examples of anaerobic bacteria that cause important human disease are *Clostridium* (causes of tetanus, botulism, and gas gangrene), *Bacteroides*, *Propionibacterium* (causes of acne), and *Fusobacterium* species. Then, just to make things more confusing, there are the facultative bacteria that can grow in both high and low oxygen concentrations; our discussion of antibiotics will be difficult enough without further considering facultative germs.

Table 5.1 illustrates the classification, using the three criteria of color, shape, and oxygen requirements, of several representative bacteria that are responsible for important human diseases. When antibiotics are developed and tested, their potencies against bacteria with each of the three criteria are quite predictive of how effective those antibiotics will be in humans with infections due to bacteria having similar properties.

As can be seen by a quick glance at Table 5.1, an antibiotic that targets gram-positive (purple) bacteria might be effective against certain causes of meningitis or pneumonia but not others. An antibiotic that is potent against anaerobic bacteria might be expected to work against some skin infections but not others. For that reason, your child's doctor might choose to use more than one antibiotic at a time when an infection is suspected until the germ causing the infection can be precisely identified.

To conclude this discussion of what bacteria are, I want to note what they are *not*. Bacteria are *not* the cause of most childhood (or adult) infections—viruses lay claim to that title. That means that the vast majority of infections don't require antibiotics—there is a lot more on that theme below and in chapters 3 and 6.

Table 5.1 Representative examples of bacteria with their important biological properties that determine antibiotic effectiveness

Bacterium	Example of disease caused	Color (Gram stain)[a]	Shape	Oxygen requirement[b]
Staphylococcus aureus	Skin infections	Purple	Round	+
Propionibacterium acnes	Skin infections	Purple	Rod	−
Bordetella pertussis	Pneumonia	Pink	Rod	+
Streptococcus pneumoniae	Pneumonia, meningitis	Purple	Round	+
Listeria monocytogenes	Meningitis	Purple	Rod	+
Neisseria meningitidis	Meningitis	Pink	Round	+
Escherichia coli	Food poisoning	Pink	Rod	+
Clostridium botulinum	Food poisoning	Purple	Rod	−
Neisseria gonorrhoeae	Gonorrhea	Pink	Round	+
Treponema pallidum	Syphilis	Neither	Spiral	−

[a]Purple, gram positive; pink, gram negative.
[b]+, aerobic; −, anaerobic.

What Are Antibiotics?

Antibiotics used to be defined as substances produced by one living thing that are toxic to other living things. With the advent of modern medicinal chemistry, most antibiotics are now chemically created in the laboratory based on structural characteristics of the target bacteria. Designer drugs, as these synthetic antibiotics are sometimes known, are often derived from microbiologists' and chemists' knowledge of the molecular structures of naturally occurring antibiotics. As an example, Alexander Fleming's miraculous penicillin has now been altered and tailored into innumerable synthetic penicillins that each have distinguishing characteristics and potencies for different bacterial germs; there is more about that in the second half of this chapter.

The term antibacterial is often used synonymously with antibiotic, although they aren't always identical—consumer product manufacturers have usurped some terminology to sell soaps, detergents, and toothpastes (see chapter 9). Other potentially confusing terms are antiseptics and antimicrobials. The latter is a broad term that includes antibiotics, as well as antiviral, antifungal, and antiparasitic medicines (see chapter 6), but the term may also appear on floor and laundry detergents (see chapter 9). Antiseptics are compounds that broadly kill bacteria or inhibit their growth but are often toxic to the patient, as well; as a result, antiseptics are limited to external application to the skin to prevent infection or, occasionally, to control infection. Examples of antiseptics are alcohol, iodine, hydrogen peroxide, boric acid, and ammonia. Depending on your age, you may remember your own parents using the antiseptic mercurochrome to treat your scrapes and cuts—the mercury content has since been recognized as a hazard, and this treatment is no longer recommended.

The discussion in this chapter addresses only true antibiotics—those compounds that exploit specific biological properties of bacteria to kill them or to limit their growth. Other germ-limiting strategies, such as antiseptics and detergents, are covered in chapter 9.

When Does My Child Need Antibiotics, and When Doesn't She?

The simplest answer to this question is that your child needs antibiotics when she is infected with a bacterium that satisfies one or more of the following criteria.

- It causes unpleasant or uncomfortable symptoms.
- It is potentially harmful to your child.
- It is potentially harmful to others who might come in contact with your child.

Strep throat (see chapter 3), due to the bacterium *Streptococcus pyogenes*, is an example of an infection that satisfies all three criteria. However, nothing is ever as simple as the simplest answer. As diagnosing strep has become much more automated and reflexive with newer tests, more and more kids are being treated for the mere presence of the germ detected by a rapid test of their throats, whether the symptoms they have are due to the germ or not; a full discussion of this conundrum is found in chapter 3. Remember, though, from chapters 1 and 2 that there is an important difference between colonization with bacteria and infection. The former means only that the germ is present, not that it is doing harm; most bacteria on and in our bodies are colonizers. Strep, on the other hand, often causes infection—but not always. Sometimes it is simply left over in your child's throat from a past infection, and the strep germ is now colonizing, not infecting.

What's important for understanding this chapter on antibiotics is to realize that both you and your doctor have roles in deciding when to use, and when not to use, antibiotics. It's up to your doctor to decide if one or more of the three criteria above are met. If they are not, it's up to you to understand the doctor's decision *not* to recommend antibiotics, and it's up to you to *not* insist on receiving them. Studies have shown that much of the antibiotic overuse in this country is due to consumer-driven preferences. In other words, the doctors prescribe antibiotics because parents want their kids to get them in the belief that their kids need them or will get better and go back to school faster with them. The same can even be extended to testing for bacteria. Parents often request a strep test whenever their kids get a cold, and doctors comply to make their patients happy (and because testing generates income). The false-positive test caused by colonizing rather than infecting strep then results in unnecessary antibiotic treatment that never would have occurred had the doctor held to his clinical instinct and judgment that this common cold was not due to strep.

That brings us to understanding when your child does *not* need antibiotics. Again, there are three simple criteria; your child does not need antibiotics when one or more of the following criteria are met.

- She is infected with a virus.
- She is only colonized with a bacterium that is not a risk for causing harm.
- She is only mildly ill, and the true germ cause of the illness is not known.

Examples of each of the no-need-for-antibiotics criteria are plentiful. Most of your kids' infections do not require antibiotics because they are due to viruses (you've read this before in this book, and you'll read it again), the first criterion above. These viral infections, discussed in detail in chapter 3, include, among many others, the common cold, some ear infections, most cases of pinkeye, the flu, bronchiolitis, croup, stomach flu, mono, chicken pox, and cold sores/fever blisters.

Colonizing bacteria that don't need antibiotics, in addition to the example of strep given above, often represent contaminants, that is, bacteria that are recovered in culture specimens from your kids but are there only because the germs accidentally found their way into the specimen being tested. The germs might have come from the skin or elsewhere, where they are harmless, but now that they are in the specimen being tested, they are erroneously considered to be true infections. These false-positive culture results include skin or fecal bacteria that can contaminate a urine culture collected by "clean catch" (e.g., urinating into a cup for an older child or a "bag" urine sample from a baby), a culture of draining fluid taken from the ear canal (where skin bacteria are plentiful), and a sputum culture from a coughing child (sputum is supposed to be secretions coughed up from deep in the respiratory tree and, in adults, may be an indicator of germs in the lung; a child rarely can produce true sputum, and what is coughed into the cup is simply spit that contains normal mouth bacteria). In more seriously ill kids for whom blood or spinal fluid cultures are required, contaminants from the skin can also be confusing, but because the child is sicker, the doctor's decision regarding holding off on antibiotics is more difficult and beyond the scope of this discussion. Here, we're really talking about kids with likely viral diseases whose bacterial cultures are positive because colonizing germs have confused the picture; antibiotics are not needed for these kids.

The best example of the third criterion for "no antibiotics needed," i.e., mild illness for which the germ cause is unknown, is fever. There is great temptation for many doctors, again prompted by the concerns of many parents, to treat fevers in young kids with antibiotics. This topic is fully covered in chapter 3, but suffice it to say here that more than 97% of fevers in young kids are *not* due to bacterial infections. Other mild illnesses that might prompt unnecessary antibiotic treatment include sinusitis when defined simply as sinus pressure or short-term colored (usually green or yellow) drainage, bronchitis (sometimes a bacterial disease in adults but rarely in children), fluid behind the eardrum without other evidence for actual infection, and most cases of traveler's diarrhea.

What Are the Risks of Antibiotics to My Child?

Early in my career as a pediatric infectious diseases doctor, I was giving a grand rounds lecture on the evaluation of kids with fever for possible bacterial infection. I presented the likelihoods and unlikelihoods, the helpful and the not-so-helpful predictive factors in distinguishing the few kids with bacterial infections (remember, only about 3%) from the many (97%) of feverish kids who had only simple virus infections. The pediatrician who cared for me as a child was in the audience. Now retired but still very much a part of the pediatric community in Denver, he came up to the front of the lecture hall to visit with me after the lecture. He proudly told me that thanks to him, I was there! I agreed that I and my family were very grateful for the good care he had given me as a child. However, that's not what he meant. He went on to tell me that while he found all of my fever predictive factors interesting and that he knew bacterial infections were rare causes of fever in otherwise healthy kids, he credited his own success rate as a doctor to one overriding practice principle that he employed: he treated every child who came to his office complaining of a fever with a shot of penicillin. As a result, he said, he never had a case of meningitis during his 50-plus years of practice!

If it was true, how bad could that be? I'm healthy and, in fact, never did contract meningitis despite all the fevers I must have had as a child. However, the naïve perspective of many in the physician and parent communities that antibiotics are harmless preventives has been long ago dispensed with—the emergence of resistant germs (see below) took care of that. Nevertheless, the reliance of both physicians and parents on antibiotics has proven to be much more difficult to combat. In this section, we'll deal with the potential risks to your child of unnecessary antibiotics; in the following section, we'll tackle the risks to society as a whole—it's there that the Andromeda strain scenario is best discussed.

The risks to your child in taking antibiotics that he doesn't need are all tied to the side effects of the medicines. Even though antibiotics have a high therapeutic ratio, meaning the benefits usually far outweigh the risks, that's true only if there are any benefits to be had at all. If antibiotics are appropriately used to treat true bacterial infections, the benefits of recovering from the infections outweigh the relatively small risks of side effects. If, on the other hand, the infection being treated with antibiotics is viral or simply bacterial colonization rather than true infection, any risks are too great and by definition outweigh the benefits—which are *none, nada,* and *nonexistent* (except for the possible placebo effects [see chapter 12]).

Antibiotic side effects can be grouped into four categories: allergic reactions, gastrointestinal upset, undesirable interactions with other medicines, and miscellaneous (sorry; I hate that category, too).

Allergic Reactions

An allergic reaction is triggered when the immune system has an overreaction to a substance that it perceives as foreign (see chapter 4), resulting in an excess production of immunoglobulin E antibodies and histamine; occasionally other immune mechanisms are involved, but all allergic reactions, by definition, are mediated by the body's own immune system. Fifteen percent or more of the population report being allergic to one or another antibiotic, yet fewer than 5% of people are actually found to be allergic when tested. The reason for the disparity is that adverse or unpleasant reactions to antibiotics, such as diarrhea, are often confused with allergic reactions by those taking the medicines. This is problematic for many reasons, the greatest of which is that patients are often denied the preferred antibiotic for the infection being treated because of the fear that they are allergic. Often the preferred antibiotic is also the one with the narrowest spectrum and therefore poses the lowest risk of creating resistant germs (see below).

True allergic symptoms are usually limited to rashes, often very itchy ones. Sometimes, itchiness occurs without the rash or persists beyond the resolution of the rash. Some allergic reactions include mild wheezing—a troubling finding, to be sure, because significant respiratory distress can be part of an anaphylactic reaction (see below), but most mild wheezing in response to an antibiotic stays mild. First-time allergic reactions typically occur days to weeks after initial exposure to the offending antibiotic, whereas the reaction can occur within minutes to a few hours after the second exposure. More severe reactions occasionally occur and may involve major body organs, such as the liver and kidney, or cause damage to circulating blood cells. Other severe reactions may result in the Stevens-Johnson syndrome, a condition in which antibodies linked to antibiotics cause severe damage to skin and mucous membranes with resultant "burns" that can be life threatening. Antibiotic allergy may trigger anaphylaxis, a condition in which the excess release of histamine is massive and affects the entire body, with tightening of the airways and difficulty breathing. In the most severe cases of anaphylaxis, dilating blood vessels cause low blood pressure and shock. The occurrence rate of anaphylaxis to penicillin, the best studied and most common of the antibiotic allergies, is only 1 in 5,000 to 1 in 10,000 exposures to the medicine.

Unless your child has had a rash or wheezing in the context of antibiotic treatment, or one of the very rare severe reactions, it's unlikely that she is allergic to the drug. However, even using the very modest 5% figure for the true incidence of antibiotic allergy means that more than 550,000 allergic reactions occur in children each year as a result of the antibiotics that are given for 11.2 million nonbacterial diseases, such as the common cold (you'll see where these numbers come from in the full discussion of the societal effects of antibiotic overuse below). Antibiotics should not be given for nonbacterial diseases, so these 550,000 allergic reactions each year are all for naught.

Gastrointestinal Upset

As many as 20% of patients taking an antibiotic will develop diarrhea as a complication of their treatment. There are three forms of diarrhea associated with antibiotic use—simple watery diarrhea, colitis, and pseudomembranous colitis. All are due to a disturbance of the balance of normally occurring bacteria in the gut. You'll recall from previous chapters that the gastrointestinal tracts of humans are home to billions of bacteria that live harmlessly within us, actually helping digest food and, equally importantly, filling the space that more dangerous germs would love to occupy if they could only gain a foothold. Antibiotics prescribed to treat, for example, an ear infection also kill the good bacteria in your child's intestines. The result is often an impaired capacity for digesting and absorbing food, and the watery diarrhea reflects that malabsorption. In more severe cases, damage to the intestines and the resulting inflammation occur because of an overgrowth of bad bacteria that fill the void left by the killed-off good germs. That condition, termed colitis, is characterized by blood and pus in the diarrheal feces. The most severe variety, pseudomembranous colitis, refers to a specific appearance of the damaged intestinal wall: a membrane-like coating over a severely inflamed layer of intestinal cells. This condition (and some cases of milder colitis and even the simple antibiotic-associated watery diarrhea) is due to an anaerobic bacterium named *Clostridium difficile* (see chapter 1) that normally lives in small and harmless quantities in the intestines. Facilitated by the antibiotic elimination of the good germs, *C. difficile* overgrows and overproduces toxins that cause profound, and sometimes even life-threatening, intestinal damage. Probiotics (good germs that live in live yogurt cultures and are sold in powdered form in health food stores) may help prevent the diarrheal complications of antibiotics (see chapter 10). Again, using the estimates for antibiotic overuse presented below, it is likely that as many as 2.5 million kids develop diarrheal complications from antibiotics prescribed for nonbacterial illnesses each year.

Diarrhea, although the most commonly experienced side effect of antibiotics, is not the only gastrointestinal complication. Many antibiotics, particularly those of the macrolide class (e.g., erythromycin, clarithromycin, and azithromycin) and the quinolone class (e.g., ciprofloxacin and levofloxacin), cause stomach irritation, nausea, and vomiting.

Undesirable Interactions with Other Medicines

Many antibiotics affect the metabolism and actions of other drugs, and vice versa. Fortunately, most of these undesirable interactions are rare—but as you've heard previously, rare becomes common when the use of antibiotics is excessive, as it is in this country. Table 5.2 lists a number of interactions of commonly used pediatric antibiotics with other medicines that children might take.

Miscellaneous Side Effects

With over 100 antibiotics available for human use, about two-thirds of which are approved for use in kids, the category of miscellaneous side effects is the most diverse and difficult to characterize. Among the most troubling and frequent (although still quite uncommon compared with allergic reactions and gastrointestinal upset) are bone marrow suppression (i.e., decreased production of blood cells), kidney damage or inflammation, hearing impairments, and nervous system effects (e.g., seizures, dizziness, and unsteadiness).

Two side effects that appear to be unique to children are worthy of special mention. The tetracycline antibiotics have long been known to stain kids' teeth. For that reason, the use of those medicines has been largely curtailed in recent years for children younger than 8 years (the age when the enamel of permanent teeth has completely formed). The degree of staining is determined by the total amount of the drug given, which in turn is a function of both the dose and the duration of treatment. Recognizing that a single course of tetracycline poses less risk than chronic treatment and that tetracycline antibiotics are the preferred choice for certain pediatric infections, there are rare occasions when a child under the age of 8 years must receive this class of antibiotic. Those rare occasions include Rocky Mountain spotted fever (a potentially fatal tick-borne bacterial infection [see chapters 1 and 9]) and a related infection called ehrlichiosis. Doxycycline (see below) is the tetracycline of choice in

Table 5.2 Interactions with other medicines of antibiotics commonly used in children

Antibiotic	Interacting drug	Effect
Amoxicillin	Tetracycline (antibiotic)	These two antibiotic classes may inactivate each other, each making the other less effective when given together.
Erythromycin	Steroids (anti-inflammatory medicines)	Increased levels and toxicity of steroids
	Theophylline (asthma medicine)	Increased levels and toxicity of theophylline
	Valproic acid and phenytoin (seizure medicines)	Increased levels and toxicity of valproic acid and phenytoin
Quinolones (e.g., ciprofloxacin and levofloxacin)	Antacids	Reduced quinolone effect
	Iron, zinc, calcium, vitamin supplements	Reduced quinolone effect
	Theophylline (asthma medicine)	Increased theophylline toxicity
Isoniazid (tuberculosis medicine)	Acetaminophen	Increased liver toxicity with combination
	Antacids	Reduced isoniazid effect
Metronidazole	Barbiturates (seizure medicines; anesthesia)	Reduced metronidazole effect
Rifampin (tuberculosis drug)	Isoniazid	Increased liver toxicity with combination
Sulfa drugs (e.g., trimethoprim-sulfamethoxizole)	Phenytoin (seizure medicine)	Increased levels and toxicity of phenytoin
Tetracyclines	Antacids	Reduced tetracycline effect
	Iron, zinc, and calcium supplements	Reduced tetracycline effect
	Phenytoin	Reduced tetracycline effect
	Theophylline (asthma medicine)	Increased theophylline toxicity
	Zinc supplements	Reduced tetracycline effect
Thienamycins (e.g., imipenem)	Theophylline	Nervous system toxicity increased when combined

these situations because it has less tooth-staining risk and is given only twice a day (versus four times for other tetracycline formulations).

The second special pediatric concern has faded substantially over the past several years. When first introduced, the quinolone antibiotics (see below) were withheld from children under the age of 18 because an animal model using the drugs showed that young animals developed cartilage destruction. Similar effects were actually seen with one of these medicines given to children in France, but that particular quinolone drug was never used in the United States. The quinolones that have been approved for use in the United States (see below) have been given to children in numerous settings and in increasing numbers—but they still are not Food and Drug Administration (FDA) approved for children under 18 years of age except under a limited number of conditions where alternative therapy is unavailable or undesirable. Once again, it is important to emphasize that to date childhood cartilage problems have not been seen with the quinolones approved for use in the United States. However, something is going on with these medicines. Adults, particularly older adults, have a higher risk of spontaneous Achilles tendon rupture when taking the quinolone antibiotics; the mechanism is unknown, as is the mechanism of the cartilage problems in animals. The FDA restrictions on the use of quinolones, even if they are now felt by many to be an overreaction to the animal data, provide us with a good excuse for not abusing these highly potent, broad-spectrum antibiotics (see the Andromeda strain discussion below).

Theories have been proffered that early exposure to antibiotics may increase your child's risk of developing allergic disorders, such as asthma, and that excessive antibiotic use in a person's younger years may increase that individual's risk of developing certain adult diseases, like breast cancer. There are some data to lend legitimacy to both of these concerns, but not enough data to move from theory to fact. First, in regard to asthma, a number of studies have reported that exposure of a child to antibiotics within the first year of life increases the risk of later development of the disease. There are many deficiencies with the methods used in these studies. The studies that found the strongest association between antibiotics and asthma were retrospective, that is, they looked back at the patient's history after the development of asthma. Recall bias is high in these types of studies, and their reliability is less than that of prospective studies that follow patients from infancy for several years to determine if children exposed to antibiotics develop asthma at a higher rate; those studies have shown a far weaker association or no association at all. Another potential flaw in these studies is that kids who are treated with antibiotics in the first year of life for, as is often the case, a respiratory infection may already be more prone to developing asthma by virtue of their respiratory symptoms. Said another way, kids who ultimately are diagnosed with asthma are more likely to be treated for respiratory infections in early childhood because asthma symptoms can mimic respiratory infections and can also be triggered by them. Hence, the association may actually be between respiratory infections in early life and asthma and may have nothing to do with antibiotics at all. This type of potentially false association of antibiotics with asthma is termed an epiphenomenon (see chapter 12). The "hygiene hypothesis" proposes another relationship between infections (or lack thereof) and allergies/asthma; there is more about the hygiene hypothesis in chapters 7 and 9.

A second possible association with childhood exposure to antibiotics that has raised recent concern is adult-onset breast cancer. A 2004 study in the *Journal of the American Medical Association* found that women with breast cancer had a history of statistically higher exposure to antibiotics over the many years prior to their diagnosis. As in all epidemiologic studies, an association does not prove cause and effect (see chapter 12)—and the association was not a very strong one in this study. Again, the possibility of an epiphenomenon is raised. When antibiotics are needed and appropriate, they should be taken. When antibiotics are not needed, as for viral infections (see above), they should be avoided. The association with breast cancer will require further study, but right now it stands as only a relationship that may or may not turn out to be true, and if true, is probably of little clinical (as opposed to statistical) consequence.

Finally, the generic question that is of greatest concern to most parents of kids taking antibiotics: is there a certain threshold of antibiotic use in childhood that, when passed, is dangerous? That is, if your child requires four, five, or more courses of antibiotics in the first 2 years of life for recurrent ear infections, for example, is that antibiotic exposure associated with any particular cumulative risk beyond the risk of each individual exposure to antibiotics already discussed above? Researchers have looked. In a large study published in 2005, teens and adults taking antibiotics for their acne for longer than 6 weeks had one long-term consequence: they had twice as great a chance of having one cold a year than those patients with acne but not on antibiotics. One cold a year! Not exactly striking data, and there is no good scientific reason to believe this is a real association (remember, colds are due to viruses, which are unaffected by antibiotics). Simply put, the answer is no; there is no compelling evidence that the number of days on antibiotics, or the number of doses taken, is related to any long-term adverse outcomes, notwithstanding the conditions and theories discussed above.

What Are the Risks of Antibiotics to Society?

Antibiotics are overused; about that there is no doubt or controversy. Overuse means that the medicines are being given for diseases that don't need antibiotics, as specified above. The numbers are staggering. Two articles in the *Journal of the American Medical Association* in the late 1990s defined the problem. Each study, one with adults and the other with kids, assessed the national prescribing patterns of physicians in private practice, looking specifically at illnesses that are typically viral, i.e., those diseases in which the patient will not benefit by antibiotic treatment. Table 5.3 summarizes these two important studies.

The important take-home messages from Table 5.3 are as follows. One-fifth of all prescriptions in the United States are antibiotics given for illnesses that are usually viral and for which antibiotics will have no benefit. One-half of all patients presenting to a doctor with a cold will get antibiotics. Although some cases of bronchitis in adults are in fact due to bacteria, virtually no cases of respiratory infection diagnosed as bronchitis in kids are the

Table 5.3 Doctors' private-practice antibiotic-prescribing patterns for adults and children

Patient population	% Antibiotics among overall prescriptions of any medicine	% of patients with the following infection for whom antibiotics are prescribed:			Millions of prescriptions in U.S./yr for these 3 illnesses
		Common cold	Respiratory infection	Bronchitis	
Adults	21	51	52	66	12
Children (≤18 yr)	21	44	46	75	11.2

result of bacterial infection—yet 4.7 million antibiotic prescriptions for bronchitis are written for children every year.

How can this be? How can physicians' actions fly in the face of everything we know about bugs and drugs? There are many answers, too many to adequately explore here. They include business and marketing issues that should have nothing to do with the good practice of medicine. They also include parental preferences, doctors' busy schedules and overflowing waiting rooms, pressures of insurance reimbursement, and the need to cover office overhead. There are the doctors' concerns regarding malpractice lawsuits should this apparent viral infection actually be a bacterial infection in disguise or be followed by a bacterial complication. Perhaps most importantly, antibiotic overuse stems from an anachronistic naiveté about the potential harm to society (as well as to the individual kids themselves [see above]) of such excess. What is that potential harm to society?

Michael Crichton's 1969 classic novel, *The Andromeda Strain,* was the story of an untreatable and universally fatal strain of virus that landed in New Mexico aboard a satellite returning to earth. Scary stuff, those satellite viruses. Viruses do, of course, scare us because of the difficulty in treating them; it is a sobering fact that no virus disease has ever been cured by treatment (see chapter 6). Fortunately most viral infections resolve on their own (see chapter 3), and for some of those that don't, antiviral medicines can contain them and limit the damage they cause (see chapter 6). Bacteria have always been less scary—at least since the antibiotic miracles of the last century. We have, indeed, become complacent and nonchalant about the risks of bacterial infection because we trust that advances in medicine will continue to stay one step ahead of bacteria—that we will always develop a new and more potent antibiotic weapon just in the nick of time to treat a new and more potent bacterium. Until recently, that infectious innocence has been borne out—more than 100 antibiotics have been developed for human use since Fleming's mold proved to be the greatest single advance in the history of medicine.

Fueled by scientific curiosity and ingenuity, as well as by the promise of commercial success and huge industry profits, new antibiotics and entirely new classes of antibiotics have appeared and been approved for use in adults and children. Many of these discoveries have been in response to a growing threat of bacterial resistance, but the discoveries have gone largely unheralded. Unlike a breakthrough new heart medicine or cancer treatment, new and more potent antibiotics are simply expected; no big deal. The threat of resistance has been below the radar, known to doctors and pharmaceutical companies, but largely unknown to the general population. That's because each new antibiotic drug discovery accomplished nothing more than humbly expanding the spectrum of bacterial germs that can be treated—a germ that became resistant to drug A and drug B was susceptible to drug C and to drug D. No problem, no panic, no press release.

Well, those halcyon days of letting the scientists worry about bacteria that are resistant to antibiotics are gone. The Andromeda strain, this time in bacterial form, has arrived—and not by satellite. Let's look first at Fleming's *S. aureus* (staph) germ—the one that revealed the power of the *P. notatum* mold's secretions to kill some of the most lethal bacteria known to humankind. As penicillin use soared in the 1940s, Fleming's staph bug began adapting in the patients who received the medicine, with penicillin-resistant strains being favorably selected over more sensitive ones by the pressure of the antibiotic (see below for how that happens). The first use of penicillin was in 1941; within 1 to 2 years, the first resistant strains were already being reported. As is true for many resistance patterns among bacteria, the first cases occurred in hospitalized patients, in whom the resistance trait rapidly spread among strains of staph bacteria. Within 6 years, 25% of all hospital strains of staph were resistant to penicillin. Resistance among community-acquired *S. aureus* strains followed shortly thereafter. The 25% resistance threshold was reached among community isolates within 15 to 20 years of the introduction of penicillin. By the early 1970s, 80% of staph strains in the United States were resistant; for all practical purposes, the resistance rate is now 100%, and it has been for many years. In a very short time, Fleming's staph bug had learned to produce an enzyme that broke apart Fleming's miraculous molecule!

Let's take a step back here. Learning is a concept that must be clarified in the context of germs. In every collection of bacteria large enough to cause infection, there are millions of individual bacterial cells. Each bacterium in that community is genetically nearly identical to every other bacterium because it is by cell division that each bacterium multiplies. The bacterium first makes a copy of all of its genes, and then, when the bacterium divides to become two, just as with human cells, each daughter cell receives a complete set of the original cell's genes. However, cell division and reproduction are not perfect. Within the teeming masses of millions of rapidly dividing bacteria, random mutations occur. A mutation is a break in a gene or a replacement of a part of one gene with a part of another. Some of the mutations are fatal, and the bacteria carrying those altered genes die off; other mutations are beneficial, and the bacteria carrying those genes have an advantage over the rest of the bugs in the pus.

Now, introduce antibiotics into the picture. When a potent antibiotic like penicillin was given to patients with *S. aureus* infections in 1941, 1942, and

1943, almost all of the staph germs in those patients were killed. *Almost* all. After enough patients were treated, and enough penicillin circulated in hospital environments, there came a patient in whom a random mutant bacterium had appeared with the ability to produce an enzyme to break apart the penicillin molecule. Had penicillin not been used, this bacterium would have been just another germ in the pus. However, because penicillin was wiping out the vast majority of the *S. aureus* germs in the same patient, the bacterium that was able to resist the antibiotic was at a selective advantage, i.e., it had a significantly enhanced ability to survive and grow in this penicillin-treated patient. Whether in this patient or in a subsequent patient who was cared for in the same hospital bed or by the same nurse (either of which may have been carrying the resistant bacterium from the original patient), the resistant germ was now established. Play this scenario over and over again for several years, and you have an emerging problem of resistance by that bug to that drug. Hence, it's probably fairer to say that rather than a bacterium simply learning resistance to antibiotics, we have taught bacteria to become resistant by our introduction of antibiotics and the selective pressure they bring—antibiotics select for resistant germs by killing off the susceptible ones.

S. aureus turned out to be an excellent student. It has learned to become resistant to several newer and more advanced antibiotics that were designed specifically to treat staph germs that had become resistant to prior therapies. Other bacteria are similarly good learners—it is far easier to list the occasional bacterial species that has *not* developed resistance to antibiotics than to list all of those that have.

Now, what about that Andromeda strain I mentioned earlier? When did we first reencounter bacterial infections that couldn't be effectively treated with anything? When did we return to the helpless state that we were in 60-plus years ago, before penicillin? The unthinkable occurred in the late 1990s. *Enterococcus faecium* and *Enterococcus faecalis,* as their last names suggest, are normal residents of the intestines in many individuals. When these germs get into places they shouldn't be, like almost all colonizing bacteria, they can cause infections—in the cases of these enterococci, urinary tract infections, infections of indwelling medical devices (such as intravenous catheters and bladder tubes), and abdominal infections, among others. These germs have become increasingly hard to treat in recent years, and by the early 1990s, an antibiotic called vancomycin (see discussion of intravenous antibiotics below) was the only choice for many of these enterococcal strains. And then, brazenly, enterococci emerged that were resistant to vancomycin—and all other antibiotics! It was only by heroic medical and surgical interventions that several patients infected with these multiply drug-resistant vancomycin-resistant enterococci (VRE) survived. Other patients weren't so fortunate and succumbed to their VRE infections, which were completely unresponsive to every antibiotic thrown their way. In the 8 years between 1990 and 1997, the occurrence rate of VRE among hospital enterococcal infections climbed from 1% to 15%. Writing in the *American Journal of Medicine* in 1997, Barbara

Murray noted, "The acquisition of resistance to vancomycin by enterococci demonstrates how an organism that usually behaves as a second-rate pathogen can become a first-rate clinical problem." By 2003, VRE comprised 28% of all enterococcal bacteria in intensive-care units, where the sickest and most vulnerable hospital patients are cared for.

Fortunately, we were pulled back from the brink just in time. Experimental antibiotics were being studied in the late 1990s that were effective against VRE, and medicine once again leaped ahead of microbiology, but not before falling behind for the first time ever since the advent of penicillin. The victory may have been short-lived. By 2002, one of the experimental antibiotics that had become more widely used against VRE, linezolid, was found to be ineffective against enterococci from seven patients at the Mayo Clinic in Minnesota. There are still two or three even newer antibiotics that have activity against the most resistant VRE—but likely not for long.

The list of bacteria that have acquired resistance to multiple different classes of antibiotics is growing longer. For most, there is still an arrow or two left in the quiver. For some, the quiver is already empty. On 4 August 2005, the following story appeared in the *New York Times*, reminiscent of another war in an earlier century.

> Troops in Iraq Bring Resistant Bacteria Home
> by Denise Grady
>
> American troops wounded in Iraq and brought back to military hospitals in the United States have unexpectedly high rates of infection with a drug-resistant type of bacteria, doctors are finding.
>
> The bacteria, *Acinetobacter baumannii*, are not unique to Iraq. They live in soil and water in many parts of the world, and had already been known to cause trouble in hospitals and on battlefields in Vietnam. They can invade wounds, the bloodstream, bones, the lungs and other parts of the body. Antibiotics can cure the infection, but doctors must use the right ones.

The story went on to say that no American soldiers had yet died of the infection but that other patients in the hospital at the same time, already very sick with other illnesses, acquired the germ and did die, either from the infection or from their underlying disease. As the article stated, infections with this germ were also seen in the Viet Nam war—the increased contact that soldiers have with soil during war time may explain the apparent increased incidence among military personnel.

A. baumannii is a germ that has natural resistance to many antibiotics, i.e., resistance that is not acquired by mutation and selection under antibiotic pressure. At least, that's the way it started out. Beginning in the United States in 1991, however, strains of *A. baumannii* began emerging that were newly resistant to antibiotics to which the germs had been exposed in hospital settings. What ensued in the United Kingdom and in other countries were dozens of deaths due to completely untreatable *A. baumannii* infections. Deaths continue, mostly in patients with already abnormal immune systems and mostly in hospitalized patients in intensive-care units. The only antibiotic choice for some of these patients is an extremely toxic drug called colistin. Will medicine again be able to leap ahead of microbiology as it did with VRE, or will the germs win this one? If not this one, how long will it be before the germs do win? This pattern of increasingly difficult-to-treat bacterial infections is being called by some the beginning of the *postantibiotic era,* i.e., the period during which antibiotics no longer work.

What can be done to fight the trend toward superbugs, or even reverse it? A lot, it turns out. We already know that we can adversely alter bacterial resistance patterns with imprudent antibiotic use (overuse). However, interventional trials in many places in the world have also shown that when a concerted national effort is made to reduce the unnecessary use of antibiotics, resistance patterns among bacteria can be altered favorably. The percentages of resistant germs in the noses and throats of kids can be lowered 10-fold or more by simply withholding antibiotics from those kids who don't need them—those with viral illnesses that otherwise create a knee-jerk prescription-writing response. Federal agencies have cooperated in recent years to issue national guidelines for appropriate use of antibiotics in hospital settings—similar guidelines in the private sector would reduce the alarming statistics presented in Table 5.3.

Then there's the role of the pharmaceutical industry. The money in antibiotics comes with new drug discoveries and with the widespread use of new antibiotics. When a promising new antibiotic is developed and found to be safe, it's tested in patients for efficacy against infections. New antibiotics are increasingly targeting multiply resistant germs, or trying to expand the spectrum of bacterial targets (there's more money in making a first-of-its-kind antibiotic than a me-too antibiotic [see below]). However, the catch-22 here is that when a pharmaceutical company comes up with a new drug that goes beyond the current weapons, treating bacteria that have become resistant to older antibiotics, the market share for that new drug is very small if it is limited only to the newly emerging resistant infections. As an example, if the only use for a new drug is to treat the several dozen patients with the potentially lethal multiply resistant *A. baumannii* strains, how profitable can that possibly be? The answer is, it's not profitable at all. New drug discovery and development are very expensive undertakings for pharmaceutical companies, and those costs must be recovered. Thus, even when a new drug is potentially the last resort and best hope for patients with very difficult-to-treat infections, the pharmaceutical company has little financial motive to limit the use of that drug to the patients who need that drug and that drug alone. Rather, the pharmaceutical company is anxious to treat all children with ear infections or strep throat with their new antibiotic, even though the current antibiotic choices for treating those infections are plentiful. Trials are conducted to show that the new wonder drug can treat the old standby infections just as well as the old standby antibiotics can, and then the new drug is marketed as a new treatment for ear infections or strep throat. You can guess what happens next. As the new, last-resort, best-hope antibiotic starts being prescribed widely for trivial infections that need only a slingshot, not this new cannon, the new cannon rapidly loses its effectiveness against the superbugs. The superbugs learn to become resistant to the new drug and spread from patient to patient. Once again, the germs leap ahead of the doctors.

Controlling the emergence of superbugs must be a partnership among doctors who understand the need for prudent antibiotic use, parents who understand that most of their kids' common infections don't need antibiotics, and the pharmaceutical industry, which must understand the vital role that new drugs can play in truly serious and life-threatening infections. Ultimately, the government will have to exert its formidable influence for all of this to work correctly. Physicians in private practice will need national guidelines, in much the same way that hospitals now have them, for prudent antibiotic use. Incentives may be necessary. The pharmaceutical industry will probably also need government incentives to withhold their newest and most promising products from the mass market applications that are most profitable. After all, the government pays subsidies to farmers to *not* grow certain crops! Finally, patients will need to be educated as to the harm—to themselves, to their kids, and to society—of antibiotic overuse.

A seminal study in Finland published in 1997 showed the rewards for prudent antibiotic use that result from such a cooperative effort. Epidemiologists and microbiologists in that country noticed an alarming rise in resistance of *S. pyogenes,* the bacterial cause of strep throat, to the macrolide (see below) class of antibiotics, which includes erythromycin and azithromycin. Between 1988 and 1990, the resistance rate rose from 5% to 13% among all strep germs studied in that country; this followed a period during which the use of those antibiotics in Finland nearly tripled, as well. National guidelines were issued at the end of 1991; public awareness campaigns targeted patients and doctors alike. The results were striking. The use of the targeted antibiotics fell by 43% between 1991, when the guidelines were issued, and 1992. This was followed by a steady fall in resistance rates among *S. pyogenes* bacteria from 16.5% in 1992 to 8.6% in 1996—a 48% drop in resistant germs correlating with the 43% decline in use of the offending antibiotics. A more highly correlated relationship between antibiotic overuse and resistance could not have been imagined—and now it was proven. These types of results have been replicated in a number of countries and communities throughout the world. Importantly, however, the use rates of antibiotics overall in Finland did not drop during the period of abstinence from the macrolide antibiotics; macrolides were simply replaced by other antibiotics in the Finnish doctors' armamentarium. There was not a concurrent campaign to teach patients and their doctors about the need to refrain from any antibiotic use when the criteria for using antibiotics (see above) are not met. What we learned from the Finns is that a societal change in policy and an increase in awareness can reverse the progression toward Andromeda strain scenarios. The next step is to implement policy guidelines that encompass all antibiotic overuse and abuse.

We are making progress in that arena, as well. Surveys in 2006 showed that many fewer patients feel that all respiratory infections or fevers should be treated with antibiotics than in similar surveys 10 years earlier. A study in Knoxville, TN, published in 2002 showed that a community-wide educational and awareness campaign resulted in an 11% decrease in prescriptions of all antibiotics to children. The word is getting out that antibiotics should be reserved for use only when truly indicated.

Should My Child Take Antibiotics To *Prevent* Infections?

Antibiotics kill bacteria, but they can also prevent bacteria from invading and causing infection. The latter effect of antibiotics results from their presence in the blood and tissues of the body. As a general rule, antibiotics should be used to treat existing infections, but there are certain circumstances under which using prophylactic, or preventive, antibiotic therapy is appropriate. They can be classified as follows:

- Exposure to dangerous bacteria
- Prevention of recurrent bacterial infections
- Dirty wounds and bites
- Certain medical, dental, and surgical procedures
- Overly susceptible patients

Exposure to Dangerous Bacteria

There are several situations in which a child (or adult) has been exposed to a potentially dangerous bacterium but has not yet been infected. Examples are discussed more fully in chapter 3 and include tuberculosis, *Neisseria meningitidis* (the cause of a rapidly fatal form of sepsis and meningitis) infection, anthrax, plague, and pertussis (whooping cough). Following exposure to any of these diseases, and select others, antibiotics are prescribed—but the definition of exposure is important. For each condition, there are specific guidelines as to what constitutes an exposure that puts someone at true risk for acquiring the disease. In the case of *N. meningitidis* exposure, for example, the American Academy of Pediatrics recommends prophylactic antibiotics in the following settings: any household exposure; child care or nursery school exposure during the week before an infected patient's illness; direct exposure to an ill patient's secretions (e.g., kissing, sharing eating utensils, or mouth-to-mouth resuscitation) during the week before that patient's illness; frequently sleeping or eating with an infected patient during the week before that patient's illness; and airplane passengers sitting for more than 8 hours immediately next to a patient who develops the infection. Although this seems to be a broad inclusion of people to receive preventive antibiotics, note that it does not include, for example, kids in the same classroom as an infected patient or kids on the same sports team, except as certain of those kids may qualify based on other criteria above.

One unusual example of prophylactic antibiotics given for potentially dangerous germ exposure is the antibiotic eye drops given to every newborn to prevent possible gonorrheal or chlamydial eye infection from a vaginally infected mother (see chapter 3).

Prevention of Recurrent Bacterial Infections

Some children are prone to developing multiple infections at the same body site—urinary tract infections and ear infections are the most common examples of this. Often, but not always, there is an anatomic explanation for this propensity—e.g., a kink in the urine collecting tubes or an unusually short and straight eustachian tube venting the middle ear (see chapter 3). Certain of these anatomic abnormalities need to be surgically corrected; others correct themselves with the growth and development of the child. Antibiotics given during infection-free periods, typically at a lower dose and less frequently than for an actual infection, have been proven to be effective in reducing the recurrence rate for both urinary tract and ear infections. However, as with all antibiotic decision making, care must be taken to prevent the emergence of resistance. Indeed, it is under the conditions of low-dose/low-frequency use of antibiotics that the selection of resistant germs is most likely. That's because slow killing of germs provides greater opportunity for bacteria to mutate and for mutants to be preferentially selected for survival (see above).

Guidelines for antibiotic prophylaxis (prevention of infection) against recurrent body site infections must be strictly adhered to; they include the following.

- Recurrence is defined as including only true acute infections (e.g., residual fluid in the middle ear does not constitute recurrence [see chapter 3]) and more than the expected number of infections for healthy children.
- A defined and limited period for prophylaxis is required—indefinite is not an option for treatment length. Either a planned surgical correction or a predefined trial period to get children through a particularly difficult period of recurrences (e.g., prophylaxis for ear infections during the winter cold and flu season, when most recurrent ear infections occur) is a potential guidepost for limiting therapy.
- The germs causing the recurrent infections should be predictable and within the spectrum of bacteria killed by the prophylactic antibiotic to be used.

Urinary tract prophylaxis has been particularly effective because the antibiotics most frequently used for that purpose—trimethoprim-sulfamethoxazole and nitrofurantoin—are highly concentrated in the bladder and do not seem to trigger the emergence of resistant bacteria as readily as do other drugs. In contrast, the antibiotics used to prevent recurrent ear infections (e.g., amoxicillin) are more likely to affect the bacteria colonizing the upper respiratory tract and select for resistant germs—hence, although effective, prevention of your child's recurrent ear infections with antibiotic prophylaxis may make the subsequent infections that do occur (breakthrough infections) more difficult to treat. On balance, for the right patient and with the right criteria satisfied (see above), antibiotic prophylaxis for recurrent ear infections remains a reasonable option.

A particularly troublesome type of recurrent infection is skin infection due to *S. aureus*. Increasingly, the strains responsible for these infections are resistant to standard antistaphylococcal antibiotics. The source of the germs, named "MRSA" (for methicillin-resistant *Staphylococcus aureus*), is often the nasal passages, where staphylococci colonize without causing infection in the nose itself but are spread from the nose to the skin by hand (or finger, as the case may be). A variety of approaches to eliminating nasal colonization with MRSA have been attempted. The most successful is the use of a topical intranasal antibiotic, mupirocin (see "Topical antibiotics" below), for a limited time. Even with this approach, recolonization occurs in a percentage of kids, and resistance to mupirocin may develop.

Dirty Wounds and Bites

Traumatic injuries that break the skin can introduce millions of bacteria into deep layers of tissue and even into the bloodstream. For most wounds, thorough cleansing and use of topical antiseptics (see chapter 9) are sufficient for removing most of the germs before they can establish a foothold and cause trouble. Depending on the nature of the wound, and the assessment by the doctor, antibiotics may be prescribed before any signs of infection appear, also in the hope of preventing the germs from establishing themselves. Doctors will also often preventively treat dirty wounds that require suturing.

The bacteria in the mouths of humans and animals are potentially vicious causes of infections; for that reason, antibiotics are also frequently given following bites (human bites are even worse than animal bites in causing infections).

Certain Medical, Dental, and Surgical Procedures

Because many procedures performed by doctors and dentists are associated with the risk of infection resulting from the procedure itself, studies have been conducted to evaluate the effectiveness of antibiotic treatment before, during, and/or following the procedure in reducing infections. While not proven to be of benefit for all invasive procedures, antibiotics have been shown to reduce the chances of infection in the types of procedures shown in Table 5.4.

Table 5.4 Types of medical, dental, and surgical procedures that warrant prophylactic antibiotics

Heart catheterization
Bladder scope (cystoscopy)
Insertion of prosthetic (artificial) material (e.g., a new heart valve or joint replacement)
Open-heart surgery
Brain or spinal cord surgery
Exploratory surgery of body cavities (especially in newborns)
Surgery to relieve blockages of the gastrointestinal tract, bile ducts, or urinary tract
Surgery of dirty and contaminated traumatic wounds
Surgery involving the gastrointestinal tract in which intestinal contents get into the wound
Dental or invasive medical procedures in patients with abnormal heart anatomy

The last category in Table 5.4, antibiotic prophylaxis for procedures in patients with abnormal heart anatomy, is to prevent the complication of bacterial endocarditis, an infection in which bacteria preferentially land and grow on abnormal heart valves or vessels. With all of these uses of prophylactic antibiotics, caution must be taken to minimize the duration of therapy to reduce the risks of resistance. For most procedures, antibiotics given 30 to 60 minutes prior to the procedure and discontinued not longer than 24 hours after the procedure are sufficient.

Overly Susceptible Patients

Certain patients, because of underlying abnormalities in their immune systems (see chapter 4), require prophylactic antibiotics to prevent severe bacterial complications of their immune deficiencies. Included are certain patients with AIDS (see chapter 3) and patients with absent or dysfunctional spleens. As noted in chapter 4, the spleen is responsible for clearing germs that have been targeted by antibodies. Patients born without spleens, or after surgical removal of the organ (e.g., following trauma and spleen rupture), are susceptible to severe and rapidly fatal bacterial infections, as are patients with sickle-cell anemia, whose spleens are dysfunctional.

Because a patient with an underlying immune problem is likely to have that condition for his entire life, prophylaxis for most such children with the other 80 to 100 forms of immune deficiency (see chapter 4) is not only impractical, but dangerous—the development of bacterial resistance under the influence of long-term antibiotics in these patients can be especially deadly.

Is My Child Exposed to Antibiotics in the Environment?

The simple answer is yes, there are antibiotics circulating in your kids' environment. The sources are both human and animal in origin; exposures from plant sources and personal hygiene/household cleaning products that contain antibiotics (see chapter 9) are probably less significant.

Human Antibiotics in the Environment

Antibiotics given to patients are only partially metabolized in the body, with some intact antibiotics excreted into the sewage system. Additionally, unused or expired antibiotics in homes, hospitals, and doctors' offices are frequently poured down the drain into the sewage system or discarded into waste containers that then find their way to landfills. When hospital effluent (the waste coming from hospitals prior to reaching the sewer system) is tested, numerous antibiotics are readily detected at levels high enough to actually treat infections. Antibiotics have also been detected in what comes out of sewage treatment facilities (after treatment), as well as in surface water and groundwater.

Animal Antibiotics in the Environment

Antibiotics have long been known to promote the growth and even extend the lives of farm animals, including poultry, cattle, pigs, and fish. Once again, though, the unmetabolized antibiotics from antibiotic-fed animals reach manure piles, are used in fertilizer, and generally make their way into the groundwater. A study published in 2004 by investigators at Colorado State University sampled dairy lagoons and manure stockpiles and found measurable concentrations of all the classes of antibiotics that had been fed to animals in the area. The antibiotics used in animals are the same, or nearly the same, as those used to treat human infections. In addition to exposure of humans to animal-excreted antibiotics through the groundwater, foods derived from these animals often have measurable concentrations of antibiotics.

A number of European countries have banned the use of antibiotics as growth promoters and life extenders in animals. The United States has not yet prohibited this practice, but public education and awareness have resulted in some improvements. In 2005, the FDA took its first ever action to ban an antibiotic from animal use; the prior approval for the potent antibiotic enrofloxacin was withdrawn after evidence emerged linking its use to the emergence of resistant campylobacter (see chapter 3) bacteria. The FDA found that resistant campylobacter infections in humans had increased substantially since the FDA first approved the animal use of antibiotics like enrofloxacin in 1995. Farming and ranching advocacy groups have argued that their constituencies are voluntarily reducing the use of antibiotics in animal feed in response to growing concerns regarding antibiotic resistance. However, they also make the point that withdrawing antibiotics from use as growth promoters in feed will increase the need to treat animals with antibiotics for actual infections they may contract as a result of not receiving the antibiotics in their feed. In other words, the antibiotics in the feed protect animals from infections that otherwise would have to be treated with the same or other antibiotics, according to the ranching industry. What is unknown is how the number of infections that result from removing antibiotic supplements from the feed would compare with the current virtually unrestricted use of antibiotic supplements in healthy animals. Additionally, the ranching industry notes that the shortened life span or reduced growth of the farmed animals deprived of supplementary antibiotics would have substantial economic impact.

Table 5.5 illustrates the scope and sources of the problem of widespread antibiotic usage (in tons!) as summarized in a 2003 review in the *Journal of Antimicrobial Chemotherapy*.

Plant Antibiotics in the Environment

Two antibiotics are used for preventing and treating bacterial diseases in agriculture; both are similar to antibiotics used in humans and are applied to plants via spraying. No studies have been done to evaluate the potential human risks from this antibiotic application, nor have studies demonstrated the extent of accumulation, if any, of these antibiotics in the surface water or groundwater.

Table 5.5 Antibiotic use patterns, human and animal[a]

Location	Yr studied	Total antibiotics used (tons)	% of total antibiotics used for human medicine	% of total antibiotics used for livestock farming
Worldwide	2002	100,000–200,000		
European Union	1996	10,200	50	50
	1999[b]	13,288	65	35
United States	2000	16,200	30	70

[a]Data derived from K. Kümmerer, *J. Antimicrob. Chemother.* **52:**5–7, 2003.
[b]May reflect tighter European Union regulations enacted between 1996 and 1999 regarding use of antibiotics in livestock farming.

Personal Hygiene and Household Cleaning Product Antibiotics in the Environment

The presence of antibiotics in commercial cleaning products has exploded in recent years, raising the specter of Andromeda strains emerging from exposure to toothpaste and bath soaps. Chapter 9 thoroughly reviews this issue and concludes that, at least for the present time, the risk of widespread bacterial resistance from these products appears to be small.

What is the evidence that environmental exposures to antibiotics may result in harm to humans? Many of the data are epidemiologic, that is, they link rises in resistance among bacteria infecting humans to usage patterns in animals. One example, campylobacter, was noted above. Another even more frightening illustration of the risk comes from Europe, where the emergence of VRE in humans has been strongly associated with the use of a vancomycin-like antibiotic in animal feed. Recall from an earlier discussion in this chapter the substantial human risk posed by VRE; vancomycin is one of the last resorts in treating those germs.

Consumers do have a choice. For a premium price at health food stores, antibiotic-free animal-derived foods are widely available. While those foods reduce the exposure of your own family to environmental antibiotics, their modest market penetration to date will do little to affect the overall societal problem of spreading antibiotic resistance pressure.

Antibiotics Simplified

From that fateful day in 1928 when Alexander Fleming returned to his London laboratory and discovered an unusual pattern of bacterial growth surrounding a growth of mold on a culture plate, antibiotics have proven to be the greatest single therapeutic advance in the practice of medicine. As you've seen in the discussions above, the ingenuity and sophisticated molecular tools employed by research scientists and pharmaceutical companies in designing ever more potent antibiotics are exceeded only by the ingenuity and sophisticated genetic transformations employed by bacteria to develop resistance to our antibiotics. As more and more elaborate strategies are conceived for killing germs, germs become increasingly adept at mutating and confounding our innovations. This means that even more arrows need to be added to our quivers; antibiotics now come in more flavors than Mr. Baskin or Mr. Robbins ever imagined and more varieties than Mr. Heinz could have dreamed of.

Much of this variety reflects the good-old American spirit of capitalism—when the FDA approves an antibiotic that has been painstakingly (and at great cost) developed and manufactured by one pharmaceutical company, typically numerous other pharmaceutical companies rush to market "me-too drugs." Me-too drugs are nearly identical antibiotics of the same class as and with chemical structures very similar to that of the FDA-approved medicine. There are many reasons for companies to develop me-too drugs—the approval process by the FDA is somewhat less stringent when the me-too drug is evaluated because many of the safety and efficacy concerns associated with a brand new class of compounds are allayed with the first drug of that class. Additionally, me-too drugs compete with their forerunners with the advantage of seeing how the original drugs of the class performed in the marketplace. Me-too drugs may have pharmacologic advantages and competitive advantages over earlier versions, e.g., a more favorable dosing schedule or a better-tasting formulation. As a result of the me-too phenomenon, for most of the drugs listed in the summary tables that follow below, there are others out there that do much the same thing; even some drugs that are listed in the tables are me-too drugs for other antibiotics also in the tables. This is job security for pharmacists.

Most of the information you need to understand antibiotics concerns orally administered antibiotics, by far the most common form likely to be prescribed for your kids. I'll also provide a brief review of intravenous and intramuscular antibiotics. Oral, intravenous, and intramuscular antibiotics are termed systemic, meaning that the antibiotics go to all parts of the body via the bloodstream. This is important, because it reminds us that just because an infection occurs in the middle ear, for example, that doesn't mean that the medicine used to treat it isn't also in contact with the liver, the kidneys, the gastrointestinal tract, and other organ systems where the antibiotic may cause toxicity. In contrast, topical antibiotics generally remain localized to the part of the body to which they are applied; these medicines will also be briefly reviewed.

Oral Antibiotics

A brief description of the classic oral antibiotics would reveal the familiar pastel elixirs that are most commonly prescribed for kids' most commonly occurring infections. Would that it were so simple! The classics have been outdone by the neoclassics and the neoneoclassic antibiotics, ever more potent weapons cleverly camouflaged in cherry and grape veneers. In reading the remainder of this chapter, keep in mind this overriding principle: the best antibiotic treatment for your child is the medicine that most specifically targets the germ or germs likely to be causing the infection. By specific targeting, I mean that the antibiotic should treat the fewest types of bacteria possible—the narrowest spectrum of germs that includes the likely offenders for the particular infection. The broader the spectrum of germs killed by an antibiotic, the greater the risk of generating resistant germs.

Antibiotic "Levels"

For simplicity and clarity, I have grouped the cornucopia of oral antibiotics into levels; this is my somewhat contrived convention that I hope will help you understand exactly what your kids' doctor is prescribing and why. My assignment of an oral-antibiotic level is based on the sophistication of the antibiotic's strategy in targeting and killing germs and the spectrum of the antibiotic's activity against bacteria (i.e., how broadly or how narrowly the antibiotic targets germs; broad-spectrum antibiotics kill germs from many different categories of bacteria, whereas narrow-spectrum antibiotics are limited in their effects to one or a few categories of bacteria). Generally (but not always), the narrowest-spectrum antibiotics are those in level 1. As the level of sophistication of the antibiotics increases, so does the spectrum of germs killed. There are times when a broader-spectrum, higher-level medicine is necessary—often that's because there may be many different kinds of germs that can cause a particular infection, and a narrower-spectrum medicine might miss the true cause. A higher-level antibiotic may also be necessary because resistance has already emerged among the bacteria causing the infection, and the level 1 medicines can no longer be depended upon to work.

Antibiotics can be classified by their chemical structures; this is the way that most doctors think of these medicines. The major chemical classifications of oral antibiotics, with examples of each grouped by my levels (of sophistication and spectrum of germs killed), are presented in Table 5.6. Not all of these medicines have been approved by the FDA for use in children, but they are available to your kids' doctors should they choose to use them. Sometimes, the only difference between a lower-level antibiotic and a higher-level antibiotic of the same chemical class is the ease of administration—higher levels of antibiotics tend to be given less frequently. For example, erythromycin, the level 1 macrolide, is given four times a day, whereas the level 2 macrolide, clarithromycin, is given only twice a day and the level 3 version, azithromycin, is given only once a day (and for half the number of days). Another important correlate of the level of antibiotic is that, almost invariably, higher-level antibiotics are more expensive.

Each of the antibiotics in Table 5.6 is presented in additional detail in Tables 5.7 to 5.10. Important to your understanding of these tables, and to how decisions regarding antibiotics are made by your kids' doctors, are the three biological properties of bacteria noted at the very beginning of this chapter—color, shape, and oxygen requirements. When a doctor identifies

Table 5.6 Major classes of *oral* antibiotics and examples of each

Classification	Example			
	Level 1	Level 2	Level 3	Level 4
Penicillins	Penicillin, ampicillin, amoxicillin	Dicloxacillin, amoxicillin-clavulanic acid		
Cephalosporins	Cephalexin	Cefaclor, cefuroxime	Cefixime, cefdinir, cefpodoxime	
Carbacephems		Loracarbef		
Tetracyclines	Tetracycline, doxycycline, minocycline			
Macrolides and ketolides	Erythromycin	Clarithromycin, erythromycin-sulfisoxazole	Azithromycin	Telithromycin
Sulfas	Trimethoprim-sulfamethoxazole			
Lincosamides		Clindamycin		
Metronidazole	Metronidazole			
Nitrofurantoin	Macrodantin			
Quinolones			Ciprofloxacin, levofloxacin, ofloxacin	Gatifloxacin
Oxazolidinones				Linezolid

Table 5.7 Level 1 oral antibiotics

Generic name	Brand name(s)	Relative susceptibility of bacteria by color[a]		Relative susceptibility of bacteria by shape[a]		Relative susceptibility of bacteria by oxygen requirement[a]		Common examples of infections treated[b]
		Purple	Pink	Round	Rod	Aerobic	Anaerobic	
Penicillin	Pen-Vee K	++	+	++	+	++	++	ST
Ampicillin	Principen	+++	++	+++	++	+++	++	ST, E, S, UTI
Amoxicillin	Amoxil; Trimox	+++	++	+++	++	+++	++	ST, E, S, UTI
Cephalexin	Keflex	+++	+	+++	+	+++	–	Skin
Tetracyclines	Minocin; Vibramycin; Doryx	+++	++	+++	++	+++	++	UTI, acne, rickettsia, Lyme disease
Erythromycin	Ery-Tab; ERYC	+++	+	+++	+	+++	–	P, pertussis, Legionnaires' disease, chlamydia
Trimethoprim-sulfamethoxazole	Bactrim; Septra; Co-Trimoxazole	+++	++	+++	++	+++	–	E, UTI
Metronidazole	Flagyl	++	++	–	++	–	++++	Vaginitis, colitis
Nitrofurantoin	Furodantin; Macrobid	++	++	+	+++	++	–	UTI

[a]The relative susceptibilities of bacteria are presented on a scale of + to ++++ and refer to the ability of the antibiotic to kill or inhibit all the bacteria of that particular category. For example, a +++ susceptibility of purple bacteria to ampicillin means that more types of purple bacteria are susceptible to that antibiotic than to penicillin, which is rated ++. An antibiotic like metronidazole being rated ++++ for anaerobic bacteria means that almost all types of anaerobic bacteria are susceptible to that antibiotic. A rating of – means the bacteria are not susceptible.

[b]ST, strep throat; E, ear infections; S, sinus infections; UTI, urinary tract infections; P, pneumonia.

Table 5.8 Level 2 oral antibiotics

Generic name	Brand name	Relative susceptibility of bacteria by color[a]		Relative susceptibility of bacteria by shape[a]		Relative susceptibility of bacteria by oxygen requirement[a]		Examples of infections treated[b]
		Purple	Pink	Round	Rod	Aerobic	Anaerobic	
Dicloxacillin		+++	–	+++	–	+++	–	Skin
Amoxicillin-clavulanic acid	Augmentin	+++	+++	+++	+++	+++	++	E, S, skin
Cefaclor	Ceclor	+++	+++	+++	++	+++	–	ST, E, S, UTI, skin
Cefuroxime	Ceftin	+++	+++	+++	++	+++	–	ST, E, S, UTI, skin
Loracarbef	Lorabid	+++	+++	+++	++	+++	–	ST, E, S, UTI, skin
Clarithromycin	Biaxin	+++	+	+++	+	+++	+	ST, E, S, P, skin
Erythromycin-sulfisoxazole	Pediazole	+++	++	+++	+	+++	–	E
Clindamycin	Cleocin	+++	–	+++	+	++	+++	Skin, vaginitis, chronic respiratory and chronic sinus

[a]See Table 5.7, footnote *a*.
[b]See Table 5.7, footnote *b*.

Table 5.9 Level 3 oral antibiotics

Generic name	Brand name	Relative susceptibility of bacteria by color[a]		Relative susceptibility of bacteria by shape[a]		Relative susceptibility of bacteria by oxygen requirement[a]		Examples of infections treated[b]
		Purple	Pink	Round	Rod	Aerobic	Anaerobic	
Cefixime	Suprax	++	+++	++	+++	+++	–	ST, E, S, UTI
Cefdinir	Omnicef	++	+++	++	+++	+++	–	ST, E, S, skin
Cefpodoxime	Vantin	++	+++	++	+++	+++	–	ST, S, P, UTI, skin
Ceftibuten	Cedax	++	+++	++	+++	+++	–	ST, E, S, UTI, skin
Azithromycin	Zithromax	+++	+	+++	+	+++	+	ST, E, S, P
Ciprofoxacin	Cipro	+++	++++	+++	++++	++++	–	UTI, S, P, dysentery, skin
Levofloxacin	Levaquin	+++	++++	+++	++++	++++	–	UTI, S, P, dysentery, skin
Ofloxacin	Floxin	+++	++++	+++	++++	++++	–	UTI, S, P, dysentery, skin

[a]See Table 5.7, footnote *a*.
[b]See Table 5.7, footnote *b*.

Table 5.10 Level 4 oral antibiotics

Generic name	Brand name	Relative susceptibility of bacteria by color[a]		Relative susceptibility of bacteria by shape[a]		Relative susceptibility of bacteria by oxygen requirement[a]		Examples of infections treated[b]
		Purple	Pink	Round	Rod	Aerobic	Anaerobic	
Telithromycin	Ketek	+++	+	+++	+	+++	–	S, P
Gatifloxacin	Tequin	++++	++++	++++	++++	++++	++++	S, P, UTI, skin
Linezolid	Zyvox	++++	–	++++	+	++	+	P, skin

[a]See Table 5.7, footnote *a*.
[b]See Table 5.7, footnote *b*.

an infection, he or she makes the decision regarding the best treatment on the basis of a number of factors:

- Location of the infection (which body site[s] is involved)
- Types of bacteria most likely to be involved (color, shape, oxygen requirements, and other characteristics [see above])
- Antibiotic history of the patient
  1. Allergies
  2. Tolerability—whether the patient has had gastrointestinal or other adverse reactions to certain antibiotics
- Advantages of a particular antibiotic
  1. Convenience of the dosing schedule
  2. Cost

The oral-antibiotic summary tables (Tables 5.7 to 5.10) are organized by levels of antibiotics. The types of bacteria targeted (by color, shape, and oxygen requirements), as well as examples of specific infections treated with each antibiotic, are included. The latter reflects today's infections that respond to the antibiotics; in earlier times, the list of infections that responded to many of these medicines would have been longer, but bacterial resistance has limited the utility of many of our antibiotics. Also note that just because an infection is listed as amenable to treatment by a particular antibiotic, that doesn't mean that antibiotic is the best choice for that infection—a particular medicine may not be potent enough or may be too potent for a particular type of infection, much as infections themselves vary in potency and risk to your child. For example, many oral antibiotics are approved for use and proven to be effective against strep throat, but only two or three antibiotics should ever be used for that infection because, remarkably, strep throat remains susceptible to good old penicillin after all these years. Hence, only the level 1 antibiotics penicillin, amoxicillin, and ampicillin should ever be used (plus erythromycin for penicillin-allergic kids) to treat strep throat—there is no reason to expose your kids or the environment (see above) to broader-spectrum antibiotics for this infection. The infections referenced in the tables are themselves discussed fully in chapter 3.

The level 1 oral antibiotics are those that, for the most part, came along first. Many are still useful for treating kids' infections; others have lost some of their utility and/or popularity.

The level 2 oral antibiotics either more precisely target specific germs (e.g., *S. aureus* by dicloxacilliin and amoxicillin-clavulinic acid) or extend the spectrum of germs covered to include additional categories of bacteria. Many of these antibiotics are derivatives of the level 1 antibiotics, but others are novel classes of medicines (e.g., clindamycin and loracarbef). Two of the level 2 antibiotics are hybrids of a common medicine with an additional chemical to enhance the effectiveness of the antibiotic (amoxicillin-clavulanic acid and erythromycin-sulfisoxazole).

Level 3 antibiotics were, until recently, the most potent and broadest spectrum (i.e., they killed the most types of bacteria) of all antibiotics. This supremacy (now usurped to some degree by the few level 4 oral antibiotics introduced in recent years, as well as by the level 4 intravenous antibiotics discussed below) is reflected in some of the brand names used to convince doctors, with prefixes or suffixes like super, omni, and max. The level 3 antibiotics, with their broad spectrum, should be reserved for those infections that cannot be treated with level 1 or level 2 antibiotics.

Until recently, the only level 4 antibiotics available to doctors were intravenous medicines for the sickest patients, who were hospitalized and for whom lower levels of antibiotic therapy had usually failed. There are now several oral level 4 antibiotics, but their use is very controversial for the reasons discussed earlier—when highly potent, broad-spectrum antibiotics are used for commonplace infections, the effectiveness of those medicines will be compromised by more quickly emerging resistance. The antibiotics in Table 5.10 should rarely if ever be used for infections that could be treated with lower-level antibiotics.

One observation you might have made from the tables is that many of the antibiotics, from level 1 through level 4, target ear and sinus infections. This is because these infections are the most frequent conditions for which antibiotics are prescribed in pediatrics. Herein lies one of the ironies alluded to earlier in this chapter—regardless of how sophisticated and potent the antibiotic, the greatest potential profit for the manufacturer lies in the widest use of the medicine. Thus, even a level 4 antibiotic is promoted for use in the most trivial of infections, because trivial infections, e.g., ear infections, are so common and so profitable. This is clearly an area that needs further cooperation between health authorities and pharmaceutical companies (see above).

Intramuscular Antibiotics

There are a few situations in which your child's doctor may decide that a shot or two of antibiotics given into the muscle (the thigh or buttocks) would be more effective or simpler than a course of oral antibiotics. Although a number of antibiotics can be given by a shot into the muscle, only two are routinely used in the outpatient setting for commonplace infections. A long-acting penicillin G preparation can be given intramuscularly as an alternative to 10 days of oral antibiotics for strep throat. The second outpatient intramuscular antibiotic is ceftriaxone, a broad-spectrum cephalosporin antibiotic that can be given as an alternative to oral medicines for ear infections. This therapy should be reserved for children for whom first-line treatment with a level 1 antibiotic (usually amoxicillin) has failed; other level 1 or level 2 oral antibiotics are also options for treatment of an ear infection that has resisted first-line therapy. The advantages of ceftriaxone given intramuscularly in this setting include one-time dosing and a broad spectrum of potential ear germs treated. The disadvantages include the pain of the shot and the broad spectrum of bacteria treated that may have nothing to do with the ear infection.

Ceftriaxone is used intramuscularly in another outpatient scenario as well—a young child with a high fever, but otherwise appearing well, may be evaluated with cultures for possible blood-borne bacterial infection (fever and bacteremia [see chapter 3]). While waiting for the culture results to return (24 to 72 hours), if your child's doctor does not believe that hospitalization is warranted, two shots of ceftriaxone given 12 to 24 hours apart may be used to cover the possibility of bacterial bloodstream infection until a definitive diagnosis can be made. The levels of this antibiotic in the blood after an intramuscular injection are comparable to those reached by intravenous therapy, which would require hospitalization.

Intravenous Antibiotics

If your child's doctor decides that intravenous antibiotics are necessary, it is likely that the infection is more serious and also likely that your child will be hospitalized. Fortunately, this brief discussion of intravenous antibiotics is well beyond what 99% of you will need to know for 99% of your kids' antibiotic experiences. Like oral antibiotics, intramuscular and intravenous antibiotics can be grouped according to my system of levels. In fact, most oral antibiotics have identical or equivalent counterparts that can be given directly into the bloodstream for more severe infections. The levels of intravenous antibiotics are, however, beyond the scope of this discussion.

Several classes of intravenous antibiotics were not described above in the discussion of oral antibiotics (in most cases because they haven't been developed in an oral form) but are frequently used for serious infections; they include:

- Glycopeptides, e.g., vancomycin, to treat aerobic purple bacteria that are either round or rod shaped (an oral form of this medicine is discussed with the topical antibiotics below—the reasons will become clear)
- Aminoglycosides, e.g., gentamicin, to treat aerobic pink rods
- Monobactams, e.g., aztreonam, to treat aerobic pink rods
- Carbapenems, e.g., imipenem, to treat aerobic purple round bacteria and aerobic pink rods and also to treat anaerobic bacteria of both colors and shapes (these are very broad-spectrum antibiotics)

For the purpose of a general understanding of the use of intravenous antibiotics, Table 5.11 presents several serious types of infections for which intravenous antibiotics are used, along with the spectrum of germs that can cause the infections and examples of specific antibiotics that might be chosen for treatment. The infections themselves are discussed fully in chapter 3.

Table 5.11 is presented to give you a very limited sampling of the difficult decisions that doctors have to make in treating serious infections. There are so many caveats to interpreting Table 5.11 that it's probably best not to interpret it at all—the clinical scenario of a serious infection in a hospitalized

Table 5.11 Select examples of serious infections requiring *intravenous* antibiotics

Infection	Bacteria involved		Oxygen requirements	Intravenous antibiotic options
	Color	Shape		
Sepsis	Purple	Round	Aerobic	Ampicillin, vancomycin, ceftriaxone
	Pink	Round	Aerobic	Ampicillin, ceftriaxone
	Pink	Rod	Aerobic	Gentamicin, ceftriaxone, azthreonam, imipenem
	Purple or pink	Round or rod	Anaerobic	Clindamycin, metronidazole, imipenem
Meningitis	Purple	Round	Aerobic	Ampicillin, vancomycin, ceftriaxone
	Pink	Round	Aerobic	Ampicillin, ceftriaxone
	Pink	Rod	Aerobic	Ceftriaxone
Appendicitis	Mixed	Mixed	Mixed	Imipenem; combination of ampicillin, gentamicin, and clindamycin
Kidney infection	Pink	Rod	Aerobic	Ceftriaxone, ciprofloxacin, imipenem
	Purple	Round	Aerobic	Vancomycin, linezolid, imipenem
Pneumonia; severe, community acquired	Unknown	Unknown	Unknown	Ceftriaxone, azithromycin, imipenem

patient is complex and defies a simplified description. The antibiotic options listed are arbitrary because other choices might also be appropriate. Many infections occur without doctors being able to identify a specific germ; doctors must arbitrarily choose a combination of antibiotics that cover the most likely bacteria based on their knowledge of the particular patient and the particular infection. Broader-spectrum antibiotics are frequently chosen in this scenario until the picture sorts itself out with time—the patient's response to therapy, as well as test results that take longer to obtain, can help to narrow the therapy after the initial hours and days. Although the intravenous antibiotics used in the hospital setting have great potential for inducing resistance and becoming a hospital-wide and societal problem, more leeway must be given for the use of these medicines in the very sick patient who presents with an infection that does not have an immediately identifiable bacterial cause.

Topical Antibiotics

A number of localized infections can be treated with antibiotics that are administered topically and have minimal to no systemic penetration, that is, they don't get into the bloodstream and circulate to other parts of the body the way oral, intramuscular, and intravenous antibiotics do. Certain infections of the skin, outer ear canal, eyes, and gastrointestinal tract fall into this category.

Skin. Antibiotic creams are available over the counter for application to minor wounds for the prevention of bacterial infection; they are reviewed in chapter 8 with other over-the-counter medicines. Prescription topical antibiotics are available and effective for mild to moderate skin infections, like impetigo, and for acne. Acne is beyond the scope of this discussion because of the myriad issues involved in diagnosing acne and the equally numerous options for appropriately treating it, both with and without antibiotics.

Impetigo (see chapter 3) is caused by staphylococcus and streptococcus bacteria (see chapter 1). Mupirocin (brand name, Bactroban), a prescription topical antibiotic (available over the counter in Canada), has been shown to be effective in the treatment of many cases of impetigo, as well as in treating minor wounds and scratches that have become infected with bacteria. A second prescription topical antibiotic, retapamulin (brand name, Altabax), was approved by the FDA in April 2007 for similar clinical uses. Alternative treatments for these mild infections are oral antibiotics (see above); more severe skin infections and those that have extended to deeper tissue layers require intravenous therapy. Once again, bacterial resistance has complicated the treatment of infections—many minor skin infections are now caused by *S. aureus* strains that are resistant to level 1 oral antibiotics (e.g., methicillin-resistant *Staphylococcus aureus*, aka MRSA; see above). Although resistant staphylococci often start out susceptible to mupirocin, resistance to this topical therapy has also developed in some patients who have been treated with the medicine or exposed to other patients who have received it. An intranasal form of mupirocin is also available to reduce or eliminate the colonization of the nose with *S. aureus;* it is nasal colonization that serves as the source for recurrent skin infections in many children.

Outer Ear Canal. Infections of the outer ear (swimmer's ear [see chapter 3]) can be treated with drops applied into the ear canal. This is in contrast to middle-ear infections, which require oral antibiotics when treatment is indicated. Although most outer-ear infections (otitis externa) arise in the skin of the ear canal (see chapter 3), occasionally a perforated eardrum with drainage of pus from a middle-ear infection (otitis media [see chapter 3]) can mimic outer ear canal infections. Topical quinolone (see above) antibiotics, such as ofloxacin drops, have become the preferred treatment of many doctors because they are less risky than aminoglycoside drops (e.g., gentamicin) in the event a perforated eardrum is the cause of the inflamed ear canal, although aminoglycoside drops are still used safely if your child's doctor determines that no perforation exists.

Eyes. Antibiotic eye drops are used in cases of bacterial conjunctivitis (pinkeye [see chapter 3]), as well as to protect the eye from infection after traumatic injury. A variety of classes of antibiotics (sulfa, macrolides, aminoglycosides, and quinolones) are available to your child's doctor. The quinolones have replaced the others in many doctors' prescribing patterns because they are well tolerated and associated with fewer allergic reactions. Most cases of pinkeye are viral or allergic and don't require antibiotics at all (see chapter 3), which is good, because kids hate eye drops.

Gastrointestinal. Vancomycin, a drug of the glycopeptide class used intravenously to treat serious purple-bacterium infections (Table 5.11), is available in an oral form that, unlike most oral antibiotics, is not absorbed into the bloodstream. For that reason, oral vancomycin is used in certain cases of colitis due to *C. difficile* (antibiotic-associated colitis [see above and chapter 3]) as essentially topical therapy of the gastrointestinal tract.

Bottom Line—Are Antibiotics a Miracle or a Menace?

They are a miracle, and if we are careful, we can limit the risk of antibiotics ever becoming a true menace to society. The clock can be turned back on the emergence of resistance, but it will require a cooperative effort by many sectors of society, both private and public—an effort of a scope and magnitude not often seen. But it's worth it.

6

Germinators II: Antivirals, Antifungals, Antiparasitics, Insecticides

The remarkable (miraculous) successes in the development of antibiotics that fight bacterial infections (see chapter 5) have led to extrapolation of those scientific techniques to the development of medicines for other types of germs. Here, too, serendipity has been coupled with scientific ingenuity. For example, one of the most widely used antifungal medicines, griseofulvin (see below), was discovered as a secretion of a common penicillium mold, a cousin of the mold that Fleming discovered to be secreting penicillin (see chapter 5). The diversity of chemical compounds available to fight viruses, fungi, and parasites, while far more limited than for bacteria, continues to expand. Unlike most antibiotics, however, some of these medicines are only partially effective and are associated with high relapse rates and side effects.

This chapter reviews the available antimicrobial (germ-killing) drugs for treating kids' nonbacterial infections to help you understand and contribute to the doctor's decisions regarding if and when your kids should be treated for those conditions. A different structure is required for the discussion in this chapter than for the review of antibiotics in chapter 5 because the number of viral, fungal, and parasitic infections for which therapy is available is so much more limited than for bacteria. This review will therefore be organized by infection rather than by drug, providing a convenient way for you to reference the treatment issues for your child once a diagnosis has been made. The infections themselves are discussed more fully in chapter 3, and nonmedicinal approaches to prevention and treatment, including hygiene and nutrition, are discussed in chapters 9 to 11.

Treatments for Viral Infections

It is ironic, as you have read repeatedly already in the preceding chapters, that although the vast majority of infections that your kids will get are viral, there are so few viral infections for which we have effective medicines. That having been said, it's also a relief that most viral infections don't require therapy because they resolve harmlessly on their own. Still, as benign as infections like the common cold, mono, and stomach flu may be, they often cause your kids substantial discomfort and inconvenience, with missed school and canceled activities. Antiviral therapy, if safe and effective, would be welcome for many of the nuisance infections that your kids seem to get repeatedly. There's no luck yet for most of these infections, but there are glimmers of progress.

Another irony is that, despite our still early progress in treating certain viruses, even those germs have so far proven to be incurable. That is, we have yet to devise a medicine that actually cures a viral infection. What we have are drugs that slow the infections long enough for your child's immune system (see chapter 4) to catch up and purge the virus from his system or relegate it to a dormant state within the system. However, unlike antibiotics, many of which definitively kill bacteria and permanently cure the patient, the antiviral compounds that have been developed to date are stabilizers or containers of the infections, not eradicators or, to use my clever term from the chapter title, germinators.

Herpesvirus Infections

Herpes simplex virus infections in children occur throughout the age spectrum, from the neonate through adolescence (see chapters 1 and 3). Neonatal infections are acquired in the birth canal and may result in severe disease and death. The first herpesvirus infection that many children get is called stomatitis (see chapter 3) and is characterized by multiple sores on the lips and throughout the mouth; this can be a very painful condition, resulting in dehydration because of the difficulty in swallowing. When stomatitis resolves, the virus remains dormant in nerve roots leading to the mouth; herpes cold sores and fever blisters are reactivations of that original mouth infection and may occur throughout childhood. Herpesvirus can also infect the eyes of kids of all ages, resulting in potentially serious complications. Adolescents are susceptible to genital herpesvirus infections, along with other sexually transmitted diseases. Herpes simplex virus is also the most commonly identified cause of encephalitis, a potentially brain-damaging infection.

There are four antiviral medicines available to treat herpesvirus infections other than those of the eyes: acyclovir, penciclovir, valacyclovir, and famciclovir. Acyclovir is approved for children and is available in intravenous form for very serious herpesvirus infections. Acyclovir is also available in a topical form that has been tested as therapy for recurrent oral cold sores and genital infections; the topical formulation does not appear to be effective in otherwise healthy patients but may be helpful when used in combination with oral or intravenous acyclovir in patients with abnormal immune systems. Topical penciclovir has been approved for kids older than 12 years of age, but studies with adolescents have been very limited. All four of these herpes medicines work because they are activated by a viral enzyme into a form that inhibits the virus' ability to reproduce itself. Because the virus' own enzyme is more effective than the patient's enzymes in activating the drug, uninfected body cells and tissues are not affected very much by the medicine, and therefore, the side effects are few and mild. These are very well-tolerated drugs.

Herpesvirus infections for which acyclovir, the most widely used of the herpes medicines for kids, has been proven to be effective and the formulation of acyclovir that is most appropriate for each condition are shown in Table 6.1.

Herpesvirus infections of the eye (herpes keratitis) can be very severe. Three topical antiviral drugs are available to children with these infections: trifluridine, iododeoxyuridine, and vidarabine. All three work by blocking the production of new viral DNA and therefore the production of new viruses. In the newborn, and in severe cases in older kids, intravenous acyclovir is given in addition to topical therapy for treatment of herpes keratitis.

Chicken Pox and Shingles

Chicken pox and shingles are two phases of the same infection caused by the varicella-zoster virus. As you recall from chapters 3 and 4, the varicella-zoster virus retreats after causing chicken pox but does not disappear. Like other members of the herpesvirus family (see chapter 1), varicella-zoster virus lies dormant in the nerve roots until a lapse in the surveillance by the immune system allows it to reactivate along a single or a couple of nerve roots, causing the painful condition of the elderly called shingles. The varicella-zoster virus is also susceptible (although not as susceptible as herpes simplex virus) to acyclovir, valacyclovir, and famciclovir; only acyclovir is approved for use in children. Although acyclovir, when started within the first 24 hours after the appearance of visible spots, has been shown to reduce the duration of illness in chicken pox and the number of spots on the skin, the reductions are too modest to recommend use of the medicine in all cases

Table 6.1 Acyclovir treatment for herpesvirus infections

Type of herpesvirus infection[a]	Formulation of acyclovir
Newborn (neonatal) infection	Intravenous
Adolescent genital infection	
First episode	Oral
Recurrent	Oral
Stomatitis	Oral
Recurrent cold sores/fever blisters	None (oral or topical acyclovir may be given in very bothersome cases; penciclovir may be given to kids older than 12 years in very bothersome cases)
Encephalitis	Intravenous
Any infection in patients with abnormal immune systems (see chapter 4)	Intravenous

[a]For a discussion, see chapter 3.

of childhood chicken pox in otherwise healthy kids. Rather, your kids' doctors may consider the use of acyclovir in the following chicken pox settings in which the disease may be more severe:

- Adolescents (and adults)
- Patients with chronic skin or lung conditions
- Patients on steroid medicines or aspirin
- Second and subsequent cases in a household

If, however, your child has an immunologic abnormality (see chapter 4), chicken pox can be a very severe and even fatal disease; therapy with intravenous acyclovir is very important in these immune-compromised kids.

Shingles is uncommon in children; it is sometimes seen during childhood when the primary chicken pox infection occurred in very early infancy. Treatment with oral acyclovir is recommended in most cases of childhood shingles; more severe cases are treated intravenously.

Influenza

Although influenza may be more severe than other respiratory virus infections, most cases resolve on their own and do not require specific antiviral therapy. However, for kids with risk factors that predispose them to more severe influenza virus infections, such as chronic lung or kidney disease, blood disorders, cancer, and heart abnormalities, use of antiviral medicines to treat established influenza virus infections or to prevent infection after exposure is appropriate. Young children (under 5 years of age, and especially those under 2 years) are also at risk for severe influenza virus infection and may be considered by their doctors for treatment. Treatment is also appropriate for a patient of any age with particularly severe influenza.

The two main groups of influenza viruses that cause infections in kids are influenza A and influenza B viruses (see chapter 3). There are four antiviral medicines available for preventing and treating influenza A, two of which are also effective against influenza B (Table 6.2). As with antibiotics and bacteria, virus germs also have learned how to become resistant to our treatments. Mutations and adaptations of influenza viruses have famously made vaccine development most challenging (see chapter 7). The same propensity for mutating has resulted in widely distributed strains of influenza A virus that are resistant to the two earliest antiviral drugs developed against this infection, amantidine and rimantidine (Table 6.2). Amantidine and rimantidine act on the influenza A virus by preventing it from shedding its outer membrane and releasing its genetic material into the host cell. Resistance to those drugs among the influenza A virus strains circulating in the United States in 2005 and 2006 was so prevalent that the Centers for Disease Control and Prevention for the first time advised that the medicines *not* be used for infection prevention during that flu season; the recommendation was extended in July 2006 to apply to the 2006–2007 season, as well. Because the two newer antiviral drugs for influenza, zanamavir

Table 6.2 Antiviral medications for treating influenza virus infections

Generic name	Brand name	Formulation	Target strain(s)	Risk of resistance
Amantidine	Symmetrel	Oral	A	High
Rimantidine	Flumadine	Oral	A	High
Zanamavir	Relenza	Inhaler	A and B	Low
Oseltamavir	Tamiflu	Oral	A and B	Low

and oseltamavir (Table 6.2), work by a different mechanism, resistance does not yet appear to be as big a problem. These two drugs act by blocking an enzyme of the virus that is needed for it to escape from one infected body cell and move on to infect another.

All four of these antiviral drugs have only modest effectiveness—they reduce the duration of illness in influenza by an average of only 1 day when treatment is begun within the first 2 days of symptoms. Generally, zanamavir and oseltamavir have been well tolerated; zanamavir is given by inhaler and causes wheezing and airway spasm in some patients, while the side effects of oseltamavir were felt to be mostly limited to stomach upset (nausea and vomiting). In November 2006, however, the Food and Drug Administration (FDA) required the manufacturers of oseltamavir to add a warning regarding possible neuropsychiatric complications in kids taking the medicine based on a few patients in Japan who developed delirium and inflicted injury on themselves. Neurological and behavioral side effects have been well known with amantidine and rimantidine in the past, but this Japanese experience was the first report of such adverse effects with the newer antivirals against influenza. Use of oseltamavir in Japan is much more widespread than in the United States, raising the possibility that a similar experience might be seen in this country if enough patients are treated. Importantly, influenza infection itself can cause neuropsychiatric symptoms. The Japanese data are undergoing analysis at the time of this writing for further clues regarding a possible cause-and-effect relationship between the drug and the symptoms (see chapter 12).

Although, as noted above, all four anti-influenza virus medicines have shown effectiveness as preventives, as well as for treatment of established infection, the best prevention is influenza vaccine (see chapter 7).

Hepatitis

In chapter 3, you read that hepatitis is due to an alphabetic potpourri of viruses, from hepatitis A virus through hepatitis E virus. To date, antiviral treatment available for the hepatitis viruses is limited to drugs directed at hepatitis B and hepatitis C viruses.

Hepatitis B virus causes acute liver inflammation and, in a significant number of patients, persists to cause chronic long-term (defined as 6 months or longer) liver disease. Chronic hepatitis B is an important risk factor for subsequent development of cirrhosis and liver cancer. Babies infected with hepatitis B virus from their mothers before or at the time of birth (perinatal infection) have a 90% risk of developing chronic infection. Although there is no medicine to treat *acute* hepatitis B virus infection (i.e., the initial phases of the disease), four drugs have been shown to be beneficial to some patients with *chronic* infection; two of the drugs, alpha interferon and lamivudine, have been approved for use in kids, whereas adefovir and entecavir have not yet received adequate study for pediatric approval. While none of the four treatments is effective in all patients with chronic hepatitis B, remission rates of up to 40% have been reported. Note the use of the term remission—recall again that cure is not part of the vocabulary of antiviral therapy.

Hepatitis C is generally a benign acute disease (i.e., the initial phases of the infection), but more than half of all patients progress to chronic infection. As with hepatitis B, chronic disease increases the risk of cirrhosis and is now the leading cause of liver transplantation in adults in the United States. Treatment options are available for chronic infections and include alpha interferon and a combined regimen of two antiviral drugs, alpha interferon (brand name, Intron-A) with ribavirin (Rebetol); the combined treatment is approved by the FDA for use in children with chronic hepatitis C. Adults treated with that combination experience a 35 to 75% response rate with improved liver function; data are more limited for kids.

Side effects for all of the hepatitis drugs can be severe, but the risks of chronic hepatitis B or C to the long-term well-being of the child usually justify coping with the side effects.

Cytomegalovirus (CMV)

Also a member of the herpesvirus family (see chapter 1), cytomegalovirus (CMV) is a very common infection, present in a dormant state in as many as 70% or more of adults. Most infections occur without causing symptoms; sometimes a mild flu-like or mono-like illness heralds the infection. Recovery is spontaneous in otherwise healthy kids, but like its cousins in the herpesvirus family, CMV is a lifelong infection. It remains dormant in blood cells and other tissues. If a child develops an immune abnormality, CMV may reactivate and cause serious disease. The other setting in which the virus causes trouble is when it is acquired from infected mothers during the birth process (see chapter 2)—most of these congenital infections are also asymptomatic, but occasionally babies can be very severely affected, with eye, brain, and other organ system damage. CMV may also cause hearing loss in infected newborns, which is often not detected for months or years. Available antiviral therapy includes four medicines, all reserved for patients with abnormal immune systems and significant CMV disease; those medicines are ganciclovir (brand name, Cytovene), valganciclovir (Valcyte), foscarnet (Foscavir), and cidofovir (Vistide). Only ganciclovir is approved for use in kids. Studies have shown some benefit of ganciclovir in treating severely affected babies with congenital infection, but there are not yet enough data to recommend it for this use. Although ganciclovir is chemically closely related to acyclovir, the side effects associated with ganciclovir are potentially much more serious than any seen with acyclovir (see above). Ganciclovir suppresses the bone marrow's production of all blood cells and can also have an adverse impact on the kidneys.

AIDS

Caused by the human immunodeficiency virus (HIV) (see chapters 1 and 3), AIDS deserves a textbook to itself—actually, there are many textbooks devoted to this extraordinary infection. In recent years, the innovative use of combinations of antiviral drugs has transformed our treatment approach to AIDS from one of desperate resignation to one of hopeful lifetime maintenance therapy. The vast armamentarium of antiviral drugs directed at HIV that has developed in the past 25 years is a testament to the power of medicine and science in the face of daunting challenge. Dozens of antiviral drugs are now available to treat AIDS, many more medicines than are available for the treatment of all other virus infections combined. This is a true case of the tail wagging the dog—frenzied research to find new treatments for AIDS has resulted in the development of new molecular techniques that have benefited all of medicine. Indeed, as an example, the highly advanced technology that is driving the Human Genome Project evolved in a greatly accelerated way because the technology was first applied to studying HIV. Additionally, antimicrobial drugs of other classes, i.e., antibiotics, antifungals, and antiparasitic medicines, have also been discovered and developed because of the secondary opportunistic infections experienced by AIDS patients with severely compromised immune systems. These other antimicrobial medicines greatly enhance our treatments of patients with many diseases other than AIDS.

A full and continuously updated listing of the moving target that is our AIDS antiviral drug list is available at http://www.aidsinfo.nih.gov, the complete AIDS website of the National Institutes of Health. Table 6.3 provides an overview of the types of antiviral medicines available against HIV and the way they work.

Hemorrhagic Fevers

There is a wide array of exotic virus infections that cause hemorrhagic fever around the world; Ebola virus (see chapter 3) is the best known among those of us in the West because of the widespread media coverage generated by the dramatic outbreaks of that infection. There is no known treatment for Ebola virus infections. Another group of viruses, termed the arenaviruses, also cause hemorrhagic fevers, including Lassa fever. This infection is amenable to treatment with ribavirin (brand name, Rebetol), and other arenavirus infections may be as well. Lassa fever is a disease mostly limited to West Africa, but it has been diagnosed in U.S. travelers to that part of the world.

Table 6.3 Classes of AIDS antiviral drugs and their mechanisms of action

Class of antiviral drugs	How they work	Example(s)[a]
Nucleoside reverse transcriptase inhibitors	Block the enzyme required for reproduction of HIV by mimicking the building blocks of DNA	Zidovudine Lamivudine
Nonnucleoside reverse transcriptase inhibitors	Block the enzyme required for reproduction of HIV by chemical means	Neviripine
Protease inhibitors	Block the enzyme needed by the virus to produce its structure	Indinavir Ritonavir
Entry and fusion inhibitors	Block the virus from getting into cells	Enfuvirtide
Combination drugs	Combine two or more drugs from the same class or from different classes	

[a]Generic names are given.

Warts

Plantar warts and genital warts are caused by papillomaviruses (see chapters 1 and 3). There are no antiviral drugs proven to be beneficial for warts, but numerous treatments based on the physical destruction of the warty growth have demonstrated variable effectiveness. These range from freezing to locally toxic chemicals and duct tape and are summarized in chapter 8.

That's pretty much it for available antiviral therapies. We clearly have a long way to go in treating virus infections before we reach the sophistication that our antibiotics have reached for bacteria, and there are many deserving viral targets waiting to be attacked.

Treatments for Fungal Infections

As with infections due to bacteria and viruses, fungal infections range from the benign nuisance variety to the severe and life-threatening level. Generally speaking, kids who are otherwise healthy, i.e., who do not have underlying disease or immune deficiency, do not get serious fungal infections. In contrast, the mild, bothersome, but not dangerous fungal infections are nondiscriminating—they may attack any child regardless of age or predisposition. Most of the nonsevere fungal infections are treated topically; some require oral medications. All of the severe fungal infections in compromised patients require intravenous therapy with antifungal drugs. Because the side effects of antifungal medicines vary greatly with the formulation (topical, oral, or intravenous), an overview of the side effects, organized by the formulation of the medicine, is presented at the end of this section.

Recall from chapter 1 that fungi grow in either yeast or mold forms, with some species able to grow as both. Approaches for the prevention and treatment of yeast and fungal infections other than with antifungal medicines, including dietary interventions and hygiene, are presented in chapters 9 to 11.

Thrush and Diaper Rash

Candida albicans is a very common yeast infection that causes white plaques (thrush) in the mouths of otherwise healthy babies, as well as inflamed diaper rashes. Candida is part of the normal flora of human mouths and skin, present in small quantities in most of us but repressed by the bacteria that colonize those areas and dominate the available growth surfaces and nutrients. When the yeast gets the upper hand, thrush or skin yeast infections become evident (see chapter 3). Antibiotic use increases the occurrence rates of both thrush and yeast diaper rashes because the bacteria that usually monopolize the growth environment are wiped out and the yeast is allowed to flourish. Antibiotics may lead to thrush in older children, as well, but when kids beyond infancy develop thrush *without* antibiotic exposure, there should be concern regarding an abnormal immune system, and an evaluation may be required.

Treatment of oral thrush in healthy babies is often not necessary, as the infection resolves on its own. If thrush persists or is particularly widespread in the mouth, treatment is with nystatin (brand name, Mycostatin) suspension given orally into the cheek pouches and by swabbing the plaques and mucous-membrane surfaces of the mouth. Candida diaper rashes in babies are treated with topical nystatin ointment (Mytrex or Mycolog). Nystatin works by punching holes in the outer membrane of the yeast, causing it to leak important minerals and sugars. Immunocompromised kids with mouth or skin yeast infections may respond to treatment with oral fluconazole (Diflucan) or itraconazole (Sporonox), but intravenous therapy is often required (see below).

Vaginal Yeast Infections

Even though yeasts in low numbers are normal inhabitants of the vagina and gastrointestinal tract, the normally acidic conditions of the vagina suppress yeast from growing to detectable and symptomatic levels. When the acidic environment is altered, for example, by antibiotics, menstruation, diabetes, pregnancy, or birth control pills, the yeast can gain the advantage over the usually predominant vaginal bacteria and cause an irritating, inflamed condition called vaginitis. There are numerous topical antifungal creams of the azole class of medicines that are effective, including miconazole (brand name, Monistat), clotrimazole (GyneLotrimin), and butaconzole (Gynazole). Oral preparations of azole medicines were mentioned above for thrush, and intravenous azole preparations are available for more serious fungal infections (see below). Azoles work by inhibiting an enzyme that fungi need to make their cell membranes.

Ringworm, Jock Itch, Athlete's Foot, and Infections of the Scalp and Nails

The pesky fungi that cause infections of the skin, scalp, and nails are called dermatophytes (because they are fond of skin [see chapter 3]). The secret to these germs' success is their ability to grow on keratin, the tough outer protein that gives our skin, hair, and nails their protective qualities. The dermatophyte fungi are formally molds (versus yeast). There are numerous species that can cause each of the itchy and scaly infections in this category.

In North America, ringworm of the scalp is most often caused by *Trychophyton tonsurans;* other species predominate elsewhere. Because the hair (and sometimes even the eyebrows and eyelashes) is involved and is very difficult to work around in trying to treat the infection topically, scalp ringworm is treated with oral antifungal medicines for several weeks. The first line of treatment is usually with griseofulvin (brand names, Grifulvin, Grisactin, and others) given orally. Griseofulvin acts by inhibiting the cell division required for fungi to multiply. Other options for scalp ringworm may be as effective as griseofulvin but have not been approved for use in kids for this purpose. They include terbinafine (Lamisil), which inhibits another enzyme (different than that acted on by the azoles) required for fungal membrane synthesis, and also the oral azoles. A medicated (selenium sulfide) shampoo is often prescribed in addition to the oral therapy; the shampoo helps prevent spread of the infection by removing the superficial germs and keeping the numbers of fungi low.

It's much easier to treat ringworm on the body, where topical therapy is very effective; again, several weeks of treatment are required. Many topical antifungal medicines are now available in over-the-counter formulations; some are available both over the counter and by prescription. Topical medicines for body ringworm include the azoles miconazole and clotrimazole (Micatin, Lotrimin, and others) and terbinafine (Lamisil). Other choices are ciclopirox (Loprox), which works by binding molecules of metal elements that fungi need for their enzymes to work, and tolnaftate (Tinactin, Aftate, and others), which inhibits the same fungal enzyme as terbinafine.

Still other fungal species, including species within the trichophyton group, cause athlete's foot and jock itch. Topical terbinafine or azole medicines treat most cases of athlete's foot; occasionally, oral griseofulvin is required for spots on the foot where the fungus is tough to eradicate, like the heel. Jock itch is treated with the same topical medicines as body ringworm. Fingernail and toenail infections with fungi are much more difficult to eradicate than are infections of the skin; fortunately, they are quite uncommon in kids. Treatment may be with topical creams or lacquers, but often oral griseofulvin or oral azoles are required—and even then success rates are only modest.

Treatment of Serious and Life-Threatening Fungal Infections

Fortunately, life-threatening fungal infections are very rarely if ever seen in otherwise healthy kids. When the immune system (see chapter 4) becomes compromised, however, because of either disease or medications, fungi become the ultimate opportunists (see chapter 1). Table 6.4 details the most commonly used antifungal medicines, organized by class of drug, in children with severe infections due to abnormal immune systems.

As noted earlier, the side effects of antifungal medicines are closely linked to the formulation of the medicine used—topical, oral, or intraven-

Table 6.4 Classes of antifungal medications for severe and life-threatening infections

Class	How they work	Examples	Formulation[a]
Polyenes	Punch holes in the outer membrane of the fungus, causing it to leak vital minerals and sugars, with death of the fungus cell	Amphotericin Nystatin	Intravenous Intravenous (not yet available)
Echinocandins	Block the formation of the fungal cell wall	Capsofungin Micafungin Anidulafungin	Intravenous Intravenous Intravenous
Azoles	Block an enzyme needed by the fungus to make its cell membrane	Fluconazole Itraconazole Voriconazole Ketaconazole	Intravenous and oral Intravenous and oral Intravenous and oral Oral
Antimetabolites	Insert themselves into fungal genetic material, blocking protein production and blocking the fungus from multiplying	Flucytosine	Oral

[a]Only the formulations used for severe disease are listed in the table. Other formulations of these medicines are available for non-life-threatening fungal infections.

ous. It is not surprising that the relationship is a direct one—the more aggressive the therapy, the more serious the potential side effects. It is also true that the intravenous, and some of the oral, treatments are reserved for patients with very severe, even life-threatening, infections; in that context, the risk of more serious side effects is justifiable.

Table 6.5 summarizes the side effects of the antifungal medicines discussed above as a function of how they are given. The theme you will notice from the table is that topical medicines' side effects are mostly limited to the skin where they are applied; most orally administered drugs not surprisingly cause gastrointestinal side effects as their major drawbacks, and intravenous medicines cause side effects of multiple organ systems as the blood carries these drugs throughout the body.

The relative paucity of antifungal medicines for serious infections, as with antiviral medicines (see above), stands in stark contrast to the panoply of antibiotics for serious bacterial infections (see chapter 5). Because we live in a market-driven world, the incentive to develop more sophisticated, safer, and more effective medicines for severe fungal infections is relatively low; that is also in stark contrast with the explosion of creams, ointments, gels, and sprays to treat the common, everyday fungi that cause athlete's foot, jock itch, and ringworm

Table 6.5 Side effects of antifungal medicines

Formulation	Name of medicine	Side effects
Topical	Nystatin	Local irritation
	Clotrimazole	Local irritation
	Miconazole	Local irritation
	Other azoles	Local irritation
	Terbinafine	Local irritation; worse if covered by bandage
	Ciclopirox	Local irritation
	Tolnaftate	Local irritation
Oral	Nystatin	Nausea, vomiting, rash
	Griseofulvin	Headache, nausea, vomiting, diarrhea, rash, numbness, dizziness, reduced blood cell count
	Terbinafine	Nausea, diarrhea, stomach ache, changes in taste sensation or loss of taste, liver problems
	Clotrimazole	Nausea, vomiting, liver problems
	Fluconazole	Nausea, vomiting, stomach ache, headache, rash, liver problems, reduced blood cell count
	Itraconazole	Nausea, vomiting, stomach ache, headache, rash, liver problems, reduced blood cell count
	Ketaconazole	Nausea; vomiting; stomach ache; liver problems; mood changes, including depression; rash
	Flucytosine	Nausea, vomiting, reduced blood cell count, liver problems, mood and behavior changes
Intravenous	Amphotericin	Fever; chills; breathing problems; headache; irregular heart beat; kidney problems; hearing loss; nervous system abnormalities, including dizziness, pains, tingling
	Capsofungin micafungin, anidulafungin	Fever, itching, headache, nausea, vomiting, diarrhea, reduced blood cell count, rash, itching, irritation of the vein with the intravenous formulation
	Fluconazole, itraconazole	Rash, liver problems, nausea, vomiting, reduced blood count

in the masses (where the money is!). As noted earlier, the AIDS epidemic, ironically, improved the market for treating severe fungal infections because patients with AIDS are, indeed, susceptible to severe infections of all types.

This unfortunate pattern of few available treatments for serious infections when the commercial market potential is small will be seen again with the discussion of antiparasitic medicines that follows.

Treatments for Parasite Infections

Recall from chapter 1 that parasites can be classified by whether they are acquired domestically (i.e., in the United States) or abroad and by whether they are simple, single-celled protozoa or complex helminths (worms). The medicines reviewed in this section are those used to treat domestic parasites, as well as those used to treat the most common wormy germs acquired by travel outside of the United States. Refer to chapter 1 to put the parasites into context and to chapter 3, where most of the common infections that these parasites cause are reviewed.

Following the format for antiviral and antifungal treatments, this section is organized by disease so that you can look up the relevant condition of concern for you and your kids; the diseases, in turn, are grouped by whether they are typically acquired in the United States or abroad. You will note significant overlap among the available therapies, as many antiparasitic medicines are effective against numerous germs. The section concludes with Table 6.6, which details the side effects seen with the most commonly used drugs, i.e., those that your kids have a reasonable likelihood of ever using.

Treatments for Parasites Acquired in the United States

The four most common domestic protozoan parasitic diseases are giardiasis, cryptosporidiosis, toxoplasmosis, and trichimoniasis (trich); a fifth protozoan disease, babesiosis, occurs much less frequently than the others and is limited to the eastern United States. The only (numerically) significant helminth (worm) infestation in the United States is pinworms.

Giardiasis. *Giardia lamblia* causes protracted diarrhea and is the most common parasite causing disease in the United States. Giardiasis is prevalent in day care centers and among hikers who drink stream water (see chapter 2). Numerous medicines are effective for treating kids with giardiasis. The three preferred drugs, all administered orally, are metronidazole (brand name, Flagyl; you may remember this medicine, which is also used as an antibiotic to treat infections by anaerobic bacteria [see chapter 5]), tinidazole (Tindamax), and nitazoxanide (Alinia). Each of these treatments has a cure rate that exceeds 80% on the first try. Other orally administered drugs that have been shown to be safe and effective in kids for giardiasis but that are usually reserved for treating worm infections include furazolidone (Furoxone), albendazole (Albenza), and mebendazole (Vermox).

Cryptosporidiosis. *Cryptosporidium parvum* is another cause of gastroenteritis that also results from contacts in day care and from contacts with contaminated water (see chapter 2); treatment is with oral nitazoxanide.

Toxoplasmosis. Toxoplasmosis is an important infection of the fetus in the womb of an infected mother. Due to *Toxoplasma gondii,* it is treated with a combination of two oral drugs, pyrimethamine (Daraprim) and sulfadiazine (Microsulfon).

Trichomoniasis. A sexually transmitted cause of vaginitis, *Trichomonas vaginalis* is treated with either oral metronidazole or oral tinidazole.

Babesiosis. Babesiosis is a malaria-like disease due to *Babesia microti* and is treated with one of two combinations, either clindamycin (Cleocin) plus quinine (Quinamm) or atovaquone (Mepron) plus azithromycin (Zithromax); all of these drugs are given orally. Clindamycin and azithromycin are antibiotics that are also used to treat bacterial infections (see chapter 5).

Pinworms. There are three effective treatments for *Enterobius vermicularis* infestations in kids, all given orally: pyrantel pamoate (brand names, Pin-X, Ascarel, and others), which is available over the counter; mebendazole; and albendazole.

Treatments for Parasites Acquired Outside the United States

As with domestic parasites, infestations acquired elsewhere in the world may be with single-celled, simple protozoa or more complex worms. The most common worldwide parasitic diseases, affecting natives and travelers alike, are amebiasis, leishmaniasis, malaria, ascariasis, hookworm, and whipworm.

Amebiasis. Amebiasis, a severe dysentery, is caused by *Entamoeba histolytica*. Once inside the body, the parasite can spread to involve major organ systems. Treatment for infestation limited to the intestines is with iodoquinol (brand name, Yodoxin) or paromomycin (Humatin), each of which remains in the intestine, where it kills the ameba. If the disease has spread beyond the intestine, either oral metronidazole or oral tinidazole is used, combined with iodoquinol or paromomycin.

Leishmaniasis. Leishmaniasis is a severe whole-body disease affecting many organ systems. Treatment is with sodium stibogluconate, a medicine that is difficult to come by and is considered to be investigational (not yet approved), or meglumine antimoniate (Glucantim), which can also be hard to access (e.g., it is not available in the United States or Canada). The antifungal drug amphotericin B (see above) is effective, as is pentamidine (Pentam).

Malaria. Malaria is an infection of the red blood cells that results in severe, potentially life-threatening total-body illness; it is caused by four different *Plasmodium* species. The species involved, the patterns of resistance to antimalaria drugs in the region, and the severity of the disease all factor into the decisions regarding which of the following medicines, or combinations of several medicines, is used: chloroquine (Aralen), quinine (Qualaquin), primaquine, atovaquone-proguanil (Malarone), mefloquine (Lariam), and artemether (Coartem). Because of the ease of contracting malaria from mosquito bites in high-incidence regions of the world, prophylactic (preventive) treatment is recommended for travelers to these areas. For that purpose, chloroquine is most often used, except when travel is to areas where resistance to that antiparasitic is prevalent. Atovaquone-proguanil, mefloquine, or doxycycline (an antibiotic [see chapter 5]) is taken preventively when traveling to such regions.

Roundworms, Hookworms, and Whipworms. Ascariasis is the most common worm infestation worldwide; it is caused by the roundworm *Ascaris lumbricoides*. Several worms belonging to *Ancylostoma* species cause hookworm disease. Whipworm is caused by *Trichuris trichiura*. All cause intestinal disease with frequent involvement of other organ systems. The treatment of choice for all three of these very common international scourges is either mebendazole or albendazole. Ivermectin (brand name, Stromectol) is a backup treatment for ascariasis and whipworm. Pyrantel pamoate is an alternative treatment for hookworm.

Table 6.6 summarizes the side effects of the more commonly used antiparasitic drugs.

Treatments for Insect Infestations

Head lice, the mites that cause scabies, and bed bugs are all insects that creep and crawl onto your kids and cause problems related to the bugs' appetite for your kids' blood and/or the bug's ability to trigger an allergic skin reaction (see chapters 1 and 3).

Head Lice. There are numerous approaches to treating head lice, more formally known as *Pediculus humanus capitis;* some are homebrew remedies involving suffocating the critters or heating them to death (see chapter 3),

Table 6.6 Antiparasitic drugs and their side effects

Medicine	Side effects[a]
Metronidazole	GI, dry mouth, tongue irritation, metallic-taste sensation, numbness/tingling, change in urine color
Tinidazole	GI, metallic-taste sensation, numbness/tingling, dizziness, headache
Nitazoxanide	GI, headache, change in urine color
Furazolidone	GI, headache, change in urine color
Albendazole	GI, headache, dizziness, fever
Mebendazole	GI, headache, fever
Pyrantel pamoate	GI
Ivermectin	GI
Chloroquine	GI; headache; rash/itching; hair loss; mood changes; neurological abnormalities, including hearing, vision, seizures
Quinine	GI; mood changes; neurological abnormalities, including hearing, vision, confusion; sweating; bleeding problems
Mefloquine	GI; mood changes; headache; neurological abnormalities, including hearing, vision, dizziness
Atovaquone-proguanil	GI, headache, altered sense of taste, sweating

[a]GI, gastrointestinal (nausea, vomiting, stomach ache, loss of appetite, and/or diarrhea).

but often medicated shampoos are required. Permethrin 1% shampoo (brand name, Nix) is available over the counter, has the lowest risk of side effects (for the most part limited to local skin and scalp irritation), and has a high rate of success. However, not wanting to be left out of the trend toward emerging resistance to standard therapies, cases of lice that resist first-line medications are being increasingly reported. A 5% solution of permethrin (brand name, Elimite), applied as a cream to the scalp overnight, is available by prescription for treatment failures; this formulation is the treatment of choice for scabies (see below). Natural extracts from the chrysanthemum flower, pyrethrins (Rid, Pronto, and others), are also available over the counter; cases resistant to these products have also been found. Like permethrin, side effects with the pyrethrins are mild, usually local skin and scalp irritation. For kids older than 6 years, prescription malathion (Ovide) lotion has been shown to be very effective; this product also has generally mild side effects but comes with some additional risks because of its alcohol content—for example, it is flammable, and your child's hair can catch fire if a hair dryer is used after an application of the lotion; the alcohol can also cause stinging of an already irritated scalp, as well as eye irritation. Ratcheting up the side effects significantly, lindane 1% shampoo (Kwell) is a prescription-only medication reserved for cases of lice that resist all other treatments. Severe side effects have been reported, including nervous system toxicity such as dizziness and seizures; several deaths have been confirmed to be linked to the use of lindane, and more than a dozen other deaths occurred while the product was being used but cannot be confirmed to be the result of the product. Lindane interacts with many other medications, resulting in even more side effects. In March 2003, the FDA issued a stern warning about the product, emphasizing that the use of lindane should be restricted to cases of lice for which all other alternatives have failed.

Scabies. Caused by the burrowing mite *Sarcoptes scabiei,* the itchy rash scabies is treated topically with insecticidal lotions and creams applied to your child's body. The first line of treatment is the permethrin 5% cream noted above as a second-line treatment for lice. Other choices include crotamiton 10% lotion (brand name, Eurax), a prescription medicine with a low risk of side effects, which are mostly limited to skin irritation. Oral treatment with ivermectin, an antiparasitic drug (see above) used as a backup treatment for numerous parasites, has also been shown to be effective in treating scabies mites. Side effects are primarily gastrointestinal—nausea, vomiting, and stomach aches. Once again, the much more toxic insecticide lindane (see above) is reserved for the most recalcitrant scabies infections.

Bed Bugs. Bed bugs, formally known as *Cimex lectularius,* infect beds and homes, not kids. The bugs, however, feed on kids and adults sleeping in infested beds and homes, causing an irritating skin reaction. Treatment, therefore, is of the home (see chapter 9), not the child, with the exception of topical medications to make the itching and skin irritation feel better.

Bottom Line: a Long Way To Go

Compared with antibiotics for bacterial infections, our progress in treating viral, fungal, and parasitic infections has been slow. The AIDS epidemic brought the need for better therapies for those infections to the forefront, and the past few years have seen a clear uptick in the number and diversity of available medicines. Concerns regarding other epidemics and pandemics of viral infections, such as bird flu (see chapter 3), highlight the deficiencies in our arsenal of weapons. We have only four drugs to treat influenza virus infections, and two of those have already been made nearly obsolete by the rapid emergence of resistant strains (see above). There are no effective medicines for two of the most vicious and prolific killers of children, rotaviruses and respiratory syncytial viruses (see chapter 3). The difficulties in developing antiviral treatments, however, have a silver lining—we have accelerated our development of vaccines (see chapter 7) to prevent infection by these ubiquitous germs; a rotavirus vaccine is now available, and respiratory syncytial virus vaccines are under development.

The lack of commercial markets for new antifungal and antiparasitic drugs continues to hamper their development, despite the fact that on a worldwide basis, infections like malaria, leishmaniasis, and helminthic (worm) infestations cause literally billions of cases of illness and death each year. People with these diseases in developing countries cannot afford medicines, so the incentive for pharmaceutical companies to discover and develop the medicines is low. Private charitable initiatives, such as those of the Bill and Melinda Gates Foundation, have spurred new research into worldwide scourges that otherwise would find limited support.

7

Vaccine Verisimilitude: Venerated, Vilified, Vindicated

Nothing stirs the passions of doctors and parents more than a discussion of vaccines. As mothers took to the streets for the March of Dimes campaign to eliminate poliomyelitis, emerging vaccines targeted some of civilization's greatest scourges. One by one, diphtheria, tetanus, whooping cough, measles, German measles, and mumps were dealt lethal or near-lethal blows. A generation has grown up not knowing the fear of these once-prevalent infectious diseases (see chapter 3). However, along with blessed ignorance came a national nonchalance. Why risk giving our kids vaccines when the diseases are gone from the headlines? The side effects, real and imagined, of vaccines took center stage as the diseases they prevented exited the theater—and some mothers again took to the streets, this time to advocate *against* vaccines. Vaccines have been blamed for sudden infant death syndrome (SIDS), mental retardation, paralytic syndromes, and, most recently, autism. If not the active ingredients of the vaccines, the preservatives have been faulted. Antivaccine advocacy groups have realized some successes in decreasing immunization rates—and the effects have already been seen, with increasing incidences of once-vanishing illnesses.

Well, who's right—the doctors or the advocacy groups? I'll try to help you sort that out, with evidence-based reviews of the issues associated with each vaccine; additionally, at the end of this chapter, I provide you with references for further reading that include the medical establishment resources and those of parent advocacy groups, both for and against vaccines.

This chapter examines the effectiveness and safety of each of the individual childhood vaccines for your own consideration and analysis. Before starting our vaccine-by-vaccine descriptions, let me give you a bit of vaccine history and background regarding how vaccines work.

How It Started

As you read in chapter 3, smallpox was a devastating total-body disease unique to humans and with a very high fatality rate. Emphasize the past tense: *was* a devastating disease. Smallpox was eradicated from the world in 1977 as the result of a heroic international vaccination campaign.

Cowpox is a disease of animals, particularly but not exclusively cows, caused by the cowpox virus. That germ is now known to be in the same family as the smallpox virus. In 18th century England, cowpox was an occupational hazard of milkmaids, women who milked cows for a living. The virus was transmitted from the cow to humans through breaks in the milkmaids' skin that came in contact with blisters on the cow. Infected humans developed a blister rash (one blister is a pock; more than one are pox) on the hands and arms; the infection remained local (at the point of contact) and typically got better on its own.

It was common knowledge on those farms in England that after milkmaids were infected with cowpox and recovered, they became immune to smallpox. That is, farm lore held that a benign, spontaneously resolving infection with cowpox made someone invulnerable to contracting a potentially fatal case of smallpox. Enter Edward Jenner, a rural physician with a Renaissance man's background in biology and natural phenomena. In 1796, 50 years before the advent of the germ theory, Jenner performed one of the first and most important clinical trials in history. He injected the pus from a milkmaid's cowpox blister into a healthy child (can you imagine what the Food and Drug Administration [FDA] would have said about that?) and then, several weeks later, injected live smallpox virus into the child's skin. Through this experiment, Jenner demonstrated that the cowpox injection protected the child from developing any reaction to the smallpox virus, either at the site of the smallpox virus injection or anywhere else in the child's body. Jenner repeated the experiment on a dozen or so other individuals, none of whom reacted to the smallpox challenge after being immunized with cowpox. The history of medicine was forever changed on those farms in England by the lessons we learned there about protection against infections.

What Are Vaccines, and How Do They Work?

The Advisory Committee on Immunization Practices (ACIP) of the Centers for Disease Control and Prevention (CDC) defines a vaccine as follows.

> A suspension of live (usually attenuated) or inactivated microorganisms (e.g., bacteria or viruses) or fractions thereof administered to induce immunity and prevent infectious disease or its sequelae. (http://www.cdc.gov/mmwr/preview/mmwrhtml/rr5515a1.htm)

Within those few words are textbooks full of information and explanations. An abbreviated discussion is presented here. Vaccines produce immunity, that is, they either directly provide protection against infection or stimulate the body's own immune system to generate protective defenses (see chapter 4). The former strategy, providing protection directly, is termed passive immunization: giving antibodies (also known as gamma globulin or immunoglobulins [see chapter 4]) from an immune person or persons (and occasionally from animals or from a preparation made in the laboratory) to a susceptible person to temporarily protect the susceptible person from an infection. In contrast, active immunization involves giving an actual infectious germ, or parts of a germ, to someone to stimulate that person's body to produce antibodies to the germ, which can produce long-term, often lifetime, immunity.

The best sources for additional information concerning immunizations are the CDC Pink Book (http://www.cdc.gov/vaccines/pubs/pinkbook/default.htm) and the American Academy of Pediatrics *Red Book.* These and other references for your further reading are provided at the end of this chapter.

Passive Immunization

As you remember from chapter 4, recovery from infection results in the formation of antibodies, protein chemicals that are made by lymphocyte blood cells and protect against future infection with the same germ. All of us have innumerable such antibodies circulating in our bloodstreams, reflecting the innumerable germs we have fought off over the years of our lives. The concentration of antibodies against any single germ is highest in the immediate days and weeks following infection; after that, the quantity of antibodies directed against that germ falls to a maintenance level that may last a lifetime. Antibodies can be collected from people via blood donation and given to other people who are in need of them. One of the most common reasons for doing this is to benefit patients with abnormal immune systems who cannot produce their own antibodies (see chapter 4). A single donor's blood will not provide enough antibodies to protect an antibody-deficient person, so the antibody-containing portions of blood from many donors are pooled and concentrated to produce sufficient antibodies. There are antibody products (immunoglobulins or gamma globulin) available that can be given by a shot either into the muscle (intramuscular) or into the vein (intravenous) of a person in need of immune boosting.

It is also possible to produce human antibody preparations that contain particularly high concentrations of antibodies against a single germ by pooling donations from selected donors who themselves have had recent infections. Remember, the highest concentrations of antibody directed against any germ are found in the immediate period following an infection with that germ. So-called *hyperimmune* globlulin is an antibody preparation that has a high concentration of antibodies directed at a specific germ of concern. Some hyperimmune globulin preparations are produced by actively immunizing donors against a germ and then collecting blood from those donors at the peak of their antibody production and pooling the collected samples.

A brief aside: you may be wondering at this point about the safety of giving people blood products from pools of human donors. You're right to wonder and even to worry. But it's okay—relax. Immunoglobulin comes from the plasma (liquid, as opposed to cell) component of blood (see chapter 2). While it's true that pooling blood donations increases the risk that one of the donors in the pool will have an infectious disease that can be transmitted in the blood, the sterilization procedures used to produce immunoglobulins from plasma are very vigorous and kill the viruses that we are worried about and can test for. There is more about the issue of blood safety, and specifically plasma safety, in chapter 2.

If you were worried about the safety of antibodies from human donors, wait till you read this. There are occasional diseases or exposures that warrant antibody treatment but for which human immunoglobulin products are not available. For those rare occasions, currently limited to diseases caused by germ byproducts (toxins) rather than the germs themselves (Table 7.1), *horse* serum is still used. These products are called antitoxins, made from blood obtained from horses that have been immunized with the toxin of concern. For yet another type of infection prevention, currently limited to respiratory syncytial virus infection of babies who also have chronic lung or heart disease (Table 7.1 and chapter 3), a laboratory-produced product called a monoclonal antibody is the best available option; this preparation is derived from a hybrid of human and mouse cells that are grown in the laboratory. This is a brave new world, indeed.

The focus in this section is on the use of immunoglobulins to passively protect otherwise healthy kids (i.e., with no deficiency of immunity) from specific infections. For most childhood infections, we are content to let the child's own immune system fight off the germ—remember that most infections are viral, and we do not have medicines to treat most viruses even if we wanted to. For certain other infections, we treat with medicines (antibiotics, antivirals, antifungals, and antiparasitics) when they are available and necessary (see chapters 5 and 6). For yet other infections, there's enough time to protect a child by giving an active immunization that stimulates the child's own immune system to protect him or her (see below). However, there are certain infections for which there may be no medicine and which are of grave enough concern that waiting for a child's own immunity, or for an active immunization to stimulate immunity, is not an option. Those infections for which we give donated antibodies are detailed in Table 7.1, along with the formulation of the immunoglobulin preparation used and the situations that warrant the use of passive immunization for each infection. The infections themselves are more fully described in chapter 3. For many of the infections that might otherwise warrant passive immunization with a shot of antibodies, *active* immunization with a routine childhood vaccine (see below) before a child can become exposed to the infection obviates the need for giving passive antibodies after exposure. Active immunization is usually the preferred method of prevention, and those conditions for which active immunization is available are also indicated in the table.

From just a brief review of Table 7.1, it's clear that it is unlikely your child will ever need passive immunization with antibodies—the conditions that warrant such therapy are very limited. This is particularly true if your kids are actively immunized with the recommended routine childhood vaccines—a convenient segue to the next section of this chapter.

Table 7.1 Passive immunization against specific infections in kids

Infection	Immunoglobulin type (brand name)[a]	When and whom to treat	Active vaccine available that may make passive immunization unnecessary
Botulism	Horse serum (Antitoxin)	As soon as possible after onset of symptoms of food- or wound-associated disease to slow the paralysis	No
	Hyperimmune IVIG (BabyBIG)	As soon as possible after infant botulism is diagnosed to reduce severity	No
Chicken pox	Hyperimmune IMIG (VariZIG)	After exposure; given only to those at risk for severe disease	Yes
	Standard IVIG	Certain at-risk patients when hyperimmune IMIG is not available	Yes
Cytomegalovirus infection	Hyperimmune IVIG (CIGI)	Certain transplant patients; perhaps to prevent infection of newborns born to infected mothers	No
Diphtheria	Horse serum (Antitoxin)	As soon as possible to patients with suspected infection	Yes
German measles (rubella)	Standard IMIG	To pregnant women (who have not received active immunization) after exposure early in pregnancy	Yes
Hepatitis A	Standard IMIG	Within 2 weeks of exposure	Yes
		Prior to travel to some areas of the world	Yes

Hepatitis B	Hyperimmune IMIG (HBIG)	Within 12 hours of birth to newborns of infected mothers	Yes
		Certain newborns whose mother's infection status is not known	Yes
		Certain infants after exposure to infected caregivers in the household	Yes
		Within 24 hours following exposure to blood or body fluids from an infected person (unless the exposed person has been actively immunized)	Yes
Kawasaki disease	Standard IVIG	As soon as possible after the diagnosis is made to reduce the heart complications	No
Measles	Standard IMIG	Within 6 days of close exposure if not previously actively immunized	Yes
Rabies	Hyperimmune IMIG (RIG)	Immediately after exposure	Yes; given along with hyperimmune IMIG after exposure
Respiratory syncytial virus	Monoclonal IM (Palivizumab)	During the winter to certain premature infants (<35 weeks of gestation) and to babies with certain chronic lung or heart conditions	No
Tetanus	Hyperimmune IMIG (TIG)	After dirty wounds in people who have not had active immunization	Yes

[a]IVIG, intravenous immunoglobulin; IMIG, intramuscular immunoglobulin; BIG, botulism immune globulin; ZIG, zoster immune globulin; CIGI, cytomegalovirus immune globulin intravenous; HBIG, hepatitis B immune globulin; RIG, rabies immune globulin; TIG tetanus immune globulin.

Active Immunization

In contrast to passive immunization, which most kids will never experience, active immunization is a form of infection protection that all of your kids should experience repeatedly throughout childhood. That said, I do not minimize the concerns raised by parents and certain parent advocacy groups regarding the possible hazards of kids' vaccines, nor do I disparage those raising the concerns. Rather, I believe the science behind the arguments of vaccine proponents and vaccine opponents should be critically reviewed and that decisions regarding vaccinating kids should be based on that science. For many of the concerns raised by parents and advocacy groups, the science has been reviewed, and the findings are summarized in this chapter.

In this section, each of the recommended routine childhood immunizations will be individually considered. First, we need a definition of recommended—recommended by whom? The CDC, through its Advisory Committee on Immunization Practices (ACIP), is the primary recommending body in the United States. The guidelines for vaccines provided by the ACIP are developed in close cooperation with the American Academy of Pediatrics and the American Academy of Family Physicians. As a result, the guidelines are almost always jointly approved (harmonized) and jointly recommended by three organizational bodies.

As with passive immunization, there are several different formulations that have been employed for active immunizations. That's because an immune response (your child's development of protective antibodies) can be produced by different types of antigens, the components of germs against which antibodies are formed. For some vaccines, whole germs are inoculated; these may be living but weakened forms of the germ or killed germs. Other vaccines use only parts of germs, not the entire organism. Still others use byproducts of germs, for example, protein toxins that are produced by the germs and can be chemically inactivated to produce toxoids; toxoid vaccines stimulate the body to produce antibody protection against the real toxin. Among the currently recommended childhood vaccines, some are designed to be given into your child's muscle, others are given just under the skin, one is given by mouth, and one is given into the nose. For all vaccines given into the muscle or under the skin, the most common side effects are localized pain, redness, and/or swelling. Many vaccines are given in combinations that protect against more than one disease; other vaccines are directed against a single infection only. All vaccines run the risk of an allergic reaction, sometimes severe; fortunately, these are extremely rare (see "General Vaccine Concerns" below).

The optimal formulation and method of delivery of a vaccine are determined by clinical studies in which the safety and protectiveness of the vaccine preparations are investigated. Most of these studies occur before a vaccine is approved and licensed for use, but increasingly, very large postmarketing studies are also being conducted to determine if previously unrecognized side effects are observed when many more children are immunized than in preapproval clinical studies.

There are currently 16 infectious diseases for which routine immunization of all otherwise healthy kids is recommended: diphtheria, *Haemophilus influenzae* type B infection, hepatitis A, hepatitis B, influenza, measles, mumps, *N. meningitidis* infection, papillomavirus infection (genital warts and cervical cancer), pertussis (whooping cough), poliovirus infection, rotavirus infection, rubella (German measles), *S. pneumoniae* infection, tetanus, and varicella (chicken pox). This chapter focuses on the vaccines; each of the diseases targeted by the vaccines is discussed more extensively in chapter 3.

The following overview of each of the routinely recommended childhood vaccines begins with a profile chart of pertinent facts about the vaccine and then focuses on its effectiveness in disease prevention and on its safety; any special concerns regarding a particular vaccine are also addressed. Following the individual reviews of all 16 vaccines, I present an overview of some of the general issues and concerns regarding active immunization.

Diphtheria

Germ causing the disease	*Corynebacterium diphtheriae* bacterium
Type of vaccine	Toxoid (inactivated bacterial toxin)
Year first recommended for use	1940s
How it is given	Into muscle
When it is given (2007 guidelines)	2, 4, and 6 months; between 15 and 18 months; between 4 and 6 years; and every 10 years thereafter
Brand name(s) of individual vaccine	Not available as an individual vaccine
Combination products (brand names) also containing this vaccine	Diphtheria vaccine is available only as a combination vaccine with: • Tetanus toxoid (DT* for kids under 7 years; Td for kids older than 7) • Tetanus toxoid and acellular pertussis vaccine 1. DTaP* for kids under 7 years (Tripedia, Infannix, DAPTACEL) 2. Tdap for kids older than 7 (Boostrix, Adacel) • Tetanus toxoid, acellular pertussis, and *H. influenzae* conjugate vaccine (TriHIBit) • Tetanus toxoid, acellular pertussis, hepatitis B, and inactivated polio vaccine (Pediarix)

*The capital D indicates a higher dose of the diphtheria toxoid component, as recommended for younger kids.

Effectiveness. Diphtheria is a severe infection of the throat in which the bacteria produce a toxin that can cause obstruction of breathing, as well as severe heart and brain effects (see chapter 3); as many as 20% of all kids with the infection in the prevaccine era died of it. In 1921, 15,520 diphtheria deaths were reported in the United States. Even before the toxoid vaccine was introduced into widespread use in the United States in the mid-1940s, the number of cases had dropped from 120,000 to 200,000/year in the 1920s to 20,000 per year because of increased awareness of hygiene and infection control. However, the drop in incidence after the introduction of diphtheria vaccine was nothing short of miraculous; the United States now sees fewer than two cases each year, and those are in unvaccinated or undervaccinated individuals. The protectiveness of the vaccine is estimated at 97% or higher.

Because there is no longer any naturally occurring infection in the United States, there is no natural boosting of immunity. Therefore, it is recommended that a combined tetanus-diphtheria immunization be given every 10 years to maintain ongoing protection against both toxin-caused diseases.

Safety. Since the diphtheria vaccine is always combined with at least a tetanus vaccine, it is difficult to tell which of the side effects are attributable to which component of the combined vaccine. As many as half of all children receiving diphtheria toxoid-containing vaccines have local reactions at the site of injection, consisting of pain, tenderness, swelling, and redness. In younger infants, this local reaction may be accompanied by fussiness and/or decreased appetite. Occasionally, kids will have a more severe local reaction (termed an Arthus reaction) with a much wider area of swelling and more pain. Fever and more worrisome whole-body reactions are uncommon but occurred more often when an older form of pertussis vaccine was routinely given with DPT; these and other pertussis-associated side effects of diphtheria-containing vaccines are discussed below [see "Pertussis (Whooping Cough)"].

Special Concerns and Controversies. Because many children receive their diphtheria vaccine combined with a pertussis vaccine, special consideration must be given to pertussis vaccine side effects. They are discussed in "Pertussis (Whooping Cough)" below.

Some diphtheria vaccine preparations contain trace amounts of a preservative called thimerosal. Concerns regarding this ingredient are discussed under "General Vaccine Concerns" below.

Because diphtheria (and other) vaccines are recommended for frequent administration during the first year of life and because sudden infant death syndrome (SIDS; crib death) also occurs (by definition) during those months, concerns have been raised about a possible association. For a discussion of this concern, see "Pertussis (Whooping Cough)" below.

Haemophilus influenzae Type B Infection

Germ causing the disease	*H. influenzae* type B bacterium
Type of vaccine	A fraction of the outer capsule of the bacterium linked to a protein carrier molecule (conjugated vaccine) that makes the vaccine more effective in stimulating protective antibodies
Year first recommended for use	1988
How it is given	Into muscle
When it is given (2007 guidelines)	Different vaccine formulations are given at different times; each product is given at 2 and 4 months and between 12 and 15 months; one formulation is also required at 6 months.
Brand name(s) of individual vaccine	ActHIB; Pedvax HIB; HibTITER
Combination products (brand name) also containing this vaccine	Available in combination with hepatitis B vaccine (Comvax) and in combination with diphtheria, tetanus, and acellular pertussis vaccines (TriHIBit; approved only as a booster fourth dose)

Effectiveness. Prior to 1988, when the *H. influenzae* type B vaccine was introduced, one of every 200 children in the United States under age 5 developed a serious infection with the germ, translating into about 20,000 cases each year. *H. influenzae* type B was the most common cause of bacterial meningitis in the United States, as well as a leading cause of pneumonia, bacteremia, sepsis, epiglottitis (a potentially fatal infection of the epiglottis, a structure near the voice box), joint infections, severe skin infections, infections around the heart, and ear infections. Kids between the ages of 6 and 18 months of age were the most often affected. As many as 600 children died each year due to *H. influenzae* type B meningitis in the prevaccine era, and survivors were often left deaf or with permanent brain damage. The vaccine has eliminated 99% of all infections due to this organism; there were 5 proven cases in 2005 among kids under 5 years and approximately 200 that could not be confirmed.

Safety. This is a very well-tolerated vaccine. Between 5 and 25% of kids have a mild local reaction at the injection site—redness, pain, and/or swelling; this resolves within 24 hours in most children. Fever is uncommon, and serious reactions are rare.

Special Concerns and Controversies. Because some children receive their *H. influenzae* type B vaccine combined with a pertussis vaccine, special consideration must be given to pertussis vaccine side effects. They are discussed below [see "Pertussis (Whooping Cough)"].

Hepatitis A

Germ causing the disease	Hepatitis A virus
Type of vaccine	Killed virus
Year first recommended for use	1995 for travelers 1999 for kids in geographic regions with high rates of hepatitis A infection 2005 for all children
How it is given	Into muscle
When it is given (2007 guidelines)	Two doses between the ages of 1 and 2 years; doses separated by at least 6 months
Brand name(s) of individual vaccine	Havrix; Vaqta
Combination products (brand name) also containing this vaccine	None for children (an adult product combines hepatitis A vaccine with hepatitis B vaccine)

Effectiveness. Protection is between 95 and 100%; however, it is unknown how long the protection lasts. Measurement of antibody levels in kids after they receive the vaccine indicates a protective level for at least 5 years. Nationally, the occurrence rate of hepatitis A has fallen dramatically since vaccine use began in the mid- to late 1990s. Previous peaks in occurrences among young children and in certain geographic regions have all leveled out since the vaccine was introduced, indicating a substantial impact of the vaccine on the disease burden in the United States.

Safety. Side effects are limited to mild local reactions at the injection site (pain, redness, and swelling). Fever, headache, and/or fatigue are uncommon. No serious side effects have been seen.

Special Concerns and Controversies. Hepatitis A in young children, typically the group with the highest occurrence rate, is usually a mild disease and often has no symptoms at all—the kids don't know they are infected, nor do their parents, teachers, or day care attendants. However, infected kids are contagious, and adult contacts often develop a much more debilitating infection (see chapter 3), although they ultimately recover, as well. There has been controversy about giving hepatitis A vaccine to kids when the major benefit is actually seen as a reduction in hepatitis diagnosed in adult contacts. The debate is further fueled due to the fact that hepatitis A, even in adults, usually gets better on its own without serious or life-threatening complications. Still, the benefit to parents of not contracting the hepatitis A infection that is circulating in their kids' day care center can be tangibly measured. Hepatitis A in adults is a very unpleasant disease (see chapter 3) and typically results in many days or weeks of missed work and disability.

It is not yet known whether booster doses will be needed, since the duration of protection with the first two doses has not yet been determined.

Hepatitis B

Germ causing the disease	Hepatitis B virus
Type of vaccine	A surface protein of the virus, produced in the laboratory
Year first recommended for use	Available in 1986
	Recommended for all infants in 1991
How it is given	Into muscle
When it is given (2007 guidelines)	At birth, between 1 and 2 months, and between 6 and 18 months
	Also given under certain circumstances after exposure to an infected person
Brand name(s) of individual vaccine	Recombivax HB, Engerix-B
Combination products (brand name) also containing this vaccine	Available in combination with *H. influenzae* type B vaccine (Comvax); the combination may not be given for the vaccine dose required at birth.
	Available in combination with DTaP and inactivated polio vaccine (Pediarix); the combination may not be given for the vaccine dose required at birth.

Effectiveness. Hepatitis B virus causes severe liver disease, including chronic (long-term) hepatitis and liver cancer (see chapter 3). It is estimated that 1.25 million Americans have chronic (long-term) hepatitis B infection, and 20 to 30% of those became infected during childhood—many at birth, acquiring the infection from their mothers. The vaccine protects 90 to 95% of kids from hepatitis B infection, and the duration of protection probably exceeds 15 to 20 years; at this time, there is no evidence that booster shots will be necessary. Between 1990 and 2004, the occurrence rate of hepatitis B in the United States declined by 75%, but among kids, the decrease was nearly 95%; the former reflects a reduction in high-risk behavior among adults, and the latter reflects the effectiveness of the vaccine.

Safety. A local reaction (pain) at the injection site occurs in 3 to 10% and temperature elevation to less than 100°F in 6% or less of kids receiving the vaccine. Occasional reports of fatigue or fussiness have been recorded. More serious reactions are very rare.

Special Concerns and Controversies. The Engerix-B and Pediarix preparations may contain trace amounts of a preservative called thimerosal. Concerns regarding this ingredient are discussed under "General Vaccine Concerns" below.

Most states require hepatitis B immunization for day care and school entry. It is up to the individual states to determine which vaccines are mandated (see "General Vaccine Concerns" below). Because hepatitis B is often sexually transmitted, some parents and religious organizations have objected to the mandating of the vaccine, preferring abstinence for prevention. However, nearly one-third of hepatitis B infections have no known sexual contact or other risk factors in their histories, and the sources of those infections are unknown.

Although the risk of chronic and severe hepatitis B infection is very high among babies born to hepatitis B-infected mothers, there is no risk at birth if mothers are uninfected. This has led some to object to the mandated immunization of all newborns, including those born to the vast majority of women who are not infected with hepatitis B. However, most kids who become infected with hepatitis B at some time during their lives are born to mothers who were not infected at the time of birth. Immunization of all babies at birth, followed by completion of the immunization series, protects against hepatitis B infection of those babies at any time during childhood and beyond.

Concerns have been raised about the possible role of hepatitis B vaccines in causing or exacerbating multiple sclerosis, a nervous system disease. Numerous large and well-controlled population studies have been conducted to assess this possible association, and no relationship has been found. The Institute of Medicine of the National Academy of Science summarized these studies, and the institute's conclusion was that hepatitis B vaccine does not cause or contribute to multiple sclerosis; the executive report of that analysis is available online (http://www.iom.edu/?ID=4705).

There have been several dozen cases of hair loss (alopecia) reported following hepatitis B vaccination; in many, but not all, of the cases, the hair grew back to normal. Further studies are being conducted to establish whether there is a relationship between the vaccine and the cases of hair loss, but with millions of vaccine doses being given, the few cases of hair loss constitute a miniscule risk even if the association is real.

Other chronic diseases have been linked to the hepatitis vaccine in isolated reports but have not been substantiated by further, more thorough study; these include diabetes, asthma, and lupus. Review of the literature also did not show evidence for an association between this vaccine and SIDS.

Influenza

Germ causing the disease	Influenza A and influenza B viruses
Types of vaccine	Components of three killed strains of virus
	Three live, weakened virus strains

Year first recommended for use	1940s for killed virus vaccines 2003 for live, weakened virus vaccine
How it is given	Killed virus vaccine into the muscle Live, weakened virus sprayed into the nose
When it is given (2007 guidelines)	Given in the fall, before flu season, to children 6 months to 5 years old (the live, weakened virus vaccine is currently approved only for kids older than 5 years). Kids older than 5 years with certain medical conditions (or living in a household with a person at risk for severe influenza) should also receive the influenza vaccine. For kids younger than 9 years who have not received prior influenza immunizations, two doses are given. Vaccine must be given every year.
Brand name(s) of individual vaccine	Killed virus vaccine: Fluzone, Fluvirin Live, weakened virus vaccine: FluMist
Combination products (brand name) also containing this vaccine	None

Effectiveness. Because the vaccine changes each year to match the anticipated circulating strains (see chapter 3), the effectiveness of the vaccine varies according to how closely it matches the actual strains that circulate each year. When the vaccine closely matches the strain causing an outbreak (meaning that the experts who predicted the most likely strain for a particular year were correct), the vaccine's protection rate is 85 to 90% among kids and adults less than 65 years old. The protection from each year's vaccine can only be assumed to last for the single year that the vaccine was designed for, but there is evidence that each time the vaccine is given, the response is improved by prior years' vaccinations. Among kids, the vaccine also reduces the rate of ear infections, because influenza predisposes kids to that illness (see chapter 3).

Safety. Local reactions (soreness or redness) occur in about 10% of kids older than 13 years who receive the killed vaccine and even less frequently in kids under that age. Fever within 24 hours of receiving the killed vaccine occurs rarely, usually in kids under 2 years of age. The live, weakened vaccine given into the nose has been associated with an increased occurrence of asthma attacks in kids under 5 years old; currently, the live vaccine is not approved for these younger kids. The killed, injected vaccine is not associated with an increase of asthma attacks or a worsening of any other chronic lung problems in kids receiving the vaccine. Live, weakened vaccine should not be given to kids with asthma or other lung diseases regardless of age. People with egg allergies may develop a severe allergic reaction to either the killed or live, weakened vaccine (see "General Vaccine Concerns" below).

Special Concerns and Controversies. The virus used to make the vaccines is grown in the laboratory using chicken eggs and should not be given to people with known egg allergies (see "General Vaccine Concerns" below).

The Fluvirin preparation contains a preservative called thimerosal. Concerns regarding this ingredient are discussed under "General Vaccine Concerns" below. The other brands available for children in the United States are thimerosal free.

The live, weakened vaccine given into the nose may result in live virus in the nose for several days to a week or more after the vaccine is received. Rarely, this live virus may cause infection in a contact of the child who received the vaccine. When this occurs, the influenza infection in the contact is typically very mild. Because live virus from the vaccine may cause infection in contacts of the vaccinated child, kids who have contact with a person who has an abnormal immune system should get only the killed vaccine.

During certain years of influenza immunization, there may have been a very slight increase in the occurrence rate of a neurological disease called Guillain-Barré syndrome (GBS), which causes paralysis (see chapter 3). The increased occurrence rate after receiving influenza vaccines is felt to be far less than the rate of severe and potentially life-threatening influenza infection that would develop in people who chose not to be immunized. The magnitude of increase of GBS cases seen during the swine flu vaccination campaign in 1976 (see chapter 3) has not been seen during subsequent flu vaccine years, and even the increased 1976 rate was not substantial (less than 1 additional case of GBS for every 100,000 vaccine doses given).

Kids on long-term aspirin treatment for arthritis or other conditions should not get the live, weakened vaccine because of the association of influenza virus with Reye's syndrome (see chapter 3). These kids should, however, receive the killed vaccine, because natural influenza infection poses a significant risk for Reye's syndrome in patients on aspirin and the killed vaccine is protective against natural influenza infection.

Influenza vaccine is specifically recommended for kids 6 months to 5 years old because those children are at greatest risk for severe disease. Although only officially recommended for kids *older* than 5 years of age if there are certain medical conditions (or high-risk household contacts), any child older than 5 whose parents wish to avoid the discomfort, inconvenience, and interruption of daily activities associated with influenza can and should receive influenza immunization. This includes kids in college not wanting to miss classes, kids on winter sports teams, and any other situations where a week or more of illness would be highly undesirable. The bottom line is that most kids should be getting this immunization regardless of age or underlying medical conditions.

In certain years, news headlines shout out stories of influenza vaccine shortages. Vaccine shortages are discussed under "General Vaccine Concerns" below.

Measles

Germ causing the disease	Measles virus
Type of vaccine	Live, weakened virus
Year first recommended for use	1963; improved vaccine in 1968; second dose recommended in 1989
How it is given	Under the skin
When it is given (2007 guidelines)	Between 12 and 15 months and again between 4 and 6 years (the second dose can be given sooner if at least 1 month after the first dose)
	Also given under certain circumstances after exposure to an infected person
Brand name(s) of individual vaccine	Although it is available as a single vaccine, only the combined vaccines are recommended.
Combination products (brand name) also containing this vaccine	MMR, with mumps and rubella; and MMRV, with mumps, rubella, and chicken pox (ProQuad)*

*At the time of this writing, a shortage of MMRV vaccine has made it temporarily unavailable (see "General Vaccine Concerns" below for a discussion of vaccine shortages).

Effectiveness. Measles is a potentially severe disease requiring hospitalization of as many as 20% of all those infected in the United States, with complications such as pneumonia, encephalitis, diarrheal dehydration, and ear infections. Prior to the introduction of the vaccine in 1963, as many as 500 patients died in the United States each year as a result of measles. In the developing world, as with many infections, measles is even more severe, with 1% of all infected persons dying due to the disease. The vaccine is very effective, with about 95% of kids becoming immune to measles after one dose of the vaccine and almost all of the remainder after two doses. The number of cases each year in the United States is now under 70 (37 in 2004, 66 in 2005, and 45 in 2006) compared with 500,000 to 4 million cases each year in the prevaccine era. However, during certain periods over the past 20 years, transient surges in the number of cases (e.g., between 10,000 and 30,000 cases per year between 1989 and 1991, including 123 deaths and more than 11,000 hospitalizations during those 3 years) occurred during times when immunization rates fell in this country and around the world (see "Herd Immunity" under "General Vaccine Concerns" below). In certain communities experiencing measles outbreaks during those years, immunization rates had dropped below 50%. With renewed efforts at immunizing all kids at 1 year of age, the annual rates of measles infections have remained at their current very low levels. Most cases in the United States now are the result of exposure abroad or exposure to others who have traveled abroad and brought home the infection; children adopted from China have been the sources of some cases. Other cases have occurred among religious-group communities where vaccination is refused. In 2005, one such outbreak occurred in an Indiana religious community, where 34 cases of measles resulted from exposure to a single case of infection; 94% of those 34 infected individuals were unvaccinated. However, a major epidemic was averted because 98% of the kids in the neighboring areas had been immunized against measles (see "Herd Immunity" under "General Vaccine Concerns" below).

Safety. Measles vaccine (and the mumps, rubella, and chicken pox vaccines in the combination products) are live, but weakened, virus vaccines; the weakened viruses can cause symptoms. The measles virus in the vaccine causes fever of 103°F or higher in 5 to 15% of kids, usually between 1 and 2 weeks after the immunization is received (that's the time period for the virus to cause infection, also known as the "incubation period"); the combination vaccine that includes chicken pox (MMRV) has a slightly higher rate of fever associated with it than with the MMR combination. A rash occurs during the same time frame in 5% of kids. Both the fever and the rash are short-lived, resolving in a day or two. Occasionally kids develop swollen glands (lymph nodes) after receiving measles vaccine and rarely swollen parotid glands (due to the mumps component of the combined vaccine); both are of short duration and get better on their own. A very small percentage of kids (1 in 30,000 or fewer) receiving the vaccine will develop low counts of one type of blood cell (platelets, the cells responsible for clotting); as with fever and rash, the occurrence of low platelet counts is much less frequent with the vaccine than with the naturally occurring measles infection that the vaccine prevents.

Special Concerns and Controversies. Because of the relatively high rate of fever associated with measles immunization, kids predisposed to having seizures with fever may have a higher risk of having a seizure after the immunization; these kids should still be immunized, however, because the risk of natural measles is greater than the risk of the vaccine.

Because autism is sometimes recognized between 1 and 2 years of age, the same time as the first MMR or MMRV vaccine is given, there has been much attention paid to a possible association between the measles vaccine and autism. The Institute of Medicine of the National Academy of Science (http://www.iom.edu/?ID=4705) and the American Academy of Pediatrics have both reviewed reports of autism occurring after MMR vaccine and have concluded that the vaccine does not cause autism or autism spectrum disorder. The issue of autism and vaccines is discussed under "General Vaccine Concerns" below.

A few studies done by a single group of researchers linking the MMR vaccine to inflammatory bowel disease (e.g., Crohn's disease) have generated substantial attention and substantial research seeking to confirm or refute a relationship. Large population studies conducted by the CDC and others, as well as large controlled studies of cases of patients with inflammatory bowel disease conducted by many different investigators, have all failed to confirm a relationship between MMR vaccine and bowel disease. Similarly, a new syndrome of inflammatory bowel disease combined with autism cannot be associated with the vaccine when large, well-controlled studies are performed.

Because the measles, MMR, and MMRV vaccines are live, weakened virus vaccines, they should not be given to kids with abnormal immune systems. However, the viruses in MMR are not contagious for contacts of vaccine recipients, so kids who have contact with others who have abnormal immune systems may still themselves receive MMR; these children should not receive the MMRV vaccine because the chicken pox component does pose a risk for contagiousness to contacts who have an abnormal immune system.

Mumps

Germ causing the disease	Mumps virus
Type of vaccine	Live, weakened virus
Year first recommended for use	1967
How it is given	Under the skin
When it is given (2007 guidelines)	Between 12 and 15 months and again between 4 and 6 years (the second dose can be given sooner if at least 1 month after the first dose)
Brand name(s) of individual vaccine	Although it is available as a single vaccine, only the combined vaccines are recommended.
Combination products (brand name) also containing this vaccine	MR, with rubella; MMR, with measles and rubella; MMRV, with measles, rubella, and chicken pox (ProQuad)*

*At the time of this writing, a shortage of MMRV vaccine has made it temporarily unavailable (see "General Vaccine Concerns" below for a discussion of vaccine shortages).

Effectiveness. Although usually a benign disease in children, but worse in adults, mumps was a significant cause of deafness prior to the introduction of the vaccine (see chapter 3); as many as 1 in 20,000 mumps cases resulted in deafness. Mumps also rarely caused inflammation of the testes (orchitis) and of the brain (encephalitis [see chapter 3]). A single dose of the vaccine is greater than 95% successful in generating protective levels of antibodies, and those antibodies appear to last for many years. However, the ability of the vaccine to protect against infection is somewhat lower than for other vaccines, and it is estimated to be between 75 and 90% protective clinically. After licensure and widespread use of the vaccine, the number of cases of mumps in the United States dropped from more than 150,000 each year to between 3,000 and 5,000 annually through the 1980s. As with measles, surges of infections broke through in certain years, particularly before mumps vaccine was required for school attendance. With required use of the vaccine, mumps cases fell to 258 in 2004 and 314 in 2005, only to bounce back to more than 6,000 cases in 2006 as a result of an outbreak occurring mostly among college students (see chapter 3), despite the fact that many of the students had been vaccinated. This further illustrates that the mumps vaccine does not appear to be as protective as other vaccines (see above) and that breakthrough infections do occur with some regularity.

Safety. The mumps vaccine is very rarely associated with any side effects; those that do occur after MMR or MMRV are usually attributable to the other vaccine components. Rare cases of testicular inflammation or parotid gland inflammation, both seen in much higher frequency with natural mumps infection, have been reported following receipt of the vaccine.

Special Concerns and Controversies. Approximately 1 of every 800,000 doses of mumps vaccine is followed by a nervous system abnormality (seizure, meningitis, encephalitis, or deafness); however, the Institute of Medicine of the National Academy of Science, on reviewing these data, could not determine that even those very rare events were the result of the vaccine (http://www.iom.edu/?ID=4705).

Because the mumps, MMR, and MMRV vaccines are live, weakened virus vaccines, they should not be given to kids with abnormal immune systems. However, the viruses in MMR are not contagious for contacts of vaccine recipients, so kids who have contact with others who have abnormal immune systems may still receive MMR; these children should not receive the MMRV vaccine because the chicken pox component does pose a risk for contagiousness to contacts with abnormal immune systems.

Concerns regarding an association between the MMR vaccine and conditions such as autism and inflammatory bowel disease are discussed above under "Measles" and below under "General Vaccine Concerns."

Neisseria meningitidis Infection

Germ causing the disease	*Neisseria meningitidis* bacterium
Types of vaccine	Portions of the outer capsules of four strains of the bacterium (polysaccharide vaccine)
	Portions of the outer capsules of four strains of the bacterium linked to a protein carrier molecule (conjugated vaccine) that makes the vaccine more effective in stimulating protective antibodies

Year first recommended for use	1978 for polysaccharide vaccine 2005 for conjugated vaccine
How it is given	Polysaccharide vaccine, under the skin Conjugated vaccine, into the muscle
When it is given (2007 guidelines)	Polysaccharide vaccine, only to children 2 years old or older with certain medical conditions that put them at increased risk for *N. meningitidis* infection; a booster 2 to 3 years after the first shot is recommended for young children who remain at increased risk for infection. Conjugated vaccine, all children older than 11 years; should be given to all adolescents not previously immunized, including all college freshmen living in dormitories
Brand name(s) of individual vaccine	Polysaccharide vaccine, Menomune Conjugated vaccine, Menactra
Combination products (brand name) also containing this vaccine	None

Effectiveness. Since the introduction of vaccines against *H. influenzae* type B (see above), the relative importance of *N. meningitidis* as a cause of meningitis and sepsis in children has increased (see chapter 3). There are several thousand cases of *N. meningitidis* infection in the United States each year, with a significant proportion of them occurring in kids younger than 2 years of age and in adolescents and young adults; two-thirds of all cases occur in people older than 11 years. The percentage of fatal cases is very high among infected kids (10 to 15%), and 10 to 20% of the survivors are permanently disabled. The *polysaccharide* vaccine is not effective in children under the age of 2 years but is protective against infection with four of the five major strains of *N. meningitidis* in kids older than 2. It is recommended only for those patients at particular risk for infection with that bacterium, for example, kids with abnormal immune systems or missing spleens or those who travel to areas of the world where the infection occurs frequently. In contrast, the *conjugated* vaccine is much more protective of younger kids but is currently only routinely recommended for adolescents (age 11 and older); at the time of this writing, the FDA was considering licensing the conjugate vaccine for kids down to the age of 2 years. Like the polysaccharide vaccine, the conjugate vaccine protects against only four of the five major strains of the germ—there is no licensed vaccine in the United States to protect against the fifth strain that commonly causes infection; that strain causes up to one-third of all cases. It is too early to determine what effect immunization of kids older than 11 years will have on the annual occurrence rate of this infection. The duration of immunity conferred by the polysaccharide vaccine is only a few years, but it is believed that the protection afforded by the conjugated vaccine will be longer. It will be several years before it is known if a booster shot is required for kids receiving the conjugated vaccine; a booster is recommended for young children who received the polysaccharide vaccine because of their particular susceptibility to this infection (see above).

Safety. The side effects for the two *N. meningitidis* vaccines are similar. Local reactions (pain and redness) occur at the injection site in 50 to 60% of kids who receive either vaccine; fever between 100 and 103°F is uncommon, occurring in 3 to 5% of kids within a week following receipt of the vaccines. As many as 60% of kids will have headache or fatigue within a week of the shot; these are mild in the vast majority of cases.

Special Concerns and Controversies. As of September 2006, 15 cases of GBS had been reported in teenagers within 6 weeks following receipt of the conjugated vaccine; two additional cases occurred in adults aged 30 and 43 years. GBS is a neurological disease that causes paralysis, and it has been associated with certain vaccines (see above and chapter 3), as well as with naturally occurring infections; many cases do not have an explanation. The CDC has not yet determined whether the cases of GBS were related to this vaccine or coincidental to it (see chapter 12).

The lack of an effective vaccine against the B strain of *N. meningitidis,* the cause of up to one-third of all cases, remains a major concern.

The relatively low occurrence rate of meningitis and sepsis due to *N. meningitidis* in this country has prompted criticism that widespread vaccination will not be cost-effective, that is, the number of cases of infection prevented will not justify the cost to society of immunizing all kids older than 11 years of age, which is the current recommendation.

Papillomavirus (Genital Warts and Cervical Cancer)

Germ causing the disease	Human papillomavirus strains 6, 11, 16, and 18
Type of vaccine	A surface protein of each of the virus strains, produced in the laboratory
Year first recommended for use	2006
How it is given	Into the muscle
When it is given (2007 guidelines)	11- to 12-year-old girls or older adolescent girls and young women who have not been previously immunized; given as a three-dose series—the second dose is given 2 months after the first, and the third dose 6 months after the first.

Brand name(s) of individual vaccine	Gardasil
Combination products (brand name) also containing this vaccine	None

Effectiveness. The human papillomaviruses cause warts, those on the hands and feet, as well as sexually transmitted genital warts (see chapter 3). Importantly, papillomaviruses also cause most cases of cervical dysplasia and cervical cancer. Papillomavirus infection is the most common sexually transmitted disease in the United States; as many as two-thirds of adolescent girls are infected in some communities studied, and most new infections each year occur in adolescents and young adults. The same papillomaviruses also cause genital cancers in men and some cases of cancers in the mouth. According to the American Cancer Society, as many as 10,000 new cases of cervical cancer are diagnosed each year, virtually all due to the strains of papillomavirus included in the vaccine. The current vaccine produces high levels of protective antibodies in nearly 100% of women studied. In prevention trials of the vaccine in 16- to 26-year-old girls and women, the vaccine was found to be 95 to 100% effective in preventing genital warts, cervical dysplasia, and early cancer changes in the cervix. The protection was against only those strains of the virus with which the women had not yet been infected; there was no effect of the vaccine on already existent papillomavirus infections. The vaccine must be given to girls and women before they have the chance to become infected, but even if they were already infected with one or more of the four high-risk strains, immunization was found to be protective against the remaining strains included in the vaccine preparation. Immunity appears to persist for at least 5 years after receipt of the vaccine, and at present, no booster shots beyond the primary series of 3 doses are recommended. Studies are continuing with men—the vaccine produces high levels of protective antibodies in men as well, but it is not yet known if there will be an effect on reducing actual infections, as has been shown for women.

Safety. As with most vaccines, local reactions (pain, redness, and swelling) at the site of injection are the most common side effects, occurring in more than 80% of vaccine recipients. Fever developed in 10% of those vaccinated, but also in 9% of those getting a placebo (sugar pill [see chapter 12]). In post-marketing surveillance, dizziness and/or fainting have been rarely reported. The Advisory Committee of Immunization Practices of the CDC recommends a 15-minute wait in the clinic after receipt of the vaccine to prevent fainting episodes.

Special Concerns and Controversies. Although the vaccine protects against the most common papillomaviruses that cause genital warts and cervical cancer, it does not protect against *all* papillomaviruses or *all* causes of abnormal cervical changes in women. Pap smears and cervical screening examinations will still be required for those immunized.

Many states are now considering mandating immunization against papillomaviruses for all girls entering middle school. It is up to individual states to determine required immunizations for school and day care entry (see "General Vaccine Concerns" below). Because papillomavirus infections are sexually transmitted, only sexually active adolescents are at risk. This has caused objections to mandated immunization against papillomaviruses by some parents and religious organizations that prefer abstinence as a preventive against sexually transmitted diseases. Indeed, abstinence is preventive against acquisition of this infection. However, studies show that 65 to 70% of adolescent girls and boys are sexually active by the end of high school and that more than 10% of 11th graders have already had four or more sexual partners. The vaccine is effective only if given *before* sexual exposure to the strains in the vaccine occurs (see above). Traditionally, mandated vaccines have been those that prevent infections that readily spread within the school community. The sexual pattern of spread of papillomavirus infections has generated debate regarding whether it satisfies the usual criteria for mandated vaccines. Tetanus immunization also does not satisfy the traditional criterion of classroom contagiousness, but it is included in the diphtheria and pertussis vaccine preparations that do protect against school contagions.

Along similar lines, some parents and religious organizations have expressed concern that immunization against a sexually transmitted disease sends the message to kids that sexual activity is expected and even permissible. Additionally, some have worried that immunized girls might be less cautious sexually because they feel immune and therefore potentially put themselves at risk for other sexually transmitted diseases and pregnancy.

Although many of the girls and women involved in the prelicensing studies of the vaccine were 16 to 26 years of age, the vaccine is recommended for use in 11- and 12-year-old girls. Some have raised concerns that the biology of these younger girls may be different and that side effects or other reactions may be more severe. However, the vaccine studies prior to licensure did also include girls as young as 9 years of age, and to date, among 11- and 12-year-old girls who have received the vaccine, the reactions appear to be similar to those of older girls and women; however, studies are continuing as the vaccine is being used (postmarketing surveillance).

Since the vaccine was marketed in 2006, a number of cases of Guillain-Barré syndrome (GBS) have been reported following immunization (GBS is a neurological disease that causes paralysis; see "Influenza" and "*Neisseria meningitidis* Infection" above and chapter 3). Studies are ongoing to determine if there is any relationship between the vaccine and this condition. A number of the GBS patients had received the vaccine against *N. meningitidis* (see above) at the same time as the papillomavirus vaccine.

The current vaccine is very expensive—almost $400 for the three doses, in addition to clinic visit charges. Some, but not all, health insurance plans cover the vaccine and clinic visits, but many adolescents in the highest-risk communities are uninsured.

Pertussis (Whooping Cough)

Germ causing the disease	*Bordetella pertussis* bacterium
Type of vaccine	Purified portions of killed bacteria (acellular; see "Special Concerns and Controversies" below)
Year first recommended for use	1940s for the whole-cell preparation no longer used (see "Special Concerns and Controversies" below) 1991 for the acellular vaccine as fourth and fifth doses of the pertussis series 1996 for the acellular vaccine as the full series
How it is given	Into the muscle
When it is given (2007 guidelines)	2, 4, and 6 months; between 15 and 18 months; between 4 and 6 years; and as a booster between 11 and 12 years
Brand name(s) of individual vaccine	Not available as an individual vaccine
Combination products (brand name) also containing this vaccine	Pertussis vaccine is available only as a combination vaccine with: • Tetanus toxoid and diphtheria toxoid vaccine 1. DTaP* for kids under 7 years old (Tripedia, Infannix, DAPTACEL) 2. Tdap for kids older than 7 (Boostrix, Adacel) • Tetanus toxoid, diphtheria toxoid, and *H. influenzae* conjugate vaccine (TriHIBit) • Tetanus toxoid, diphtheria toxoid, hepatitis B, and inactivated polio vaccine (Pediarix)

*The capital D indicates a higher dose of the diphtheria toxoid component, as recommended for younger kids.

Effectiveness. Pertussis begins as a mild upper respiratory tract infection but progresses in stages to a severe coughing disease that, in young infants and children, can cause death due to exhaustion and oxygen deprivation (see chapter 3). Older children, adolescents, and adults have less severe disease but can harbor the germ and pass it on to babies and young children. Following the introduction of vaccines, the annual occurrence rate of pertussis in the United States fell from 175,000 to 15,000 by 1960 and to 3,000 or fewer cases each year between 1980 and 1990. Since that time, the rate of pertussis in the United States has been increasing, and in 2004, there were nearly 26,000 cases, the largest number since 1959. There has also been an age shift, with many cases now occurring in older children. At least some of the explanation for the increase in the number of cases of pertussis rests with a decrease in the immunization rate of babies and infants due to parental concerns regarding short-term side effects and long-term safety. In the 6-year period between 1990 and 1996, more than half of the cases of pertussis occurred in kids who were unvaccinated or incompletely vaccinated. Another factor in the failure to sustain pertussis rate declines is the relatively low protective power of the vaccines—a protection rate of only 80 to 85% (compared with 95 to 100% for many of the other vaccines discussed in this chapter).

Safety. Local reactions (pain, swelling, and redness) at the injection site occur in approximately 30% of kids after each of the first three doses of the series of DTaP shots; the rate is higher after the fourth and fifth doses, and rarely, more severe local reactions, such as severe swelling, may occur with the booster doses (in 2 to 3% of vaccine recipients). Occasionally, a sterile abscess develops at the injection site—an accumulation of fluid that is not infected and goes away on its own. Fever of 101°F or higher occurs in about 5% of children after vaccination, with lower rates of drowsiness, irritability, and loss of appetite. Much higher fevers, to 105°F, and more serious whole-body reactions (seizures, increased or decreased muscle tone, and prolonged crying) are much rarer with the acellular vaccines than with the whole-cell vaccines that are no longer used in this country (see below), but those more-severe reactions still occur approximately once every 10,000 or more doses. The Tdap preparation of pertussis vaccine used for older children and adolescents has a higher rate of local pain at the injection site (about 60%) but is otherwise comparable to DTaP in rates of redness, swelling, and fever; Tdap has not been associated with the severe whole-body reactions that rarely occur with DTaP, likely because the pertussis vaccine component is in a lower dose in Tdap.

Special Concerns and Controversies. Among all licensed vaccines, none have been more clearly linked to unpleasant and unsettling side effects than pertussis vaccines. Until 1996, a whole-cell pertussis vaccine was used in the United States. As the name implies, this whole-cell formulation contained all of the components of the killed pertussis germs and caused many more bad reactions and side effects than are seen with the current purified acellular preparations. The lingering bad reputation of the older whole-cell vaccine continues to limit parents' comfort and compliance with recommendations for pertussis immunization.

Pertussis vaccines, both the old whole-cell preparations and the current acellular products, are associated with an array of rare but disturbing side effects, in addition to the local reactions they cause. These whole-body side effects include a dramatic drop in muscle tone and/or decrease in consciousness ("hypotonic-hyporesponsive episodes" [HHE], also referred to as "collapse"), very high fever (105°F or higher), and crying episodes lasting for

3 hours or more. While none of these reactions has been associated with long-term harm, they are obviously and appropriately of concern to parents. Importantly, all of these severe reactions are significantly less common (occurring 1/5 to 1/10 as often) with the current acellular vaccine preparations than with the older, and no longer used, whole-cell pertussis vaccines. Seizures rarely (0 to 26 cases for every 100,000 doses of the vaccine given) complicate both the whole-cell and acellular pertussis vaccines; some studies have shown a reduction in the occurrence rate with the acellular preparation, whereas other studies have not shown a decrease.

In addition to the side effects noted above that have been proven to be due to pertussis vaccines, there are several severe medical conditions that have been alleged to be due to these shots, as described in the paragraphs below.

Brain injury. Most troubling among the alleged associations is serious brain injury resulting in permanent damage. Individual cases of children who had an early bad reaction to the old whole-cell pertussis vaccine, for example, seizures or HHE (see above), and went on to have long-lasting abnormal nervous system development or long-term neurological problems put everyone on alert. The difficulty in determining the true cause of individual cases of brain damage, however, is substantial. Many neurological abnormalities are first diagnosed during the same age window as that for immunizations; establishing a cause-and-effect relationship between any factor and any end result requires large, controlled trials to eliminate the possibility of epiphenomena, or coincidences (see chapter 12). With very rare events, such as brain damage in children following an immediate bad vaccine reaction (which is also a rare event), the population size that must be studied to rule out a coincidence is huge. Several such huge studies have since been performed, looking cumulatively at hundreds of children who had bad immediate neurological reactions to the vaccine (seizures, HHE, etc.)—none of those studies have found long-term or permanent brain damage in those children at a rate that exceeds that in a control population of kids the same age. While everyone is hesitant to say absolutely that the old whole-cell pertussis vaccine never caused permanent brain damage, it is safe to say that in the large and well-controlled studies that have been done, there is no evidence for such an association. Of course, that issue is somewhat moot in the United States now that the whole-cell vaccine preparation is no longer used here. A very large, 10-year Canadian study of children with neurological problems has been conducted looking for any possible association of permanent brain damage with the new acellular vaccines that are now used in the United States, and no relationship has been found.

Infantile spasms. Infantile spasms are a particularly severe form of epilepsy that results from abnormal brain function. At one time this condition was also suspected to be a reaction to whole-cell pertussis vaccines; studies done in the late 1970s and early 1980s refuted this association. There is no evidence that the newer acellular pertussis vaccines cause infantile spasms.

Sudden Infant Death Syndrome (SIDS). Because pertussis (and other) vaccines are recommended for frequent administration during the first year of life, and because SIDS (crib death) also occurs, by definition, during those months, concerns have been raised about a possible association between the two events (pertussis vaccine administration and SIDS). Indeed, by sheer concurrence of timing alone, cases of SIDS may occur within hours, days, or weeks of routine childhood immunization. How can one distinguish a true association between vaccines and SIDS from a coincidence of timing? Once again, epiphenomena, or coincidences, plague our interpretation of many scientific and medical observations (see chapter 12). To answer questions of vaccine safety, such as the possible relationship between SIDS and vaccines, the Institute of Medicine of the National Academy of Science established an Immunization Safety Review Committee. Their study of this subject evaluated all available research and heard from experts from around the world who came to present data before the committee. Published in 2003, the Institute of Medicine report rejected an association between childhood vaccines, including those containing a pertussis component, and the occurrence of SIDS. The executive summary of the Institute of Medicine report is available online (http://www.iom.edu). Of note, the incidence of SIDS in the United States has steadily fallen since successful educational efforts regarding safe sleeping positions for babies have been widely accepted and guidelines have been followed. This fall in SIDS incidence has occurred despite the increase during the same period of childhood immunization rates, further attesting to the lack of an association between SIDS and vaccines.

Poliovirus Infection

Germ causing the disease	Poliovirus strains 1, 2, and 3
Type of vaccine	Killed viruses; strains 1, 2, and 3
Year first recommended for use	1955; improved version in 1988
How it is given	Under the skin or into the muscle
When it is given (2007 guidelines)	2 months, 4 months, between 6 and 18 months, and between 4 and 6 years
Brand name(s) of individual vaccine	IPOL
Combination products (brand name) also containing this vaccine	With DTaP and hepatitis B (Pediarix); only approved for the first hree doses tof the polio vaccine series

Effectiveness. Poliomyelitis was a dread paralytic disease in this country through the first half of the 20th century (see chapter 3). Vaccines introduced in the 1950s and 1960s eliminated naturally occurring poliomyelitis from the United States in 1979. There continued to be a few cases of poliomyelitis associated with a live, but weakened, virus vaccine (see "Special Concerns and Controversies" below). In 2000, the live, weakened vaccine was replaced entirely with the killed virus vaccine used today; there have been no further cases of vaccine-associated disease. In 1952, before the vaccines, the number of cases of paralytic poliomyelitis in the United States peaked at 21,000. Worldwide, fewer than 2,000 cases were reported in 2006 thanks to international collaboration for worldwide polio eradication. Immunity after the routine four-dose series lasts for many years, and boosters beyond the four-dose series are not currently required. However, exclusive use of the killed vaccine in this country is relatively new (since 2000), and long-term protection studies have not been done. The previously used live, weakened vaccine produced lifelong immunity (see below).

Safety. The vaccine is very safe, with only mild local reactions at the injection site reported. No serious side effects have been seen.

Special Concerns and Controversies. There was a major change in polio vaccine policy beginning in the late 1990s. Although the live, weakened vaccine was very successful in eliminating the infection from the United States during its use from 1963 through the mid-1990s, that vaccine was replaced beginning in 1996 with a strategy that gave the first polio immunizations with the killed vaccine, followed by the last two doses with the live, weakened vaccine. In 2000, the killed vaccine replaced all doses of the live version. The reason for the change is that the live vaccine caused polio in an average of 8 to 10 people each year—most with abnormal immune systems who could not handle the vaccine virus even though it had been weakened.

The live, weakened vaccine had several advantages that have been lost with the current exclusive use of killed vaccine. The live vaccine was given by drops orally, whereas the killed vaccine is given by shot; parents and, of course, kids prefer oral vaccines. Additionally, because the vaccine virus actually grew in the intestines of vaccinated kids, it conferred "herd immunity," meaning it protected those who were not themselves immunized because the live virus was excreted by those who were immunized (see "General Vaccine Concerns" below); viruses and other germs in the feces of young kids spread around very effectively (see chapter 2). Finally, the live, weakened vaccine produced "intestinal immunity," antibodies secreted in the intestines (see chapter 4), in addition to the antibodies in the blood that both types of vaccine produce. The intestinal antibodies are more effective at protecting against natural infection and limiting excretion of contagious natural virus in the feces; this is not an issue any longer in this country (where there is no natural infection), but it is the reason (along with the aforementioned herd immunity) that live, weakened vaccine still has an important role in the global polio eradication programs.

The viruses used to make polio vaccines are grown in the laboratory in cell cultures; some of those cells are derived from monkeys. In 1960, a newly discovered monkey virus named simian virus 40 (SV40) was found to be contaminating the killed polio vaccines that were being manufactured in the late 1950s and early 1960s; the live, weakened vaccines used in the past were not contaminated with SV40. Vaccines made since 1963, including the currently used killed vaccine, are free of SV40. Because SV40 has been found in several human cancers in recent years, the concern regarding possible contamination of childhood vaccines with this virus has resurfaced. The presence of SV40 in human cancers does not prove that the virus causes those cancers, and it is known that SV40 can occur in humans independent of vaccines. It is not known precisely how humans become infected with SV40, but spread from person to person is a possibility. Since 1963, poliovirus vaccines have been free of SV40; between 1955 and 1963, however, nearly 100 million people received poliovirus vaccines that may have been contaminated with SV40. The estimates are that one-third or fewer of those vaccine recipients actually received contaminated vaccine, and it is unknown if any human cancers or other adverse outcomes resulted from contaminated vaccine.

Rotavirus Infection

Germ causing the disease	Rotaviruses
Type of vaccine	Live cow rotaviruses into which portions of five strains of human rotaviruses have been inserted in the laboratory
Year first recommended for use	2006
How it is given	Orally (into the mouth)
When it is given (2007 guidelines)	2, 4, and 6 months
Brand name(s) of individual vaccine	RotaTeq
Combination products (brand name) also containing this vaccine	None

Effectiveness. Rotaviruses are the leading cause of gastroenteritis in the United States and around the world (see chapter 3). In the United States, infections with these germs usually occur in the winter and early spring, although cases occur in late fall in some parts of the country. In the United States alone, rotaviruses cause more than 2.5 million infections each year, resulting in more than 500,000 emergency room and physician office visits and 50,000 to 70,000 hospitalizations each year (most for dehydration), as well as several dozen deaths. Almost all kids have at least one rotavirus infection within the first few years of life. In studies done prior to licensure, the following measures of vaccine effectiveness were found:

- Reduction in severe gastroenteritis, 98%
- Reduction in hospitalizations, 96%
- Reduction in emergency room visits, 94%
- Reduction in physician office visits, 86%
- Reduction in any gastroenteritis, 74%

Safety. As the only oral vaccine currently recommended in the United States, this is the only vaccine without injection site side effects. In placebo-controlled trials (see chapter 12), there were very slight increases (1 to 3%) of the following symptoms in babies within 6 weeks of receiving the vaccine compared to those receiving placebo: vomiting, diarrhea, runny nose, and ear infection. There was no difference in the percentages of kids developing a fever when vaccine was compared to placebo.

Special Concerns and Controversies. There was an earlier oral rotavirus vaccine using similar technology but crafted in the laboratory with monkey rotaviruses (versus the cow rotaviruses in the current vaccine) hosting the portions of human viruses. This earlier vaccine was licensed for use in the United States in 1998 but was voluntarily withdrawn from the market by the manufacturer in 1999 because of a 20-fold increase in the risk of a severe intestinal complication called intussusception occurring within the first 2 weeks of receipt of the vaccine. The risk was still low overall, estimated at only 1 in 10,000 vaccine recipients. Intussusception is a form of obstruction of the intestine that may require surgery and can even cause death; the condition also occurs naturally in babies of the same age as those receiving the vaccine, but the rate among vaccine recipients was higher than would be expected to occur naturally. Importantly, the time period identified when the highest number of cases clustered, 1 to 2 weeks after receipt of the vaccine, suggested a direct cause-and-effect relationship with the vaccine. In contrast, large prelicensing studies of the current vaccine showed no increase in the rate of intussusception among vaccine recipients compared with placebo controls. Those cases that did occur did not cluster in any period after vaccination, suggesting that they were naturally occurring cases of intussusception rather than vaccine related. However, the risk associated with the earlier vaccine became most apparent in postmarketing surveillance—reports that occurred after the vaccine was already in widespread use in many more patients than were studied prior to licensure. Postmarketing surveillance of the current vaccine is being carefully conducted to search for any increase in the rate of intussusception. In the 15-month period between March 2006 and June 2007, the reported number of cases of intussusception in rotavirus vaccine recipients was no greater than that expected in unimmunized infants.

Because the rotavirus vaccine contains live viruses, it should be avoided for children with known abnormalities of the immune system (see chapter 4). The vaccine can be given to babies who have contact with an older child or adult who has an abnormal immune system. It is felt that the benefit of the protection afforded by the vaccine against a potentially severe naturally occurring rotavirus infection being introduced into the home by a naturally infected baby is greater for the contact person with an abnormal immune system than the risk to that person from the vaccine virus in the feces of an immunized baby in the home.

Rotavirus infections are seasonal (see above), and the vaccine has only been studied sufficiently to be proven safe in a narrow age group window:

the first dose must be given between 6 and 12 weeks of age, and the final dose must be completed by 6 months of age, with the three doses ideally separated in time by 4 to 10 weeks. This makes the timing of vaccine administration tricky. Your child's doctor will try to start the vaccine series before the annual winter-spring peak season and to fit all three doses into the age windows required.

Because the vaccine is given orally, concern has been raised regarding the possible effects of breast milk antibodies in reducing the benefit of the vaccine; no such reduction has been seen, so the vaccine should have no effect on breast-feeding considerations, and vice versa.

Babies who have *moderate* to *severe* gastroenteritis at the time they are due for the vaccine should not receive the vaccine until their symptoms are resolved; *mild* cases of gastroenteritis probably do not disqualify babies from receiving the vaccine at the time of their visit to the doctor (see "General Vaccine Concerns" below). Other gastrointestinal conditions that certain babies might have could also be a reason not to give the vaccine.

Some have expressed concern regarding the bovine origin of the vaccine. The vaccine is constructed using a cow rotavirus as the carrier of portions of human rotaviruses that the cow virus acquires by being mixed with human rotaviruses in the laboratory. This is the only current vaccine that is directly derived from an animal virus and recalls the dramatic success of Jenner in preventing smallpox by using the cowpox virus as a vaccine (see above). It has also raised concerns about potential risks from cow products (see "General Vaccine Concerns" below).

Kawasaki disease (KD) is an inflammatory condition of young children that has characteristics of an infectious disease, but a germ cause has not been found (see chapter 3). Several cases of KD have been reported within 6 weeks of receipt of the rotavirus vaccine. As with other purported associations, KD's naturally occurring age range overlaps with the age range of vaccine recipients. Although the occurrence rate of KD in kids after receiving the rotavirus vaccine does not exceed the expected rate of naturally occurring KD, the occurrence of KD has been added to the package label of the vaccine. This reflects an abundance of caution in an era of intense vaccine scrutiny. There is no evidence to date to suggest that the rotavirus vaccine causes KD.

Rubella (German Measles)

Germ causing the disease	Rubella virus
Type of vaccine	Live, weakened virus
Year first recommended for use	1969; improved current vaccine in 1979
How it is given	Under the skin
When it is given (2007 guidelines)	Between 12 and 15 months old and again between 4 and 6 years old (the second dose can be given sooner if at least 1 month after the first dose)

	Also given under certain circumstances after exposure to an infected person
Brand name(s) of individual vaccine	Although available as a single vaccine, only the combined vaccines are recommended.
Combination products (brand name) also containing this vaccine	MR, with mumps; MMR, with measles and mumps; MMRV, with measles, mumps, and chicken pox (ProQuad)

*At the time of this writing, a shortage of MMRV vaccine has made it temporarily unavailable (see "General Vaccine Concerns" below for a discussion of vaccine shortages).

Effectiveness. Rubella vaccine is given for one very specific reason—to prevent pregnant women from becoming infected with rubella virus and passing the infection on to their unborn babies. When babies are infected in the womb early in pregnancy as the major organs are forming, up to 90% develop congenital rubella virus infection. Congenital rubella syndrome is associated with severe neurological and other birth defects (see chapter 3). Rubella in children and adults, if not acquired in the womb, is usually a benign infection—only the developing organ systems of the fetus are particularly susceptible. In 1964 and 1965, before the 1969 introduction of the vaccine, an outbreak of 12 million cases of rubella occurred in the United States (see chapter 3), with 20,000 resultant cases of congenital rubella syndrome; 11,600 of those babies were born deaf, 3,580 blind, and 1,800 mentally retarded. Following vaccine introduction, the number of cases of rubella fell dramatically. By 1983, there were fewer than 1,000 cases of rubella each year, and since 2001, there have been fewer than 25 cases of rubella infection in the United States each year. By combining childhood immunization with the targeting of vaccine delivery to adolescent girls susceptible to rubella, the rate of congenital rubella syndrome has fallen dramatically as well—only two cases were reported in 2001, none in 2004, and a single case in 2006.

Safety. Although reactions are rare in children, about one-fourth of adolescent girls and young women develop joint pain within 3 weeks of receiving rubella vaccine, and 1 in 10 have arthritis symptoms (joint warmth, swelling, redness, and/or pain). The symptoms are temporary. The other side effects seen after rubella immunization are usually due to the measles component of the MMR combined vaccine (see above).

Special Concerns and Controversies. Because the mumps, MMR, and MMRV vaccines are live, weakened virus vaccines, they should not be given to kids with abnormal immune systems. However, the viruses in MMR are not contagious for contacts of vaccine recipients, so kids who have contact with others who have abnormal immune systems may still receive MMR; these children should not receive the MMRV vaccine because the chicken pox component does pose a risk for contagiousness to contacts with abnormal immune systems. Occasionally, a recently immunized mother has been shown to pass the vaccine virus on to her breast-fed baby, probably through the breast milk. Illnesses in babies have been very mild (transient rash), and breast-feeding is not a reason not to immunize a susceptible woman, and vice versa—recent immunization is not a reason to withhold breast-feeding.

Importantly, pregnant adolescents or women should not receive the rubella vaccine. This is a theoretical risk only—there have been no cases of congenital rubella syndrome resulting from accidentally giving vaccine to someone who didn't know she was pregnant, but the severe impact of naturally occurring rubella on the fetus warrants the extra caution.

Concerns regarding an association between the MMR vaccine and conditions such as autism and inflammatory bowel disease are discussed above under "Measles" and below under "General Vaccine Concerns."

Streptococcus pneumoniae Infection

Germ causing the disease	*S. pneumoniae* bacterium
Types of vaccine	Portions of the outer capsules of 23 strains of the bacterium (polysaccharide vaccine) Portions of the outer capsules of 7 strains of the bacterium, linked to a protein carrier molecule that makes the vaccine more effective in stimulating protective antibodies (conjugated vaccine)
Year first recommended for use	1977 for the polysaccharide vaccine with 14 strains; replaced in 1983 with the current 23-strain formulation 2000 for the conjugated vaccine
How it is given	Polysaccharide vaccine, under the skin or into the muscle Conjugated vaccine, into the muscle
When it is given (2007 guidelines)	Polysaccharide vaccine, only to certain children over the age of 2 years at high risk for infection with *S. pneumoniae* Conjugated vaccine, at 2, 4, and 6 months and again between 12 and 15 months.
Brand name(s) of individual vaccine	Polysaccharide vaccine, Pneumovax 23 Conjugated vaccine, Prevnar
Combination products (brand name) also containing this vaccine	None

Effectiveness. *S. pneumoniae* is the leading cause of bacteremia, bacterial meningitis, and ear infections in children (see chapter 3). The highest occurrence rate of serious infections with *S. pneumoniae* prior to the introduction of the current conjugated vaccine was in kids 2 years old and younger; the estimated attack rate was 188 cases for every 100,000 kids in that age group. Since the conjugated vaccine was introduced, the rate of serious infections in kids under 2 years old has fallen by about 75%. The current conjugated vaccine protects against only 7 of the 90 strains of *S. pneumoniae,* but those 7 strains were selected for the vaccine because they cause more than 80% of cases of bacteremia and bacterial meningitis and two-thirds of all cases of ear infections. The 23 strains in the polysaccharide vaccine cause nearly 90% of serious *S. pneumoniae* infections, but the vaccine is not effective in kids under 2 years old. When tested for its protective ability against clinical infections in large studies prior to licensure, the conjugated vaccine reduced the occurrence rate of all serious *S. pneumoniae* infections by 89%, that of all cases of pneumonia by 11%, and that of all ear infections by 7%. The diagnosis of the precise cause of pneumonia is difficult, with bacterial pneumonia more likely to cause larger abnormalities on chest X rays. The conjugated vaccine prevented three-fourths of all larger pneumonias compared with kids who were not vaccinated. Although the polysaccharide vaccine came to be known as a pneumonia vaccine when given to elderly adults, that is a misnomer; it does not protect against pneumonia due to this germ.

Safety. As with other vaccines that are injected, local reactions are the main side effects of both the polysaccharide vaccine (approximately 40% of recipients) and the conjugated vaccine (approximately 15% of recipients). Fever may occasionally occur following receipt of the conjugated vaccine. Severe reactions have not been seen.

Special Concerns and Controversies. Since the introduction of the conjugated vaccine, it has been plagued by manufacturing-related shortages (see the discussion of vaccine shortages under "General Vaccine Concerns" below).

As noted above, the conjugated vaccine now routinely given to infants in the United States contains only 7 of the 90 strains of *S. pneumoniae,* but those 7 strains account for the vast majority of serious infections in kids. There has been concern since the licensure of that vaccine that other strains of the bacterium would emerge to take the place of those protected against by the vaccine. Evidence for that phenomenon has begun to accumulate. In April 2007, a study among Alaskan kids reported that while the occurrence rate of serious *S. pneumoniae* infections fell by nearly 70% in the years immediately after the vaccine was introduced, the rate has now climbed again, and the increased number of infections is due to strains of *S. pneumoniae* not included in the vaccine. Newer conjugated vaccines with more strains of *S. pneumoniae* are under development.

Tetanus

Germ causing the disease	*Clostridium tetani* bacterium
Type of vaccine	Toxoid (inactivated bacterial toxin)
Year first recommended for use	1940s
How it is given	Into muscle
When it is given (2007 guidelines)	2, 4, and 6 months; between 15 and 18 months; between 4 and 6 years; and every 5 to 10 years thereafter, depending upon individual circumstances, such as potential exposures and dirty wounds
Brand name(s) of individual vaccine	Although it is available as a single vaccine, only the combined vaccines are recommended.
Combination products (brand names) also containing this vaccine	Tetanus vaccine is available as a combination vaccine with:

- Diphtheria toxoid (DT* for kids under 7 years; Td for kids older than 7)
- Diphtheria toxoid and acellular pertussis vaccine
  1. DTaP* for kids under 7 years (Tripedia, Infannix, DAPTACEL)
  2. Tdap for kids older than 7 (Boostrix, Adacel)
- Diphtheria toxoid, acellular pertussis, and *H. influenzae* conjugate vaccine (TriHIBit)
- Diphtheria toxoid, acellular pertussis, hepatitis B, and inactivated polio vaccine (Pediarix)

*The capital D indicates a higher dose of the diphtheria toxoid component, as recommended for younger kids.

Effectiveness. Tetanus (also known as "lockjaw") is due to a toxin produced by the *C. tetani* bacterium. The toxin paralyzes muscles throughout the body (see chapter 3). Neonatal tetanus results from nonsterile births and is protected against by immunization of adolescent girls and women. As with diphtheria, declines in the occurrence rate of tetanus occurred even before the introduction of the vaccine in the 1940s. Tetanus immunization dramatically reduced the number of cases of tetanus seen among troops in World War II. For the past several years in the United States, there have been 20 to 40 cases of the disease annually, mostly in adults and almost always in those unvaccinated or not appropriately boosted. The vaccine's protectiveness declines with time; most vaccine recipients are protected for 10 years, but some become potentially susceptible sooner. The recommendations for booster immunizations account for that decline in immunity. Routine boosters are given every 10 years, but for a dirty wound or for a potential exposure that might occur out of reach of a health center (e.g., travel abroad or wilderness camping), boosters are given every 5 years rather than every 10.

Safety. Local reactions at the injection site are common (50% or more) with all tetanus toxoid-containing vaccines. In younger infants, this local reaction may be accompanied by fussiness and/or decreased appetite. Occasionally, kids will have a more severe local reaction (termed an Arthus reaction) with a much wider area of swelling and more pain. Fever and more worrisome whole-body reactions are uncommon but were more common when the older form of pertussis vaccine (whole cell) was routinely given with diphtheria and tetanus vaccines (DPT); these and other pertussis vaccine-associated side effects of a tetanus toxoid-containing vaccine are discussed above under "Pertussis (Whooping Cough)." Two serious neurological reactions have also been associated with tetanus immunizations—both Guillain-Barré syndrome (GBS) (see above and chapter 3) and brachial neuritis (an inflammation of a nerve in the arm) are extremely rare.

Special Concerns and Controversies. The possibility that tetanus toxoid-containing vaccines cause severe and long-lasting reactions has been raised, but the main culprit in these reactions has been suspected to be pertussis vaccine, which is frequently given along with tetanus and diphtheria vaccines. The debate concerning a role for tetanus toxoid-containing vaccines in causing permanent brain damage, SIDS, infantile spasms, etc., is presented in "Pertussis (Whooping Cough)" above.

Some tetanus vaccine preparations contain trace amounts of a preservative called thimerosal. Concerns regarding this ingredient are discussed under "General Vaccine Concerns" below.

Varicella (Chicken Pox)

Germ causing the disease	Varicella-zoster virus
Type of vaccine	Live, weakend virus
Year first recommended for use	1996
How it is given	Under the skin
When it is given (2007 guidelines)	Between 12 and 15 months and between 4 and 6 years
Brand name(s) of individual vaccine	Varivax

Combination products (brand name) also containing this vaccine	With MMRV (ProQuad)

*At the time of this writing, a shortage of MMRV vaccine has made it temporarily unavailable (see "General Vaccine Concerns" below for a discussion of vaccine shortages).

Effectiveness. Although a benign and limited illness in most children, chicken pox can be severe. Prior to vaccine use, 4 million cases of chicken pox occurred in the United States each year, 11,000 patients were hospitalized, and there were more than 100 deaths due to the germ. In the 10 years between the introduction of the vaccine in 1995–1996 and 2004, population studies have shown an 80 to 90% drop in the occurrence of chicken pox among children. The death rate in the United States due to chicken pox declined from 145 cases each year to 66 cases each year in the early years (1999 to 2001) after the introduction of the vaccine according to a 2005 study published in the *New England Journal of Medicine;* an even further decline is expected as the use of vaccine has become even more widespread. A parallel study published in the journal *Pediatrics* found a 74% reduction in the U.S. hospitalization rate attributable to chicken pox in the same period compared with the preimmunization era (before 1996). The chicken pox vaccine protects against developing any chicken pox infection in 70 to 90% of kids; the protection rate against moderate to severe infection is 90 to 100%. If infection develops in an immunized child, the child is much less likely to have fever than an unimmunized child, and the contagiousness of an immunized child who develops naturally occurring chicken pox is significantly less than the contagiousness of a chicken pox-infected unimmunized child. However, infections continue to occur at a low rate even among kids who have received the vaccine (these cases are typically very mild). The likelihood of such a breakthrough infection (i.e., infection despite being vaccinated) increases with the length of time since the vaccination, as does the likelihood of developing a more severe case of chicken pox. In 2006, the CDC recommended a second (booster) dose of the chicken pox vaccine for kids between the ages of 4 and 6. Supporting the need for the second dose, a 2007 study published in the *New England Journal of Medicine* found that the chances of vaccine failure and breakthrough infection after a single dose rose nearly 13-fold between the first year and the eighth year after immunization, and the chances of severe disease in those kids doubled when 5 years had elapsed since immunization. Of concern, however, is a preliminary 2007 CDC report of an Arkansas school chicken pox outbreak occurring in fall 2006 that included a number of kids who had received *two* doses of the varicella vaccine. Although most of the cases were in children who had received only one vaccine dose and cases were mild in all kids, the failure of two doses to protect against infection was surprising.

Safety. Reactions at the site of the vaccine, including redness, swelling, and pain, are seen in 20 to 25% of kids; they are usually mild. Approximately one-third of children have these local reactions after the second dose of the vaccine. A few vaccine recipients (less than or equal to 5%) develop sores looking like chicken pox at the injection site, and a few kids have several scattered chicken pox sores on their body (also less than or equal to 5% of vaccine recipients). Fever occurs in 10 to 15% of children who receive the vaccine but in many cases may be unrelated to the vaccine (e.g., a concurrent viral illness). Serious reactions to the vaccine are very rare.

Special Concerns and Controversies. Very rarely (only a few cases have been proven), a child who develops sores looking like chicken pox after receiving the vaccine has been suspected of transmitting chicken pox from the vaccine to someone else. This is of primary concern only if there is a person with impaired immunity (see chapter 4) with whom a vaccinated child may come in contact. In that situation, contact with the person who has abnormal immunity should be avoided until the spots on the vaccinated child resolve.

Because this is a live, but weakened, virus vaccine, it should not be given to certain patients who themselves have abnormal immune systems.

Chicken pox in adulthood is a much more severe disease than in childhood. Some have feared that the vaccine given in childhood would shift the peak occurrence of the infection from kids to adults as the immunity from the vaccine waned with age. Indeed, the protection rate does drop with time, from 97% in the first year after receipt of the vaccine to 81% in the seventh or eighth year following immunization. These data are based on a single dose of the vaccine; the recently recommended second dose is expected to improve the duration of protection. A 10-year study following kids who received two doses showed that the incidence of chicken pox was cut by more than two-thirds compared with kids followed for 10 years after only a single dose. Again, chicken pox cases in both groups of kids who had received vaccine, either one dose or two, were generally very mild. This suggests that if there is a shift from children to adults as the age group most susceptible to chicken pox, the adult disease will not be of the severity we fear in unimmunized adults.

The chicken pox vaccine is a weakened live-virus vaccine with a virus that is known to establish a dormant state after both natural infection and immunization. This dormant virus can reactivate and cause a painful nerve root infection called shingles (see chapter 3). The occurrence rate of shingles after chicken pox immunization is very low—lower than the risk of shingles after natural infection with the virus. Additionally, vaccine-associated shingles is milder and less painful than shingles after natural infection.

Shingles is typically a disease of the elderly, a reactivation of their own naturally acquired chicken pox infection from when they were children. Some have raised the concern that giving the weakened chicken pox virus as a vaccine to children will shift the age of onset of shingles to younger adults. There is no evidence yet to suggest this will occur, but the vaccine has been given to children in the United States for only about 10 years; it will take longer to know if these children are at risk for earlier onset of shingles as adults. (Incidentally, a vaccine to prevent shingles has recently been approved for adults older than 60 years; the vaccine is similar to the chicken pox vaccine given to kids but contains a higher dose of the weakened virus; it reduces the occurrence rate of shingles by about half.)

One of the hypothesized reasons that shingles doesn't occur until an advanced age is that repeated exposures of young adults to children with chicken pox boost the immunity to the virus and help retain the virus in the dormant state. The theory is that older people have fewer boosting exposures to young kids with chicken pox, and immunity naturally wanes with increasing age. This raises the concern that as chicken pox in young children is prevented with the vaccine, perhaps adults of all ages will have fewer natural boosts to their immunity and will therefore reactivate their shingles more readily. In fact, at least one study published in 2005 in the journal *BMC Public Health* has found a near doubling of the occurrence rate of shingles in adults since the introduction of routine chicken pox vaccination in kids. It's not the kids getting the shingles, it's the adults in the community—perhaps (perhaps!) because they are less often exposed to kids with natural chicken pox infection and adult immunity is less boosted. It remains to be seen whether this is a proven association or a coincidence (see chapter 12).

General Vaccine Concerns

With the previous sections of this chapter as a reference for each of the 16 currently recommended childhood vaccines, we are now ready to tackle some of the more generalized concerns that apply to vaccines as a whole. In this section we'll review the following:

- Sick on the day immunizations are due—should your kids still get them?
- Allergic and anaphylactic reactions
- Thimerosol (mercury-containing) preservative and neurological abnormalities
- Potential cumulative risks of multiple vaccines
- Do vaccines contain cow products that may cause mad cow disease?
- Do vaccines cause autism?
- The National Childhood Vaccine Injury Act (NCVIA), the Vaccine Adverse Event Reporting System (VAERS), and the National Vaccine Injury Compensation Program (NVIC)
- Mandatory vaccines for school and day care entry
- Vaccine shortages
- Herd immunity

Sick on the Day Immunizations Are Due—Should Your Kids Still Get Them?

The two most common worries in this situation, ironically, are really not of significant concern at all: vaccines are not dangerous to sick kids, and vaccines still work (i.e., kids respond to them by forming the desired protective antibodies) even when your child is sick. In answer to another frequently asked question, there's no problem with giving a vaccine when a child is on antibiotics.

However, there is still a dilemma that parents and doctors face when a sick child shows up for shots. The yin and the yang of this dilemma are as follows—deferring vaccines during minor illnesses (such as the common cold, an ear infection, or a low-grade fever) increases the likelihood that kids will fall behind on their shots. It's often difficult to promptly get back to the doctor's office, for example, if your child misses his or her 12- to 15-month immunizations and isn't otherwise due to return for another clinic visit for 6 to 12 months. For some kids, deferring vaccines when the kids are actually in the office may mean never getting those shots. That's the yin. The yang is that we don't want the side effects of a vaccine—which can include fever, fussiness, and decreased appetite—to cloud our abilities as parents and doctors to monitor the course of the child's concurrent illness. In other words, if your child has a fever before getting the shot and the fever goes higher and is accompanied by loss of appetite after the shot, was it the vaccine, or should we be concerned that your child's illness is getting worse?

The compromise solution is that kids should get their shots if they have only a minor illness, even if that minor illness comes along with a low-grade fever. If, however, the doctor is concerned (i) that your child's illness is more than minor, (ii) that it may develop into something more serious, or (iii) that the fever may rise higher, vaccines should be put off until another visit. The next visit should be scheduled while your child is in the office and not more than a week or two later to improve the chances that it will actually occur. This is clearly a judgment call by the doctor and by you as parents. If you promise to promptly come back for a rescheduled vaccination visit, the threshold for deferring vaccines can be lowered—if your child is feeling ill and you're feeling overwhelmed, and the last thing you need is a fussy baby after a vaccination, talk with your doctor and assure him or her that you'll be back. Remember, the hard line on giving shots to kids even when they have minor illnesses is meant to guarantee that you will do your job as parents and immunize your kids. You are not putting your child at significant risk if you wait a week or two to make up the missed shot(s); longer than that and the chances go up that your child may slip through the cracks and fall further behind and that your child will encounter the naturally occurring infection that the missed vaccine would have protected against.

Beyond the general approach to receiving or deferring vaccines when your child is ill, certain specific vaccines have their own special considerations in the context of a sick child. For example, the new rotavirus vaccine (see above) should not be given when your child has a moderate or severe case of gastroenteritis, but a mild case does not preclude receiving the vaccine. A pregnant adolescent (I know, pregnancy is not an illness, but lots of people are just "sick" about a pregnant teenager, so I include this discussion in the current paragraph) should not receive the MMR vaccine or the varicella vaccine (see above). Kids with certain illnesses that require long-term aspirin treatment should not get varicella vaccine or the live influenza vaccine (the killed influenza vaccine is okay). Kids with asthma should not get the live influenza vaccine, either (again, killed vaccine is okay).

Allergic and Anaphylactic Reactions

From the earlier discussion in this chapter, you know that vaccines are prepared using many different techniques. It is also true that vaccines contain a myriad of different ingredients, including both germ components (i.e., those germs, or parts of germs, included in the vaccine to provoke the body to form antibodies) and nongerm components (carrier chemicals, preservatives, diluting fluids, etc.). You would be right to assume that the different preparations and ingredients allow for ample triggers of allergic reactions. However, allergic reactions, and especially severe allergic reactions (e.g., anaphylaxis), are extremely rare. Some of the implicated causes of allergic reactions and the vaccines in which they are found are shown in Table 7.2. Below are a few comments about the table.

- If a *severe* allergic reaction (e.g., anaphylaxis, where there is a total-body reaction that may include breathing difficulty and low blood pressure) develops immediately following receipt of any vaccine, that vaccine should never be given to your child again, even if the exact triggering component of the vaccine is determined. Hives and swelling in the mouth or throat following a vaccination are also reasons never to give that vaccine to your child again.
- In contrast, mild or moderate allergic reactions, such as local redness, pain, and swelling, are not reasons to avoid the same vaccine when it is next due to be given.
- Although measles, mumps, MMR, and MMRV vaccines are prepared in laboratory cells that come from chicken embryos, kids with egg allergies very rarely have adverse reactions to these vaccines because there aren't many actual egg proteins present in the cells that grow the vaccine

Table 7.2 Vaccine components implicated in rare allergic reactions

Vaccine component implicated	Vaccines in which component is found
Germ components	Diphtheria, tetanus, pertussis, perhaps others
Egg proteins	Measles, mumps, MMR, MMRV, influenza (both killed and live preparations)
Antibiotics	Polio, measles, mumps, rubella, varicella, MMR, MMRV
Thimerosal preservative	Diphtheria, tetanus, DtaP, hepatitis B (Engerix), 5-vaccine combination DTaP-hepatitis B-polio (Pediarix), influenza (killed)
Gelatin	Measles, mumps, MMR, MMRV, varicella
Yeast proteins	Hepatitis B

strains of virus. In contrast, influenza vaccines are actually grown in chicken eggs themselves, not laboratory cells with chicken origins, and should be avoided for kids with known egg allergies.

- Most vaccines for kids are now entirely thimerosal free. The few with thimerosal listed in the table contain only trace amounts. Concerns regarding thimerosal extend beyond allergic reactions; they are discussed separately below.
- Very rarely, kids with severe allergy to latex rubber may react to vaccines that come with rubber stoppers or to syringes with rubber plungers. If your child has a severe latex allergy, you should discuss this with your doctor before vaccines are given.
- Also very rarely, certain kids are very allergic to a wide variety of different triggers. The CDC, in its Pink Book on immunizations (http://www.cdc.gov/vaccines/pubs/pinkbook/default.htm), provides an exhaustive list of all the minor and trace substances found in each vaccine. Read this list with the right perspective—even though some of the contents of vaccines sound exotic and very chemical, so do the contents of most foods we eat and drinks we drink. Preservatives and stabilizers are in most things our kids consume, and vaccines are no different. If your child has a specific and severe allergy to one or more triggers (e.g., monosodium glutamate, present in both the varicella vaccine and the live influenza vaccine), however, this resource from the CDC can be a big help in sorting out which vaccines may pose a unique risk to your child. As noted above, the Pink Book is also an extraordinarily helpful reference for vaccine information in general.

Thimerosol (Mercury-Containing) Preservative and Neurological Abnormalities

Thimerosol prevents germs, such as bacteria and fungi, from growing in open vials of vaccine—it is a preservative. For that reason, thimerosal was (and in some cases still is [see below]) routinely used in vaccines produced in multidose vials (vials that are opened and used for numerous patients in the doctor's office until all doses are administered; this is in contrast to single-dose vials, which are opened, the contents are given as a single shot to a single patient, and then the empty vial is discarded). In recent years, thimerosol has received lots of scrutiny, not only as a potential cause of allergic reactions (see above), but also because its mercury content has provoked fears regarding mercury toxicity and brain damage in kids receiving vaccines. Virtually all childhood vaccines are now entirely thimerosal free. A few vaccines contain such a small trace of residual thimerosal (following a thimerosal removal process) that the CDC, the American Academy of Pediatrics, and the FDA advise that such vaccines be considered, for all intents and purposes, to be thimerosal free. Table 7.3 summarizes the vaccines that are entirely thimerosal free, those that are virtually thimerosal free, and those few that still contain thimerosal used as a preservative; as a caveat, this is a moving target—manufacturers are working to make all vaccines entirely

Table 7.3 Thimerosal contents of current childhood vaccines

Thimerosal content	Vaccine (brand names)[a]
None	• DTaP (Infanrix, DAPTACEL) • *H. influenzae* type B • *H. influenzae* type B-hepatitis B combined vaccine (Combivax) • Hepatitis A • Hepatitis B (Recombivax) • Influenza (FluMist, Fluzone pediatric prefilled syringe) • MMR • MMR-varicella combined vaccine (ProQuad) • *N. meningitidis* (Menactra) • Papillomavirus (Gardasil) • Poliovirus (IPOL) • Rotavirus • *S. pneumoniae* • Tdap • Varicella
Virtually none	• DT single-dose vial • DTaP (Tripedia) • DTaP/*H. influenzae* type B combined vaccine (TriHIBit) • DTaP/hepatitis B/poliovirus combined vaccine (Pediarix) • Hepatitis B (Engerix-B) • *N. meningitidis* (Menomune single-dose vial) • Td
Present (in concentration of 1 part per 10,000)	• DT multidose vial • Influenza (Fluvirin, Fluzone multidose vial) • *N. meningitidis* (Menomune multidose vial) • Tetanus (alone, not in combination vaccine)

[a]If specific brands are not named, all available brands for kids in the United States are included.

thimerosal free, and some of the content of this table may be outdated by the time you read it.

The concerns and allegations regarding mercury toxicity as a result of thimerosal exposure in childhood vaccines have centered on childhood neurological and developmental problems, such as attention deficit/hyperactivity disorder (ADHD), speech and language delays, and autism/autism spectrum disorder. The CDC and the National Institutes of Health commissioned the Institute of Medicine Immunization Safety Review Committee of the National Academy of Science to review this issue. The report of the committee was issued in 2001 (http://www.iom.edu/?ID=4705) and was based on all published medical and scientific literature pertaining to a possible association between thimerosal exposure and neurological and developmental abnormalities. In addition, the Institute of Medicine report reviewed research in progress that had not even been published yet in case there was work being done that might find results other than those that had already reached publication. The Institute of Medicine report noted that although thimerosal is 50% mercury by weight, the chemical form of mercury in the preservative is different than the form known to cause neurological problems from mercury poisoning; secondly, the Institute of Medicine concluded that even the neurological problems caused by mercury of the form *not* found in the vaccine do not resemble autism or ADHD. However, because a child receiving all the recommended doses of vaccine in the years prior to the current thimerosal-free (or virtually thimerosal-free) era could have had more mercury than was considered a safe level, the Institute of Medicine committee concluded that there was biological plausibility to the theory that thimerosal could be causing problems. Biological plausibility means that there is enough science known about mercury to make thimerosal exposure a possible risk to kids. In subsequent reports issued by the Institute of Medicine committee, they criticized their own use of the term "biological plausibility" because they became aware that parents and medical practitioners were overinterpreting the term's meaning—it was meant only to suggest that a theoretical scientific basis could be imagined, not that there was any actual, real-life evidence for an association.

The committee then tackled the more important question—did this possible, theoretical risk ever translate into *real* risk; that is, did any kids develop autism, ADHD, or speech and language delays as a result of vaccine exposures? The answer is that there is insufficient evidence to suggest that thimerosal-containing vaccines have caused or contributed to the aforementioned problems, but there is also insufficient evidence to entirely rule out a role for thimerosal. In other words, a complete review of all the published and unpublished research could not establish thimerosal as a problem or definitively conclude that the preservative was without risk. Beginning in 1999, on the recommendations of the U.S. Public Health Service and the American Academy of Pediatrics, manufacturers began making vaccines that are entirely thimerosal free (Table 7.3). As of 2001, all recommended routine childhood immunizations were available in entirely thimerosal-free preparations. The risk of the preservative has never been established, but the issue has been greatly alleviated by the new products.

An ironic outcome of the removal of thimerosal from childhood vaccines has been that the removal has been seen by some as an admission of harm or risk associated with thimerosal. The prompt action by the government and pharmaceutical manufacturers in producing thimerosal-free vaccines was in response to mounting public concern based on an entirely theoretical possibility of mercury toxicity from the preservative, i.e., there was not, and still is not, actual evidence to suggest that thimerosal is linked with any human neurological or developmental abnormality. Thimerosal-containing vaccines are still used in many parts of the world, and research regarding any possible adverse consequences of their use is still under way.

Potential Cumulative Risks of Multiple Vaccines

While surveys of parents regarding immunizations reveal that nearly 90% recognize the importance of vaccines and support giving them to their children, approximately one-quarter of all parents believe that kids get too many vaccines and that multiple vaccines can weaken the immune system.

It is not surprising that some parents are growing vaccine wary or vaccine weary. Within the first 6 months of life, infants may receive as many as 18 doses of immunizations (15 as shots and 3 orally) directed at preventing eight different infectious diseases. By 18 months of age, kids receiving the currently recommended regimen of vaccinations will receive about 25 vaccine doses (the 18 doses mentioned earlier plus 7 more) to prevent 14 different diseases (the 8 mentioned earlier and 6 additional infections). That sounds like a lot. It certainly seems that this must represent a huge potential burden on kids' immune systems—until you consider the naturally occurring "immunizations" against infections that kids encounter every day in day care or school, on the soccer field, and even in their own homes (see chapters 2 and 9). Recall from chapter 4 that it is not unusual for kids to have a dozen or more viral and bacterial infections every year—and those are the ones we know about because the kids have symptoms. In addition, kids are exposed to, and their immune systems react to, dozens if not hundreds of other infectious challenges every year. Then there are the noninfectious immune stimuli our kids encounter—almost everything they eat, drink, inhale, and touch causes immune system recognition. Indeed, our immune systems are capable of responding to millions (perhaps billions) of antigens, substances in the environment that must be interpreted by the body as friend or foe (see chapter 4). Whether friend or foe, your kids' immune systems react to these challenges without becoming overwhelmed or incapacitated. Now, in that context, the 14 infections that we immunize against in the first 18 months of life seem like a pittance to an immune system that is challenged millions or billions of times over a lifetime. However, those 14 infections, if not immunized against, are potentially life threatening.

Parents' concerns regarding overimmunization have focused on three potential adverse outcomes: (i) increased susceptibility to infections (i.e., vaccines depress the immune system and make a child more likely to be sick more often), (ii) increased risk of autoimmune diseases (i.e., vaccines confuse or misdirect the immune system, prompting a child's body to attack its own tissues), and (iii) increased risk of allergic disorders, such as asthma (i.e., vaccines confuse or misdirect the immune system to overreact to common everyday substances and become allergic to them). Increases in the occurrence rates of autoimmune and allergic disorders over the past 2 decades have generated numerous theories that seek to causally link environmental or other factors with the rising number of these diseases. Vaccines are among the factors that have been hypothesized.

The question of vaccines causing immune system harm, confusion, or misdirection has understandably been subjected to a great deal of scientific scrutiny. The Institute of Medicine Immunization Safety Review Committee addressed this issue, reviewed all the relevant medical literature, and concluded that multiple immunizations do not pose a risk to kids for increased infections (i.e., a depressed immune system) or for type 1 diabetes (an autoimmune disease in which the body's overactive or misdirected immune system attacks its own pancreas). Regarding allergies such as asthma, the committee felt that there was insufficient evidence to prove or disprove an association. The committee did not recommend any changes or further review of the immunization schedules for infants and children on the basis of concern regarding immune system health or on the basis of any other concerns.

The Hygiene Hypothesis. Over the years, some have worried that we are creating an environment that is too clean (yes, you read that right). Despite the clamor over environmental pollution and its impact on global warming and human health, there are those who are also concerned that we are raising our kids in a too-sterile environment. The hygiene hypothesis holds that our soaps, detergents, toothpastes, etc., contain too many antigerm substances, reducing the opportunity for our kids to naturally acquire immunity to germs in the environment. That aspect of the hygiene hypothesis is discussed more fully in chapter 9. With regard to vaccines, proponents of the hygiene hypothesis argue that our use of vaccinations to protect kids against what otherwise might be naturally occurring infections does not allow kids' immune systems to develop normally and therefore results in either autoimmune or allergic diseases. The Institute of Medicine's aforementioned report on multiple vaccinations concluded that there is no evidence to suggest that protecting kids from the 14 infectious diseases in the current childhood immunization regimen is making those kids too clean for their own good. Once again, the contribution of 14 vaccine-preventable infections (compared with the thousands of other immune stimuli, both infectious and noninfectious, our kids receive each year) to immune system development is miniscule.

Finally, vaccines are increasingly available in a combined form that reduces the number of injections that kids need to receive. Examples are the addition of polio vaccine and hepatitis B vaccine to the DTaP vaccines and the addition of varicella (chicken pox) vaccine to the MMR vaccines. These combinations are good for your kids—they mean less pain and lower risk of local reactions. The combined vaccines work just as well as the vaccines given individually and pose no known additional harm from cumulative exposure to vaccine components (see above). At the time of writing, a new filing for approval had been made to the FDA for a combination vaccine against diphtheria, tetanus, pertussis, polio, and *H. influenzae* type B; the CDC's Advisory Committee on Immunization Practices has recommended that the FDA license this vaccine (brand name, Pentacel), which has been widely used in Canada for many years.

Do Vaccines Contain Cow Products That May Cause Mad Cow Disease?

There are a number of vaccines that use cow products in the manufacturing process. These products include blood serum from cows, used to grow viruses in the laboratory, and gelatin, which is derived from skin and connective tissue of cows in the rendering process. The human version of mad cow disease (see chapter 3) is thought to occur because of consumption of the brain or spinal cord matter of cows infected with an abnormal protein called a prion (see chapters 1 and 3). There has been no link between other cow parts and human mad cow disease; most people consume, and feed their kids, cow-derived products daily. As noted in chapter 3, the human form of mad cow disease failed to reach the epidemic proportions that were projected, and fewer cases are being seen now than 10 years ago. There is no evidence, either biological or clinical, that vaccines pose any risk for human mad cow disease; surely with increasing immunization rates in recent years, any such association would have resulted in a parallel rise in mad cow disease, not the declines that have actually been observed.

Finally, the new rotavirus vaccine is constructed in the laboratory using a cow rotavirus to carry portions of human rotavirus strains. There is no association between cow rotaviruses and mad cow disease in cows or humans.

Do Vaccines Cause Autism?

As the diagnoses of autism and autism spectrum disorder have increased in recent years, there has been great attention to potential environmental sources to explain the rise. Two potential vaccine culprits have been alleged, and both have been studied extensively without finding a link to autism. The concerns center on the MMR vaccine and thimerosal-containing vaccines. Thimerosal and its possible relationship to neurological or developmental (such as autism) abnormalities have been discussed above.

The initial concerns regarding a possible link between the MMR vaccine and autism were generated by a report in 1998 of 12 children in England who developed autism within a few weeks of receiving the MMR vaccine; these children also had intestinal symptoms. The investigators hypothesized that the measles virus in the live, weakened MMR vaccine damaged the intestines, allowing toxins to escape from the intestines into the blood, reach the brain, and cause the developmental problems of autism. A subsequent study by the same investigators identified evidence of measles virus in more intestinal biopsy specimens taken from kids with autism than from kids without autism. These published reports generated enormous media attention, and an instant controversy was born. As with most controversies, new perspectives emerged with time, and data were more critically assessed. Many other experts reviewing those two initial studies identified numerous and serious flaws in the way in which those investigations were conducted and in the interpretation of the results. Among the flaws were the following.

- A very small number of patients were studied.
- The studies failed to compare the initial 12 patients with any of the millions of kids immunized during the same time frame who did not go on to develop autism.
- Overlap of the timing of MMR vaccination with the timing of the natural development of autism meant that kids diagnosed with autism were very likely to have recently, and coincidentally, received an MMR vaccine (see chapter 12).
- There were no kids studied who developed autism without having received the MMR vaccine in order to look for comparable intestinal symptoms or laboratory findings.
- The laboratory testing of intestinal biopsy specimens used a technique that has a propensity to produce false-positive results, meaning results that are erroneous because the test is too sensitive.
- It was not clear from the papers whether the kids in the intestinal-biopsy studies were appropriately matched for other characteristics that could have influenced the results.
- It was not disclosed whether the investigators examining the biopsy samples were themselves unaware of the kids' vaccine histories (blinded), a necessary precaution to prevent bias on the part of those doing the study (i.e., to prevent the natural tendency of people to find what they believe they should be finding rather than finding what's actually there [see chapter 12]).

In addition to finding flaws in the original studies and results, many lines of evidence have since accumulated to refute the association between measles vaccine and autism. They include a study from Denmark of more than 500,000 kids comparing those who received MMR vaccine with those who did not—the occurrence rates of autism in the two groups of kids were identical regardless of vaccine status. Other studies have looked at large numbers of home movies made of kids *prior* to their diagnosis of autism and *prior* to their MMR vaccination, as well as home movies of other kids matched for age and other factors who did not ultimately get diagnosed with autism. Expert autism investigators who didn't know whether the kids they were watching in the home movies went on to be identified as autistic were able to very accurately predict from the kids' behavior in the movies which kids would and which would not be diagnosed with autism—and all of those telltale behaviors occurred before MMR was given to any of the kids. Thus, if a child received the MMR vaccine a month after the home movie was taken and was diagnosed with autism 2 months after the home movie, there might be concern that the vaccine had something to do with the development of autism (or perhaps that the home movie caused autism, an epiphenomenon [coincidence] just as plausible, if timing alone is used, as the association with vaccines [see chapter 12]), but the movies prove that the signs of autism in those kids preceded the vaccine and preceded the actual time of diagnosis of the disorder. Finally, studies of identical twins and of toxic exposures that occur during pregnancy that are known to cause autism all suggest that autism is a condition that develops prior to birth and is subsequently diagnosed when kids' behaviors are more easily assessed, often around the time that vaccines are coincidentally given.

The Institute of Medicine of the National Academy of Science (http://www.iom.edu/?ID=4705) and the American Academy of Pediatrics have both reviewed all of the literature regarding autism and the MMR vaccine and have concluded that the vaccine does not cause autism or autism spectrum disorder.

The National Childhood Vaccine Injury Act (NCVIA), Vaccine Adverse Events Reporting System (VAERS), and the National Vaccine Injury Compensation (NVIC) Program

Necessarily, the approval of new vaccines for use in children and adults is based on large, but not enormous, studies of otherwise healthy individuals. The vaccines are assessed for safety and effectiveness, followed by FDA review and, if warranted, FDA approval. At that point, the vaccine is licensed; recommendations for its use are then determined by the CDC's ACIP, in collaboration with the American Academy of Pediatrics and the American Academy of Family Physicians (see above). Once the vaccine is put into use, however, the number of vaccine recipients is, to use the term from a few sentences ago, enormous. As millions of children are immunized, occasionally unexpected side effects, and/or benefits, may be appreciated that were not evident in the prelicensure trials of the product. One such example was discussed above: an earlier version of a rotavirus vaccine was voluntarily recalled after postmarketing surveillance—the collaborative effort between the pharmaceutical industry and the government that looks for unexpected events after a vaccine has already been licensed and put into widespread use—revealed an increased risk of an intestinal disease called intussusception.

The mechanisms for postmarketing surveillance of the type that detected the unanticipated side effect of that earlier rotavirus vaccine are well established and, in fact, legislated. In 1986, the NCVIA was approved by Congress. At the time, the primary purpose of the legislation was to assure an adequate supply of childhood vaccines—a growing number of lawsuits based on alleged damages from immunizations threatened to drive pharmaceutical companies out of the vaccine business. Sympathetic juries, confronted by neurologically damaged kids, were making large awards for injuries despite, in some cases, minimal medical or scientific basis for the claims.

The NCVIA established sources of government funds to compensate those truly injured by vaccines and established guidelines for determining which events are vaccine related. Under that legislation, doctors are mandated to keep precise and permanent records of all childhood vaccines given, including, of course, the date the vaccine was given, but also the vaccine's manufacturer and lot number. Doctors are also required to report all possible side effects (adverse events) experienced by patients following an immunization. The latter is accomplished through VAERS. Finally, claims may be made by injured parties under the NVIC Program—a no-fault system by which valid claims may be resolved. The system is funded by taxes paid by the manufacturers of the vaccines.

As an outgrowth of the NCVIA, VAERS has provided an extensive, indeed massive, system for identifying unanticipated vaccine effects. However, although massive, VAERS is also *passive*, meaning that the system depends upon the voluntary reporting of adverse events by patients to their doctors and on the compliance of doctors in reporting the events to the government. Shortly after the NCVIA was enacted, the CDC expanded its postmarketing efforts to include an *active* reporting system in which large populations (more than 10 million patients) are followed after receiving their vaccines with regular and periodic phone surveys seeking information on possible vaccine-related events; this system is known as the Vaccine Safety Datalink Project. Information about these and other government-sponsored vaccine safety initiatives can be found online (http://www.cdc.gov/vaccines/vac-gen/safety/default.htm).

If you believe your child has experienced a serious side effect of immunization, you should immediately report it to your child's doctor to trigger the recording and reporting functions of the NCVIA. The types of severe injuries that may occur with each vaccine, the time frame required for associating them with a vaccine, and the potential for receiving compensation for those injuries are provided on the government's Vaccine Injury Table (http://www.hrsa.gov/vaccinecompensation/table.htm). A cautionary note before reading this table—I wouldn't blame you if your reaction to seeing the table was to never vaccinate your child again. It's a horrific-sounding list of bad reactions, including anaphylactic shock (see above), encephalopathy (brain dysfunction), nerve damage near the vaccine injection site, long-term arthritis, bleeding disorders—and death! However, it is very important to point out that the reactions listed in that table, and eligible for government compensation, are extraordinarily rare—and that is an understatement of their true rarity.

Many other possible and actual vaccine side effects should be reported to the government under the VAERS system; reporting to VAERS does not mean that a medical condition occurring in the days, weeks, or months following a vaccine is due to the vaccine, but it is a way of gathering data which, cumulatively, could indicate a problem to those at the FDA and CDC who monitor these reports. Occasionally, such events are severe enough and provide enough evidence to justify inclusion on the Vaccine Injury Table; intussuception following the old rotavirus vaccine is one such example. Usually, however, the VAERS-reported events do not cause lasting harm and do not require, nor are they eligible for, financial compensation. These less serious reactions are important to track, however, because they provide a means for setting priorities in new vaccine development. As an example, the old DPT vaccine using whole-cell pertussis (see above) caused many, many more undesirable side effects (the vast majority of which caused no lasting damage) than the newer acellular-pertussis-containing vaccines. The impetus for developing the acellular pertussis vaccines was the large body of accumulated VAERS data linking the old vaccines to those many side effects.

The data gathered under VAERS are publicly available (all personal identifiers are of course redacted). This is a double-edged sword. On the one hand, readily accessible information on vaccines fulfills a parent's right to know about any and all events that may or may not be related to a vaccine. Remember, however, that VAERS does not prove an association of an adverse event with a vaccine; it only reports two chronologically related occurrences: immunization and the development of a medical condition. Sometimes, such as for a local reaction at the site of an injection, the two occurrences are almost certainly related. Other times, because vaccines are given frequently during childhood and because a variety of childhood diseases occur during the age range of vaccine administration, there may be no relationship between a vaccine and an event—a true coincidence (see chapter 12). Overinterpretation of the VAERS data, for example, has prompted some parent advocacy groups to claim that there are "hot lots" of vaccine—certain lot numbers associated with a higher rate of bad reactions. That is a flawed interpretation of the VAERS database. The size of any vaccine lot cannot be determined from the VAERS data. Clearly, a very large lot of vaccine from a manufacturer will have more reported side effects than will a smaller lot of vaccine from the same or another manufacturer. The FDA monitors VAERS reporting with the information on lot sizes; the FDA is empowered to remove any bad lot of vaccine should one be identified. Such a mandated recall has never been necessary to date.

Mandatory Vaccines for School and Day Care Entry

Although the federal government, through agencies like the FDA and the CDC, is responsible for approval, licensing, and monitoring of vaccines, the immunization requirements for day care and school entry are determined by each state. The use of mandatory immunizations prior to school entry is the single most important factor in achieving the high rates of childhood immunization, and the low rates of vaccine-preventable infections, that exist in the United States today. Initially enacted to control and ultimately eliminate smallpox in the United States, school entry vaccine laws have helped contain national outbreaks of polio, measles, and other infections readily spread in the schools and from the schools to the rest of the community. School entry vaccine requirements protect the kids in the schools, and because those kids are frequently the reservoirs of community-wide infection, the protection extends much further than the classrooms (see "Herd Immunity" below).

While all states have mandatory school entry vaccines, not all have mandatory day care immunization laws. For your own state, you should check with your child's physician, with your state health department, or with the Immunization Action Coalition (http://www.immunize.org/laws/), a private organization that works in collaboration with federal and state governments and major medical organizations around the country.

The designation of a vaccine as mandatory can have its own repercussions and controversy—see, for example, the discussion above regarding the new vaccine to prevent human papillomavirus infections that cause genital warts and cervical cancer. Similar controversy exists regarding the mandatory designation for hepatitis B immunization in infancy (see above). Additionally, a mandatory vaccine must be available to and provided for uninsured kids; it also must be provided to government-insured kids in the Medicare and Medicaid programs. These insurance issues make the determination of a vaccine's required versus nonrequired status for day care and school entry both a public health issue and a government budget issue.

Colleges and universities are increasingly requiring up-to-date immunizations for entry. The immunizations must include all childhood vaccines, as well as those now recommended for adolescents (see individual vaccines above). This should not be surprising; the close quarters of dormitories and fraternity/sorority houses, as well as the intimacy of person-to-person contact in those settings, mimic those of day care centers (much to the chagrin of college students' parents!) (see chapters 2 and 9).

Exemptions to mandatory immunization can be obtained by parents for special medical circumstances in all 50 states, as well as for religious objections in 48 states (not available as of January 2007 in West Virginia or Mississippi) and personal or philosophical reasons in 20 states (http://www.vaccinesafety.edu/cc-exem.htm). If an outbreak of a vaccine-preventable disease occurs, kids who have received exemptions from receiving that vaccine may be asked to stay home from school or day care until the outbreak is over.

Vaccine Shortages

Influenza vaccines, recommended annually, have been in short supply in certain years, prompting headline-grabbing pictures of long lines of the elderly waiting for the few available doses. In recent years, numerous other vaccine shortages with greater impact on children have occurred, albeit with less dramatic press coverage. These childhood vaccine shortages have forced temporary changes in the government's recommended schedules, as well as the establishment of prioritization guidelines for distributing vaccines. The affected vaccines have included the DTaP, MMR, varicella, *N. meningitidis* conjugate, *S. pneumoniae* conjugate, and hepatitis A vaccines. The reasons for the shortages are multifactorial and include breakdowns at individual manufacturers' production lines, contamination of certain production processes, and unanticipated high demand. National agencies have appointed task forces to identify ways of guaranteeing uninterrupted vaccine supply and preventing shortages. One issue of great concern is that for a number of recommended childhood vaccines (polio, MMR, varicella, rotavirus, *N. meningitidis* conjugate, *S. pneumoniae* conjugate, and human papillomavirus) there is only a single manufacturer—an interruption in that company's operations can cause a severe shortage.

When shortages develop, it is the responsibility of the physician and of the parents to make sure that any delayed or deferred immunizations are made up after the shortage is relieved.

Herd Immunity

This section, despite its name, has nothing to do with the mad cow disease issue discussed above. Vaccines protect those who receive them, and when immunization rates in a community are high, vaccines also protect those who are not immunized; a high rate of immunization in a community limits the spread of infection and the exposure of susceptible people to the infection. Here's the way it works. Let's assume that child A develops chicken pox. He exposes child B, who has had the varicella vaccine. When child B now has contact with child C, who has not had the vaccine, child C is protected by child B's immunity—that is, because child B was vaccine protected and did not develop chicken pox after exposure to child A, child C will also not develop chicken pox, at least not with this exposure. Now, alternatively, if child A exposes child D, who has not had the varicella vaccine, child D will likely develop chicken pox, and when child D comes in contact with the same non-immunized child C who did not get infected earlier, child C now develops the infection. The more child B children there are in the community, i.e., those who have had the vaccine, the lower the chances that the child C kids will ever contract the infection even if child C kids don't get the vaccine.

Because of the herd immunity effect (herd refers to the mass of people who benefit by the immunization of the majority, but not necessarily all, of the susceptible individuals in the herd), smallpox was eradicated worldwide and polio has been eliminated from the United States and the Western Hemisphere. Measles, mumps, German measles, and most other vaccine target germs have also been kept at bay (but not yet eliminated) thanks to herd immunity. When immunization rates fall for those infections, case numbers increase, reflecting the lower level of herd protection. Perhaps the most striking drop in herd immunity (and rise in number of infections) in recent years has been that of pertussis. As people became more concerned about the side effects of the old whole-cell pertussis vaccines (see above), they stopped immunizing their kids. Herd immunity fell, and case numbers rose throughout the United States and worldwide.

The beneficial herd immunity effects of vaccines on the wider community, including those not immunized, appear to be realized with the newer vaccines, as well as with the classics. In 2002, a study published in the *Journal of the American Medical Association* found that the occurrence rate of hepatitis A among adults in a California community fell with the increasing administration of hepatitis A vaccine to kids. Similarly, a 2005 study published in the same medical journal showed that pneumonia due to *S. pneumoniae* bacteria in elderly adults fell concomitant with the widespread use of the newly recommended *S. pneumoniae* conjugate vaccine for children. The exposure risks of the whole herd, including adults, went down when a high percentage of kids became immune thanks to these newly introduced vaccines.

Parting Shot

Forgive the pun. Vaccines are good for kids, good for society, and safe. Clearly, there are caveats attached to all of those statements, but the caveats are minor and the overriding principles remain true. The National Commission on Preventive Priorities ranked seven widely employed societal disease prevention strategies on a scale of 1 to 10, with 10 being judged the most effective. The commission also estimated the percentage of the population receiving each prevention initiative. The top three most effective strategies, each receiving a 10, were aspirin for the prevention of heart attacks, tobacco screening and intervention, and childhood immunizations. Aspirin therapy and tobacco screening/intervention are received nationally by 50% and 35% of the population at risk, respectively, whereas childhood immunizations reach more than 90% of the target population. This makes kids' vaccines the most widely available and most beneficial disease prevention strategy in this country. Investigators writing in the *Archives of Pediatric and Adolescent Medicine* in 2005 calculated the financial savings directly accruing from 10 childhood vaccines (i.e., the disease burden prevented by immunizing against diphtheria, tetanus, pertussis, measles, mumps, rubella, *H. influenzae* type B infection, polio, hepatitis B, and varicella) at nearly $10 billion annually. When the days of missed work and school are factored in, the annual savings are more than $40 billion. There is $16.50 saved by society for every $1 spent on vaccines.

We are a healthier nation and world today because of childhood immunizations. Ask your kids' grandparents and great-grandparents; there aren't many in those great generations who will argue against vaccinating kids. They remember what it was like in the days before we gave shots.

Further Reading

Advisory Committee on Immunization Practices of the CDC. 27 February 2007. Child and Adolescent Immunization Schedules. http://www.cdc.gov/vaccines/recs/schedules/child-schedule.htm.

American Academy of Family Practitioners. 2007. Immunization Resources. http://www.aafp.org/online/en/home/clinical/immunizationres.html.

CDC. National Immunization Program Vaccine Safety. http://www.cdc.gov/vaccines/vac-gen/safety/default.htm.

CDC. February 2007. Epidemiology and Prevention of Vaccine-Preventable Diseases. (Pink Book.) http://www.cdc.gov/vaccines/pubs/pinkbook/default.htm.

Immunization Action Coalition. http://www.immunize.org/index.htm. (Educational site funded by both government and private sources.)

Institute of Medicine, National Academy of Sciences. 15 March 2007. Immunization Safety Review. http://www.iom.edu/?ID=4705.

Johns Hopkins University Institute for Vaccine Safety. 22 June 2007. http://www.vaccinesafety.edu/. (Educational materials.)

National Network for Immunization Information. 2007. http://www.immunizationinfo.org/about/index.cfm. (A private, not-for-profit affiliation of numerous medical societies, including the Infectious Diseases Society of America, the Pediatric Infectious Diseases Society, the American Academy of Pediatrics, the American Nurses Association, the American Academy of Family Physicians, the National Association of Pediatric Nurse Practitioners, the American College of Obstetricians and Gynecologists, the University of Texas Medical Branch, the Society for Adolescent Medicine, and the American Medical Association.)

National Vaccine Information Center. http://www.909shot.com/. (A private, not-for-profit consumer advocacy group; the most vocal and prominent organization among vaccine skeptics or opponents, with an emphasis on examining vaccine safety.)

Pickering L. K., C. J. Baker, S. S. Long, and J. A. McMillan (ed.). 2006. *Red Book: 2006. Report of the Committee on Infectious Diseases,* 27th ed. American Academy of Pediatrics, Elk Grove Village, IL.

8

Over-the-Counter or Over the Top: Symptom Relief

These days, supermarket and drugstore aisles can look more like the public library, with parents studying label ingredients and trying to interpret the nuances of claims on the boxes of their kids' medicines. It's a large public library, too. There are more than 100 classes of over-the-counter (OTC) drugs (each class of drug refers to a specific function of the medicine, e.g., nasal decongestants, antihistamines, laxatives, and antacids are each a separate class of drug). The 100-plus classes of drugs, in turn, include more than 800 active ingredients that have been incorporated into more than 100,000 OTC drugs. Our discussion will be limited to the OTC medicines that are used to treat the symptoms of infection—still an ambitious list. There are hundreds of symptom relief products that are marketed for a host of viral diseases ranging from colds and flu to diarrhea and stomach flu.

This will be a short chapter—this book is about the prevention and treatment of your kids' infections, and most of the medicines discussed in this chapter neither prevent nor treat infections. Rather, they treat (or purport to treat) the symptoms caused by germs, which may (or may not) make your kids feel better. However, none of these medicines will make the germ go away faster than it would if the medicines were not used. Unfortunately, though, for many infections, especially those due to viruses, symptomatic relief is the only treatment available, so familiarity with OTC symptom relievers can be useful.

Regulation of OTC Medicines

All of the symptom relievers discussed in this chapter are regulated by the Food and Drug Administration (FDA), albeit by a different process from the one used to regulate prescription medicines. For most OTC medicines, the FDA publishes monographs that provide guidelines for the entire class of medicines (e.g., nasal decongestants or antacids). Each FDA monograph covers the relevant active ingredients for that class of medicines, as well as the doses, formulations (how the medicine is given—topical, oral, pill, liquid suspension, etc.), labeling (everything from the claims made on the box and in advertisements to the fine-print enclosures you find inside the box), and the testing required to prove safety and effectiveness. If a manufacturer can demonstrate that all the requirements of the monograph covering their OTC medicine have been met, further FDA approval procedures are not required. A smaller number of OTC products were first approved by the stricter and more formal New Drug Application process required for prescription drugs and then subsequently received approval from the FDA to be sold without a prescription.

It is important to understand that OTC products may become available to children, with FDA consent, on the basis of safety and effectiveness studies conducted only with adults; in those cases, kids' doses are approximated based on adult doses, with adjustments for size and weight. There are inherent problems with this method of determining pediatric treatments because kids are not just little adults; their metabolisms are different, and the effects and side effects of a medicine in kids may not be predictable based on adult experience.

The involvement of the FDA in approving OTC drugs stands in stark contrast to the route by which dietary supplements, "nutriceuticals", vitamins, minerals, probiotics, prebiotics, and herbal remedies reach the shelves of the supermarkets and health food stores. Those mysterious substances almost entirely (and sometimes dangerously) bypass FDA scrutiny, as described in detail in chapter 10. It's to that chapter that you should turn now if you're looking for information about zinc, echinacea, vitamin C, or products that mix and match these ingredients and other herbs (for example, those effervescent tablets that are supposed to protect your kids from the common cold and treat colds when they develop).

In this chapter, we'll cover the FDA-regulated OTC drugs available to treat the following germ-induced symptoms:

- Fever
- Nasal congestion
- Runny nose and runny eyes
- Cough
- Diarrhea

We'll also briefly discuss the use of OTC products for:

- Cuts and scrapes
- Warts

OTC products to treat fungal infections, such as athlete's foot, jock itch, and vaginitis, are discussed in chapter 6.

Fever

Not all fevers need to be treated (see chapter 3), and not all treatment needs to be with drugs. Tepid baths, cool drinks, and reducing layers of clothing can make kids with fever feel better. It's because fever often makes kids feel crummy that the market for antipyretics, as fever-reducing medicines are known, is so large. As a bonus, the antipyretics are also effective pain relievers. Hence, if an antipyretic is given for the fever caused by an ear infection, for example, the medicine will also reduce the ear pain; ditto for a sore throat or headache accompanied by fever.

There are two antipyretics commonly used for kids: acetaminophen (brand names, Tylenol and others) and ibuprofen (brand names, Motrin, Advil, and others). In a review published in the *Archives of Pediatrics and Adolescent Medicine* in 2004, investigators assessed the results from 10 different studies comparing acetaminophen with ibuprofen. Both medicines were found to be effective at reducing fever (and relieving pain), but ibuprofen offered a more sustained reduction of fever—more kids remained without fever at 4 and 6 hours after ibuprofen treatment than with acetaminophen. There was no difference in side effects between the two medicines in this analysis of multiple studies. (A 2007 study showed ibuprofen to also be more effective than acetaminophen in the relief of kids' pain from trauma, such as broken bones and sprains, but pain relief outside the context of infections is beyond the purview of this book.)

Because both ibuprofen and acetaminophen are effective in reducing fevers and act by different chemical mechanisms, it has become common practice to alternate dosing of the two, allowing treatment every 2 to 3 hours with one or the other of the medicines; surveys have shown that as many as half of pediatricians recommend this practice to their patients. Most experts, however, discourage alternating treatments—there are few data to support either the safety or greater effectiveness of alternating medicines versus the proper use and dosing of a single fever-reducing drug, and there is a risk that confusion in dosing will result in too much medicine being given; each drug can have serious complications if given in excess. Additionally, each drug comes in different dosing formulations (i.e., different amounts of the medicine in different volumes of liquid or pill sizes), further complicating the safe use of alternating schedules.

Finally, it is very important to take this opportunity to remind you that aspirin should never be used for fever reduction (or pain relief) in children. Aspirin therapy has been found to be strongly associated with the development of Reye's syndrome (see chapter 3) in children with infections, particularly influenza and chicken pox. Aspirin has very few uses in children any longer, and those kids who do need aspirin have unique and specific reasons for needing it—the vast majority of kids should grow up never receiving aspirin for anything. The National Reye's Syndrome Foundation provides a list of many OTC products that contain aspirin or aspirin-like chemicals (salicylates) and that should therefore be avoided for kids (http://www.reyessyndrome.org/aspirin.htm).

Nasal Congestion

Decongestants work by reducing the blood flow to the nose and thereby reducing swelling in the nasal passages (the fluid that causes swelling is the result of heavy blood flow to an area of the body where cells that fight infection and/or inflammation have been summoned [see chapter 4]). Decongestants can be given by two different routes, orally (by mouth) and topically (into the nose).

Oral Medicines for Nasal Congestion

If you have tried to purchase decongestants lately, for yourself or for your kids, you may have felt like the prime suspect in a TV crime drama. In 2006, federal legislation mandated that the most widely used ingredient in nasal decongestants, pseudoephedrine, be pulled from the open shelves and stocked behind the counter; the purchaser must now be screened, and quantities are limited with each purchase. Like the hoops everyone has to jump through at airports, this drugstore adventure is brought to you by the bad guys who managed to find a way to take advantage of the rest of us. Pseudoephedrine, it turns out, can be used to make methamphetamine, and home chemists were buying large quantities of the cold remedies containing pseudoephedrine for their meth labs. The most widely used decongestant products have now been reformulated using a different ingredient, phenylephrine, which has returned the familiar-looking decongestant packages to the open shelves but without the pseudoephedrine.

The use of decongestants for kids is complicated by three facts: (i) while *pseudoephedrine* products have been shown to have some beneficial effects for adults with nasal congestion due to the common cold, there are few, if any, data to support the use of pseudoephedrine products for kids; (ii) the data supporting the effectiveness of *phenylephrine* products (the ingredient that replaced pseudoephedrine in many OTC decongestants) in adults are far less compelling than the evidence for pseudoephedrine, and studies with kids using this substitute ingredient are entirely lacking; and (iii) the side effects of decongestants in kids can be serious and include high blood pressure, hyperactivity, and sleep disturbance.

Topical Medicines for Nasal Congestion

Products containing oxymetazoline are marketed for kids older than 6 years of age for direct application into the nose. As with decongestants given by mouth, topical oxymetazoline acts by restricting blood flow and swelling in the nose. There are no data to prove that this treatment is effective in kids, and the side effects can be serious (including blurred vision, sleep problems, and irritability). The medicine can be given for only a few days, after which worse congestion (rebound congestion) may be an additional side effect.

Combination Products

Many OTC symptom relievers for the common cold come in combinations—decongestants with antihistamines, cough medicines, fever and pain relievers, etc. These products magnify the risk to kids by not only combining the intended beneficial effects, but also combining the side effects. As a general rule, you should avoid using combination products for kids. There are two reasons for this: (i) even most of the single symptom-relieving products have not been proven to be effective for kids and (ii) if you're going to treat symptoms, you should treat only those symptoms that your child actually has; combination products often include medicines that are unnecessary for the particular illness your child is experiencing. There is a possible exception to this general rule (see the discussion of combinations of decongestants and antihistamines under "Runny Nose and Runny Eyes" below).

The Bottom Line

If you choose to treat your kids' symptoms of the common cold with decongestants, be aware that there is little if any evidence to prove that it will make a difference, that the risk of side effects is very real, and that combination products should be avoided. There is one more factor to consider. Every year, hundreds of kids overdose on OTC common-cold medicines, either because they are inadvertently given too much by their parents or because the medicines taste good and are not properly stored out of the reach of young kids. A group of prominent pediatricians and child health advocates in Maryland have filed a petition with the FDA to ban all OTC common-cold products for kids under 6 years old. They note that in 2004 alone, 900 kids in Maryland under the age of 5 years overdosed on OTC common-cold medicines. In response, the FDA issued a statement in March 2007 that it was reevaluating the use of OTC common-cold medicines for children.

Runny Nose and Runny Eyes

Antihistamines target runny noses and runny eyes by blocking the effect of histamine, a chemical produced by the body in response to some infections and allergies. Histamine causes an increase in blood flow and the accumulation of fluid and other chemicals that perpetuate the runny nose and runny eyes. In addition to runniness, itchiness is another symptom of histamines that is also blocked by antihistamines. The most prominent side effect of antihistamines is sleepiness— in fact, this may be a beneficial side effect in the eyes of some parents whose kids aren't sleeping well because of their cold symptoms. Studies examining the benefits of antihistamines in relieving symptoms of the common cold have found the following.

- Antihistamines alone are ineffective for children and adults.
- Antihistamines in combination with decongestants are ineffective for young children.
- Antihistamines in combination with decongestants are modestly effective for older children and adults.
- Sleepiness (sedation) is the most common side effect of antihistamine-containing products for children and adults. The antihistamines discussed here are all so-called "first generation" medicines and include diphenhydramine, brompheniramine, and chlorpheniramine, among others. Another group of antihistamines, the second generation, do not cause sleepiness; these newer antihistamines are widely marketed for relief of seasonal allergies. However, they are entirely ineffective against

symptoms of the common cold or other infections of the respiratory tract. Second-generation antihistamines include loratadine (brand names, Claritin and Alavert) and cetirizine (brand name, Zyrtec), among others.

Antihistamine-containing products should not be given to young children (under 6 years of age). The modest benefit of combination antihistamine-decongestant products for older children must be weighed against the side effects of each class of medicine in the combination; sleepiness is the most significant antihistamine side effect. Adolescents should not drive a car when taking antihistamine-containing products.

Cough

Two types of medicines are most commonly used for cough symptoms: expectorants and antitussives. Expectorants thin the secretions in the airways and loosen the cough, helping to get the mucus up and out; the most widely used expectorant is guaifenesin. In contrast, antitussives are cough suppressants—they inhibit the cough reflex. Dextromethorphan is the most common cough suppressant in kids' medicines, but there are others, as well.

Expectorants should not be used for dry coughs because there is no mucus to thin; cough suppressants should not be used for wet coughs because coughing is an important mechanism for clearing out the secretions. Despite their different effects and the different types of coughs targeted by each medicine, the many, many commercial products that contain these ingredients do not distinguish one from the other in their packaging or marketing—coughs are coughs in the eyes of the manufacturers. This confusion, intentional or unintentional, over which type of cough medicine is appropriate for which type of cough is moot for kids, however. There's no evidence that *either* type of cough medicine ingredient works better than placebo in kids with coughs and common colds. This one's easy; your kids don't need, and probably won't benefit from, cough medicines. Side effects also occur, as do overdoses (these medicines are also tempting for young kids who find open bottles on the countertop). A prominent group of cough experts issued guidelines in January 2006 recommending that kids under the age of 15 years not be treated with OTC cough medicines.

Many combination common-cold products contain cough medicines (see the comments about combination products in "Nasal Congestion" above). Cough drops help keep the throat moist and may ease the tickle behind some older kids' coughs, but cough drops are a choking hazard for younger kids.

Diarrhea

Besides the discomfort and inconvenience of diarrhea, the main concern associated with this symptom is the potential for dehydration, particularly in babies and young kids. When your child develops diarrhea, work with his or her doctor to tailor an age-appropriate fluid replenishment plan to prevent dehydration.

While there are several types of OTC medicines that slow or stop diarrhea, none are routinely recommended for kids, and some can be dangerous. For example, diarrhea due to certain bacteria (see chapter 3) may be worsened, and undesirable whole-body effects can occur, when a medicine such as loperamide (brand name, Immodium and others) is used. Loperamide slows the movement of feces through the bowel—this slowed motility of the intestines can let the bacteria and their toxins do more harm. Other OTC antidiarrheal medicines contain aspirin-like chemicals (salicylates) that should be avoided with kids (see the discussion of aspirin under "Fever" above); the FDA required all manufacturers of salicylate-containing antidiarrheal medicines (e.g., bismuth subsalicylate; brand names, Pepto-Bismol, Kaopectate, and others) to remove directions for use in children (under the age of 12) from the packaging and labeling of those products; many experts feel that even kids older than 12 should not be treated with salicylate products. Still other antidiarrhea medicines contain lead, which is also bad for kids. Binding agents have shown no benefit over temporary age-appropriate changes in diet as recommended by your child's doctor.

The bottom line (a pun of sorts) is, don't use antidiarrhea medicines for your kids; work with your children's doctor to modify their diet and attend to their fluid needs.

Cuts and Scrapes

Kids will be kids. They will get repeated cuts and scrapes as they do what they should be doing, playing outside, roughhousing, and generally messing around. The basic treatment for minor cuts and scrapes is the simple, but thorough, washing of the wound with soap and water (see chapter 9). Obvious debris should be removed, with flowing water or even tweezers if necessary, and then the wound should be covered with a clean bandage and allowed to heal. Most pediatricians and family doctors recommend applying a topical antibiotic ointment or cream to wounds after washing them and before bandaging them. There are few data to prove that this is more effective than simple washing, but the FDA, in approving topical antibiotics for first aid, has cited a small study showing that topical antibiotics resulted in fewer subsequent wound infections than placebo. Experience generally seems to support the modest benefit with minimal risk of this type of intervention. There are more than a half-dozen OTC antibiotic ointments and creams approved by the FDA for this purpose. The application of a thin layer of one of these products over cuts and scrapes also helps prevent the bandage or gauze from sticking to the wound, a gesture your kids will appreciate. There has been debate about whether these topical treatments speed wound healing—data are inconclusive, but if there is such a benefit, it's a subtle one.

It's okay to use topical antibiotic medicines as first aid, both for their potential to prevent infection and for their lubricating action on the bandage (see above). Allergic reactions and hypersensitivity to topical antibiotics occur, and you should immediately stop using the creams or ointments if you see redness or swelling. Contact your doctor. Those symptoms can mean either a reaction to the medicine or a spreading infection at the site of the injury that may require more aggressive therapy.

Warts

Warts are caused by papillomaviruses (see chapter 3). Although there is no effective antiviral treatment for papillomaviruses, the warts themselves can be treated by physical means, both in the doctor's office and at home. The two traditional strategies have been freeze therapy and chemical therapy. Recently, duct tape has also been shown to be somewhat effective in wart removal, representing an entirely different type of "counter" (the hardware store counter) for OTC therapy.

OTC chemical wart treatments for home use contain salicylic acid, a chemical related to aspirin, which is locally irritating and removes layers of the wart with each application until the wart itself is gone; this can take many weeks or months. In addition, salicylic acid use for kids poses a risk for Reye's syndrome (see "Fever" above). There are no data on the actual risk of topical aspirin-like chemicals causing Reye's syndrome, but if possible, they should be avoided for kids, particularly when a child has the flu or chicken pox. In the doctor's office, a stronger and more blistering chemical treatment can be given, which sometimes works faster than the home therapies.

Doctors use liquid nitrogen to freeze the superficial layers of the wart, causing them to peel off over time; repeated treatments gradually reduce the wart until it's gone. Home freeze therapy is also now available for wart removal; thorough comparisons of home freeze treatments with home chemical treatments have not been done.

Protocols for the use of duct tape include a 5- or 6-day continuous application of tape to the wart, followed by removal of the tape, soaking, scraping, and reapplication of the tape. Like other methods, this approach may take weeks or months for complete wart removal.

Most warts resolve on their own (see chapter 3) and, if they are not in a painful location (such as the bottom of the foot), needn't be treated. Warts are contagious, however (see chapter 3), both from person to person and from one location on a child to another on the same child, so care should be taken to wash the hands after touching the wart.

It's Counter-Intuitive

As with everything else, your common sense and intuition must prevail when you stand before the dizzying array of colorful packages—and even more colorful package claims—on the supermarket and drugstore shelves. Just because a product is available, and even FDA approved, as are the OTC drugs discussed in this chapter, that doesn't mean it's appropriate for your child. Although you don't need a prescription to buy these medicines, your child's doctor is still a valuable resource in determining the appropriateness of any treatment for your kids. For most symptoms of mild infections, no therapy at all is needed. For others, nondrug treatments, like rest, fluids, good nutrition, and loving attention, suffice (see chapters 10 and 11). If you decide to use a medicine to treat the symptoms caused by germs, treat just the symptoms that are most bothersome for your kids—combination products targeting multiple symptoms compound the risks of the medicines and are rarely, if ever, needed. As unnecessary as most OTC drugs are, dietary supplements, herbal remedies, and vitamin or mineral supplements are usually even more unnecessary; remember, holistic, herbal, and homeopathic do not necessarily translate into healthy (see chapter 10).

III. WEAR YOUR BOOTS IN THE RAIN

9

Sanitary Sanity: Personal, Household, and Community Hygiene

Congratulations, dear reader, you have now reached the "punch line" chapters of this book. Having mastered the names and addresses of offending germs, the routes that germs take to get into us, the most common and most important diseases that germs cause, the defenses our bodies mount against them, and the vaccine and medicine antidotes available to us, you are now ready for reconciliation. It is time to reconcile medicine's sophisticated molecular approaches to disease prevention with Mom's low-tech, simple interventions—time to reconcile the scientific truths about germs and our war against them with the myths and near myths that we grew up believing about infections.

This chapter begins the reconciliation with a discussion of hygiene. When your mother said to wash your hands before dinner and after using the bathroom, you groaned and then usually stomped over to the sink to do what she said. You then grew up to train your own kids the same way your mother trained you. Now, we'll learn the scientific basis for Mom's advice, as well as highlight the many new nuances that personal hygiene has taken on in recent years—from the antibacterial-soap controversy to the realization that sharing is a *bad* thing for kids (what would your mother think about that?!).

After dissecting personal hygiene, we'll move on to hygienic household habits—how can you make your home a safer (from germs) place for your kids? We'll do a room-by-room analysis. And, finally, we'll take a look at community-wide hygiene issues, including the quality of the air our kids breathe and the water they drink, and how they influence our kids' germ exposures. Also in the section on community hygiene, we'll take a special look at precautions you should take for your kids in high-risk germ locations ("hot zones"), like day care centers, schools, petting zoos, doctors' office waiting rooms, and college dormitories.

Finally, we'll tackle the hygiene hypothesis, a theory that says our kids are too clean. Yeah, right.

Personal Hygiene

In Japan, the custom of ojigi, or bowing, substitutes for handshaking as the traditional greeting. Bowing is a sign of respect, with great significance attached to the details—the inclination and depth of the bow say things about social class, gratitude, or remorse. Although perhaps it is unintended as such, ojigi is also a brilliant infection prevention strategy, since bowing bypasses the hand-to-hand contact proven to be the foremost risk factor for transmitting the majority of viruses and bacteria (see chapter 2). Incidentally, although infection prevention may be an *inadvertent* benefit of bowing, many Japanese *intentionally* act to prevent spreading their own germs by wearing a mask when they have a cold—a courtesy for the protection of those they might expose!

Hand Washing

When you hear (or read) an infectious-disease specialist like me discussing hand washing, a severe obsessive-compulsive disorder probably comes to mind. We doctors are, indeed, hooked on hand hygiene. That's because study after study for decades has proven hand washing to be the single most effective intervention in reducing the spread of infection. From the bathroom to the classroom, from the kitchen to the office, from the playground to the hospital ward, hand washing protects kids of all ages—and even saves lives.

The Evidence. As far back as 1822, a French pharmacist showed that hand washing with a chlorine-based solution could eliminate odors from hands, and he went on to recommend that doctors in contact with a patient who had an infectious disease should treat their hands with a chlorine wash before moving on to the next patient. However, it is Ignaz Semmelweis who should be known as the father of hand washing. In 1846, he demonstrated that when obstetricians washed their hands (a radical concept at the time and only accomplished in the context of Semmelweis' research study!) between their dual responsibilities of performing autopsies and delivering babies, women in labor developed fewer fatal infections. The field of obstetrics has, of course, evolved (autopsies are no longer part of the job description!), but the hand-washing lessons taught by Semmelweis a century and a half ago changed medicine forever; hand hygiene guidelines have become a nearly biblical mainstay of every hospital in this country.

Guidelines are good, but they don't guarantee compliance. The Centers for Disease Control and Prevention (CDC) reports that despite the existence of well-displayed wall charts and codified guidelines, nearly 2 million patients in the United States acquire an infection in hospitals each year, 90,000 of whom die as a result. The reason for these persistently high rates of hospital-acquired infection: only 40% rates of adherence to hand-washing guidelines! Not surprisingly, the CDC's response to these findings has been recommendations for increased attention to hand washing or, in their words, "More widespread use of hand hygiene products that improve adherence to recommended hand hygiene practices will promote patient safety and prevent infections."

Like everything else in a constantly modernizing world, hand washing has become more complicated in recent years, with new products on the market and new issues to consider in how our kids should wash and with what. The hand-washing discussions below address the growing menu of hand-washing products (simple soap, bar versus liquid soap, antiseptic hand washes, and antibiotic-containing hand washes), the optimal timing and techniques for your kids in washing their hands, and finally a not-so-encouraging look at how well adults are behaving as role models for hand hygiene.

Simple Soap. Simple soap is not antimicrobial, that is, it doesn't kill germs. Rather, it cleans hands mechanically by detergent action, lifting and washing away dirt and organic material and the associated germs that are not very sticky on the skin. Evidence for the effectiveness of hand washing, using simple soap, in reducing the rates of kids' infections comes from many sources in the United States and abroad. In U.S. day care center studies, the rates of diarrhea were reduced by 25 to 50% and the rate of vomiting illnesses was reduced by two-thirds when hand-washing protocols using simple soap were introduced. In U.S. school-based studies, hand washing with soap and water alone reduced total absence days by 25% when all infectious diseases were included, by 21% for respiratory infections, and by 57% for gastrointestinal infections. Similar results were obtained in a Canadian school study, where a hand-washing and hygiene education program resulted in 22% fewer absences, 25% fewer doctor visits, and 86% lower prescription and over-the-counter medicine use.

Outside the confines of school and day care, an Indonesian study showed that education about personal hygiene in general, combined with providing simple soap and hand-washing instructions, resulted in a nearly 90% community-wide reduction in diarrhea and a 45% decrease in skin and eye infections among children and adults. A similar approach in the United States produced similar results—75% reduction in shigella dysentery (see chapter 3) during a community-wide epidemic. Combining hand washing, free soap, and hygiene education reduced diarrhea by 40% in a Thailand community and by more than 60% in a Bangladesh community study. Extrapolated to an international scale, it has been estimated that worldwide institution of better hand-washing practices could result in saving more than 1 million lives each year by reducing deaths due to diarrhea and other communicable diseases. For this reason, the World Health Organization (WHO) has said that routine hand washing with simple soap and water is "the most important hygiene measure in preventing the spread of infection."

Implementation, however, is another matter, and saving 1 million lives a year is easier said than done. In many parts of the world, the infrastructure to provide even the clean water necessary to make hand washing useful is missing, as is government commitment to public health. Closer to home, estimates of compliance with hand-washing recommendations among kids in U.S. schools are in the 10% range, proving that you can lead a kid to water, but you can't always make him wash.

Bar versus Liquid. Soaps themselves can become contaminated with germs on the bars or on the liquid dispensers, and since simple soap isn't antimicrobial, the germs can live in the soapy goo for a long time, potentially becoming a source for transmitting infections from one child to another. Studies have shown that liquid soap in a dispenser is less likely than bar soap to become contaminated with the very germs we're trying to prevent kids from spreading to each other; bar soaps clearly harbor more bugs. Having said that, what's missing are the studies that actually show fewer infections in households using liquid soap than in those using bar soap. Hospitals long ago abandoned bar soap for use by health care personnel, and the safety rationale and theoretical advantages of liquid soap prompt many to recommend its use in households as well, even if the studies showing fewer infected kids haven't been done.

If your household uses bar soap, it should be placed on a drain stand rather than in a soap dish to minimize pooling of contaminated soapy water. Liquid dispensers, although more sanitary than bar soap dishes, are not immune to contamination and require their own strategy. Refillable dispensers should not be topped off; when empty or nearly empty, they should be thoroughly washed, both the container and the pump, preferably in the dishwasher. Routine washing of the pump surface between refills should be performed with a clean wash rag or disposable paper towel and with a disinfectant solution (see "Household Hygiene" below) when the other bathroom fixtures are cleaned—once a week, at least.

Antiseptic Hand Washes. Okay, hand washing is good, and simple soap appears to work just fine according to all of the studies mentioned above, but if simple soap is good, are other hand-cleansing agents better? For the answer to that question, we turn to many studies of health care workers in hospital and clinic settings, where the spread of infection from patient to patient via the hands of caregivers has been recognized since our hero Semmelweis first called attention to it on his obstetrical wards in the middle of the 19th century. In hospital settings, the goals of hand washing are threefold: (i) to remove the germs acquired from one patient (transient flora) before the health care provider moves on to the next patient, (ii) to keep the health care provider's personal germs (resident flora) off of his or her patients, and finally, (iii) to provide some ongoing antigerm activity between washes to minimize the stickiness of new germs from any source.

Hand-washing substances that do kill germs (as opposed to simply washing them away) are called antiseptics and are said to have antimicrobial activity. Among the most effective and widely used are alcohol-containing compounds. These kill germs, including bacteria, fungi, and some viruses, and the killing is quicker than with other substances because it results from the almost immediate breakdown of germ proteins on contact with the alcohol. In recent years, alcohols have been incorporated into hand rubs, rinses, foams, and gels that don't require water. Alcohols evaporate quickly, leaving little ongoing killing power behind, but other antiseptic agents that linger longer, like hexachlorophene, chlorhexidine, and iodine-containing products, are slower acting than alcohols and can be more irritating.

There is a paucity of direct comparison studies between any of these antiseptic compounds and simple soap. The few studies available measure the relative germ reduction abilities in laboratory-controlled experiments as opposed to real-life situations in day care centers, schools, or the workplace. In these laboratory experiments, all antiseptic solutions are more effective than simple soaps at removing both transient flora and resident flora. Alcohols are 3 to 5 times more effective than simple soaps at removing resident germs from hands and as much as 10 times more effective with transient germs; 4% chlorhexidine solution is 2 to 3 times more effective than simple soap on resident bacteria and slightly more effective than soap on resident bacteria.

Alcohol sanitizers contain a moisturizing agent, which makes them very well tolerated by the users (except those who have cuts and scrapes, which burn for a few seconds after use), and alcohol products cause little irritation or drying, whereas chlorhexidine (and even simple soaps) can cause drying, irritation, and allergies. The result is that in hospitals, compliance with handwashing guidelines increases with the use of alcohol-based solutions but decreases with soaps and other antiseptics. For these reasons, the CDC guidelines for hospital hand hygiene recommend alcohol-based hand rub use between patient contacts for all health care workers. Although alcohol-based hand sanitizers have been slow to find commercial consumer success outside of hospital and clinic settings, they are now increasingly being marketed for home use by major consumer product manufacturers. Chlorhexidine and other chemical antiseptic products remain virtually nonexistent on store shelves.

Triclosan (Antibiotic) Products. In contrast to the slow emergence of alcohol-based washes in the consumer product industry, one of the most popular consumer ingredients *outside* the hospital setting is triclosan, which blocks the synthesis of fats in germ membranes. Although technically an antiseptic (because it kills germs), triclosan is also an antibiotic in the sense that it acts specifically against unique properties of bacteria (see chapter 5). Depending on the concentration of triclosan in a cleansing solution, the enhanced antigerm action over simple soap can range from negligible to 100-fold. A 0.2% triclosan solution, for example, has no advantage over simple soap, whereas a 1% solution is better than soap. Although triclosan reduces germ counts in the laboratory more than simple soap, there are no studies demonstrating that triclosan-containing products actually reduce personal or household infections. The majority of household illnesses are due to viruses, and the effects of triclosan and other antiseptic products on viruses have been less well studied. A recent study of households randomly assigned to use either antibacterial or nonantibacterial hand-washing, cleaning, and laundry products showed no difference over a 1-year period in the occurrence of symptoms typically attributable to viral illnesses. This is not surprising, since antibacterial products aren't designed to kill viruses (see chapter 5). However, the absence of a benefit in reducing *viral* symptoms likely predicts that overall household health will not be greatly benefited by the use of antiseptic products, because viral illnesses are so much more common than those due to bacteria. Studies examining the effect of antibiotic (like triclosan) hand-washing and cleaning products on household *bacterial* infections, such as strep throat, have not been conducted to date.

Despite the fact that triclosan's benefits in reducing actual infections are theoretical and based on laboratory germ counts, the antibiotic can now be found in as many as *three-quarters of all liquid soaps* marketed to households in the United States—and triclosan use in commercial products has not been limited to soaps. The National Institutes of Health provides a list, shown in Table 9.1, of personal hygiene and cosmetic products containing triclosan; many, if not most, of these are marketed to parents for the protection of their children, as well as themselves; at least a couple are marketed for pets!

Later in this chapter, additional triclosan applications are discussed, including its incorporation, under the trade name Microban, into kitchen countertops, children's toys, linens, and bathroom fixtures. It is not at all clear why triclosan, of all possible antiseptic compounds, has found such widespread use, but the fact that it has become so ubiquitous raises issues of safety and resistance. Safety concerns, as usual, have ranged from the reasonable to the hysterical. Skin irritation and allergic reactions to triclosan do occur; chemical conversion of triclosan to the pesticide dioxin appears to be of theoretical, but not practical, concern. The theory that overcleaning our kids will cause them to be too hygienic and therefore more prone to allergies and asthma when exposed to new environmental challenges (the "hygiene hypothesis") is discussed at the end of this chapter.

Whenever widespread exposure to antibiotics is considered, so too must be the development of antibiotic resistance. Resistance, as you recall from chapters 5 and 6, results from the conditioning or acclimating of resident germs to their environments. When antimicrobial substances are part of that environment, germs that are naturally or mutationally resistant to those substances have a selective advantage and outgrow the sensitive bugs. Initial concerns about triclosan, in particular, have been largely resolved regarding resistance—it does not appear that the use of triclosan and other antiseptic products in the household, even continuously for a year, results in the emergence of resistant germs. However, recall that it is also not proven that triclosan-containing products are in any way more advantageous than nonantibiotic cleaning products in reducing personal or household infections.

The absence of evidence suggesting an advantage of antiseptic soaps, and the theoretical risks of increasing resistance (although also not actually observed to date), prompted a Food and Drug Administration (FDA) advisory panel to recommend in October 2005 *against* these products for home use in favor of simple soap.

Table 9.1 Personal hygiene and cosmetic products that contain Triclosan

Brand name	Category	Form	% Triclosan
Revlon ColorStay LipSHINE Lipcolor Plus Gloss, Dazzle	Personal care/use	Liquid	
Softsoap Fruit Antibacterial Liquid Hand Soap, Fresh Picked Raspberry	Personal care/use	Liquid	
Revlon ColorStay LipSHINE Lipcolor Plus Gloss, Blast	Personal care/use	Liquid	
Clean and Smooth Kitchen Antibacterial Liquid Hand Soap	Personal care/use	Liquid	0.3
Softsoap Fruit Antibacterial Liquid Hand Soap, Orchard Fresh Peach	Personal care/use	Liquid	
Colgate Total Toothpaste	Personal care/use	Paste	
Dial Liquid Antibacterial Soap, Original Formula	Personal care/use	Liquid	
Aveeno Therapeutic Shave Gel	Personal care/use	Aerosol	
Bath & Body Instant AntiBacterial Hand Gel-Freesia	Personal care/use	Liquid	0.3
Softsoap Gentle Antibacterial Body Wash with Vitamins	Personal care/use	Liquid	
Suave Liquid Hand Soap, Antibacterial, with Light Moisturizers	Personal care/use	Liquid	
Shield Deodorant Soap Bar, Surf Scent	Personal care/use	Solid	
Suave Liquid Hand Soap, Antibacterial, with Extra Aloe	Personal care/use	Liquid	
Clearasil Daily Face Wash	Personal care/use	Cream	
Right Guard Sport, Deodorant Aerosol, Original	Personal care/use	Aerosol	
Gentle Antibacterial Body Soap with Moisture Beads	Personal care/use	Liquid	
Softsoap Extra Moisturizing Antibacterial Liquid Hand Soap	Personal care/use	Liquid	
Revlon ColorStay LipSHINE Lipcolor Plus Gloss, Electric	Personal care/use	Liquid	
Old Spice Red Zone Antiperspirant & Deodorant	Personal care/use	Semi-solid	
Softsoap Fruit Antibacterial Liquid Hand Soap, Juicy Melon Scent	Personal care/use	Liquid	
Revlon ColorStay LipSHINE Lipcolor Plus Gloss, Solar	Personal care/use	Liquid	
Clean and Smooth Antibacterial Liquid Hand Soap	Personal care/use	Liquid	0.3
Softsoap Antibacterial Liquid Hand Soap with Light Moisturizers	Personal care/use	Liquid	
Colgate Total Toothpaste, Fresh Stripe	Personal care/use	Paste	
Softsoap Liquid Antibacterial Body Soap	Personal care/use	Liquid	

(Continued on following page)

Table 9.1 Personal hygiene and cosmetic products that contain Triclosan *(Continued)*

Brand name	Category	Form	% Triclosan
Softsoap 2 in 1 Antibacterial Hand Soap Plus Moisturizing Lotion	Personal care/use	Liquid	
Noxzema Triple Clean Antibacterial Lathering Cleanser	Personal care/use	Tube	0.3
pHisoderm Antibacterial Skin Cleanser	Personal care/use	Pump, spray	
Imina Lathering Facial Cleanser	Personal care/use	Cream	
Old Spice High Endurance Stick Deodorant, Original	Personal care/use	Solid	
Lever 2000 Deodorant Soap Bar	Personal care/use	Solid	
Suave Deodorant Soap, Antibacterial	Personal care/use	Solid	
Lever 2000 Soap Bar, Antibacterial	Personal care/use	Solid	
Right Guard Sport, Clear Stick Deodorant, Original	Personal care/use	Solid	
Old Spice Classic Stic-Original Scent	Personal care/use	Solid	
Right Guard Sport, Deodorant Aerosol, Fresh	Personal care/use	Aerosol	
New Vaseline Brand Intensive Care Antibacterial Hand Lotion	Personal care/use	Liquid	
Softsoap Aquarium Antibacterial Liquid Hand Soap, Clear	Personal care/use	Liquid	
8 in 1 Perfect Coat Select Antibacterial Deodorizing Shampoo	Pet Care	Liquid	0.03
Pet Gold 3 in 1 Antibacterial Shampoo for Dogs	Pet Care	Liquid	

Timing of Hand Washing. Well, now that we've established that hand washing is good and that simple soap is more than sufficient, how and when should your kids wash their hands? Vigorously and often. Recall from chapter 2 that germs get on kids' hands from contacts with other people, animals and birds, and inanimate objects. The kids then touch their eyes, noses, or mouths, and the germ is introduced onto their mucous membranes, where it can infect cells locally and/or spread throughout the body. Hence, the most effective way of preventing transmission of germs is to teach kids not to touch their eyes, noses, or mouths before a thorough hand washing. Good luck with that! Kids' fingers explore their faces constantly. Plan B, then, is to minimize the opportunities for germ transmission from other kids to our kids' hands and from our kids' hands to their own orifices. This is somewhat more realistic and not unfamiliar to those of us who ourselves grew up with parents. Table 9.2 lists the top 10 most important hand-washing moments for kids, not necessarily in order of significance.

Table 9.2 Top 10 most important hand-washing moments

1. After playing with a sick friend or sibling (or after handling things that a sick child might have handled, as in the doctor's waiting room [see "Doctors' Offices" in the text])
2. After using the bathroom (use the hand towel to turn off the sink and open the bathroom door [see "Hand Drying" in the text])
3. Before eating
4. After high-fiving the opposing team at the end of a sports competition (or any other mass-handshaking event, like the receiving line at a bar mitzvah or graduation)
5. After recess
6. After school or day care
7. After playing with animals or in areas where animals hang out
8. After playing outside
9. After blowing the nose or coughing into the hands (Although your kids cannot give themselves an infection by contact with their own secretions, this is a very considerate gesture that protects other kids from the germs on your kids' hands.)
10. Before bedtime

One way to accomplish the ambitious goal of hand washing at many of these 10 opportune moments is to keep a box of alcohol-containing hand wipes in the family van; these require no towel for drying, as the moist alcohol quickly evaporates from hands. When you pick your kids up from school or practice, have them use the wipes and discard them in the car's trash bag—easy and efficient hand hygiene.

Technique for Hand Washing. The best hand-washing technique to teach kids recognizes that the single most important determinant of effective hand washing is time—how much time your child spends at the sink. The difference between removal of bacteria during a 10-second wash (longer than most of our kids spend at the sink) and a 4-minute wash (longer than any of us spend at the sink) can be as much as a several thousandfold reduction in germs. Studies timing kids' hand washing find that the water is typically running for less than 5 seconds and kids leave the sink with their hands dripping wet (undoubtedly to save the additional time needed to find the towel on the rack). If you can push that time under the water up to 15 or 20 seconds, or even 30 seconds if you're feeling ambitious, the effectiveness in reducing the germ load is dramatically increased, regardless of the actual technique the kids use. That's because, out of sheer boredom alone, if your child has to stand at the sink for 15 seconds, she'll do something useful with her hands—wetting, rubbing, lathering—just to pass the time.

This is where a musical interlude is helpful. Teach your young kids to sing Twinkle, Twinkle Little Star or the ABCs (this makes your job easy, because both songs are sung to the same tune—a bit of pediatric trivia). Those songs take exactly 20 seconds (so does Mary Had a Little Lamb, if your child is the creative type who craves variety). By the time the song is done, it's time to dry—almost as important as the wash itself (see below). The wash method is pretty intuitive; lots of water, followed by lots of rubbing with the soap to create a good lather (include the wrists, between the fingers, and around the nails), followed by lots of water again. It's impossible to do all those "lots ofs" in 5 seconds; hence, the importance of the musical interlude.

Adults Behaving Badly. Our mothers and theirs before them taught all of us about hand washing. They were right, of course, as the foregoing paragraphs illustrate. Since the value of hand washing has been known and taught for generations, it's reasonable to ask how effective all of that nagging has been. The American Society for Microbiology's Washup.org has surveyed and spied on Americans' hand washing for the past 10 years. Adults are asked by phone about their hand washing, and then other adults with similar demographics are secretly observed at the sinks (or walking past the sinks, as is too often the case) in public bathrooms. The general findings are as follows: women wash more than men, men lie more than women, baseball fans have dirtier hands than museum attendees, adults rarely wash their hands after coughing or sneezing into them, and, over the 10-year study period, adults are washing slightly less today than in the past.

Now here are the specifics. Phone surveys in 2005 found that 91% of adults said they always washed their hands after using a public restroom (94% of women, 88% of men); this compares with 95% of adults in 2003 (97% of women, 92% of men). Direct observations in 2005 of more than 6,000 adults in public restrooms at six public attractions in four major cities indeed found that 90% of women washed—but only 75% of men! Phone surveys also revealed that women always washed more than men after using the bathroom at home (88% versus 78%), after changing their baby's diapers (82% versus 64%), after petting a cat or dog (50% versus 34%), before handling or eating food (82% versus 71%), and after handling money (27% versus 14%). Perhaps most telling about our need to educate better regarding preventing infections, only 32% of adults always wash their hands after coughing or sneezing into them (39% of women versus 24% of men). If adult compliance with commonsense hand washing is so poor, clearly the challenge of educating our kids' generation is even greater.

Hand Drying

Hand washing is important, but it's only part of the story. Remember the lesson from chapter 2, that moisture is a potent breeding ground for all germs. Wet hands transmit germs much more efficiently than dry hands. A fascinating review by the FDA several years ago investigated the optimal hand hygiene for food handlers, and the results apply to everyone. Regardless of the washing technique, there is a strong inverse correlation between how dry hands are and how many germs can be recovered from them (and passed on to the next victim)—the dryer the hands, the fewer the transmissible germs. In most of the studies reviewed by the FDA, towels, both paper and cloth, were more effective in drying hands and eliminating germs than were electric air dryers. It is thought that the physical rubbing of hands on towels further reduces hand germs from levels obtained by washing alone; the damp hands left after your kids electric air dry (or drip dry) are likely to get rubbed on their clothes—providing yet another inanimate source of germs.

Moreover, wet hands mobilize bacteria, viruses, and fungi so that even with the reduced germ levels afforded by effective hand washing, those that do remain are more able to drip, creep, and seep onto inanimate surfaces, food, and unsuspecting friends. Thus, both for the sake of your kids (to eliminate even more germs than by washing alone) and for the sake of those who come in contact with the residual puddles on your kids' hands after they wash, drying is critical. Again, there are no secret methods here—clean towels and thorough rubbing. Tell your kids to sing another verse while drying or that it's icky to leave the bathroom with hands that are sticky (and wet).

After your kids dry their hands, they should use the hand towel to open the bathroom door (to avoid the germs left there by the last user who forgot to wash or dry).

Cough and Sneeze Etiquette

Recognition that respiratory infections can be spread by droplets, as well as by direct hand-to-hand contact (see chapter 2), reminds us of proper cough and sneeze etiquette to protect those near your kids from the germs your kids spray in the 3-foot radius around them (see chapter 2). It's been my experience that teaching kids these coughing and sneezing manners is futile; kids are just way too distracted to pay attention to these niceties.

However, you're welcome to try. Here's what your mother taught you: cover your mouth with your hand or a tissue when you cough or sneeze. That remains pretty good advice with a few additions and modifications. Today's advice for kids is sneeze or cough into your sleeve or elbow (this acknowledges that snotty hands are the most effective way for kids to spread respiratory germs; kids' shirts or elbows become more infective with today's etiquette, but shirts and elbows are less likely to find the hands of another susceptible child than are your child's hands), wipe up any secretions with a tissue, dispose of the tissue right away, and then wash your hands.

How many kids have you seen do this correctly? For that matter, how many adults (see the adult spy studies and surveys above)?

Toothbrushes, Toothpaste, Towels, Toiletries, Tissues, and Teacups

Sharing Personal Hygiene Items. Sharing is bad. Although it's counter to everything we have tried to teach our children from crib to college, don't let your kids share anything that goes in their mouths or on their faces. This precaution is particularly relevant when any of the potential sharers are sick. It's better for your kids not to brush when they forget their toothbrush and toothpaste for the sleepover party than for them to borrow from a friend. Personal toothpaste tubes are a good idea even at home. Damp cloth towels, and even pillows that haven't had a linen change between users, can transmit germs. Kids shouldn't drink out of the same cups, soft drink cans, or water bottles as other kids (or adults). Make sure your kids take their own water bottle to their sports activities so they don't have to drink from the community urn that many coaches bring to the sidelines (often forgetting the paper cups, prompting the kids to patiently wait their turns to put their mouths directly on the spigot!). If teammates tend to borrow your kids' water bottles because their parents haven't yet read this book, have your kids put an extra water bottle in their sports bags as the community bottle for borrowers. A facial tissue is meant to be disposable; it should be tossed after each use, or at the very least, after any single user has used it up and before the next snotty or stuffy nose picks it up from the bathroom counter. Not sharing combs and hairbrushes reduces the risk of contracting head lice and ringworm (see chapter 3). To prevent potentially devastating infections with blood-borne germs, adolescents should never share face or leg razors with anyone. (There are many other things that adolescents shouldn't share [see below].) Other items not to share, especially with sick siblings and friends, are discussed under "Household Hygiene" below.

Antibacterial Toothpastes and Mouthwashes. It has been proven that antibacterial toothpastes and mouthwashes can help prevent and control gingivitis, an inflammation of the gums, which develops in part due to overgrowth of mouth bacteria; recent interest in gingivitis has included a theoretical role in adult heart disease (see chapter 3). In 1997, the FDA approved the first antibacterial toothpaste, Colgate Total, for the treatment and prevention of gingivitis. Because of its triclosan ingredient, Colgate Total has been approved only for children older than 6 years of age. As of 3 August 2005, the most recent product labeling approved by the FDA, use in children under 6 years old should only be under the direction of their doctor or dentist. This does not mean that the product is unsafe for younger children, only that safety and efficacy have not been proven in that age group.

Although they have been proven beneficial in reducing gingivitis, a noncontagious disease, there is no evidence that oral antibacterial products prevent any infections that are spread by oral secretions, such as strep throat, the common cold, pertussis, influenza, or tuberculosis.

Toothbrush Sanitizers and Toothbrush Care. After reading about the panoply of germs that normally colonize our kids' mouths, it's reasonable to wonder whether their toothbrushes are sanitary and safe—and that's exactly what the marketers of toothbrush care products hope you'll wonder. Over the past few years, a number of sanitizers, sterilizers, and protective carrying cases have appeared on the scene as aids to good oral health. There are three toothbrush truths: (i) toothbrushes do harbor many germs, from viruses to bacteria; (ii) commercial sanitizers do reduce the counts of germs on toothbrushes; and (iii) there is no evidence to suggest that the use of sanitizers (ultraviolet light or chemical) reduces infections or improves oral health.

Remember, unless your kids violate the important rule against sharing (above), the germs on their toothbrushes are mostly their own. Although it seems intuitive that cleaner toothbrushes might be less likely to promote gingivitis (which also comes from the child's own germs) in kids who are prone to that inflammatory condition of the gums (see above), there are no data to support that intuition.

Table 9.3 presents a condensed version of the recommendations from the American Dental Association (ADA) for toothbrush care.

There is one more commonsense recommendation not listed by the ADA: flushing the toilet vaporizes lots of germs into the air (see below) that shouldn't be on a toothbrush—store your toothbrushes far away from the toilet.

Hugs and Kisses (and Beyond)

Kissing. As noted earlier in this chapter, hand-to-hand contact is the most important risk factor for transmitting many viruses and bacteria. That's only true, however, because hand shaking, hand holding, and hand handling of inanimate objects are so much more prevalent than kissing. Kissing mouth to mouth, of course, very effectively bypasses the need for hands to transmit mouth germs from person to person. Some households, and some cultures, routinely kiss mouth to mouth among family members and friends. Ironically, some of the most ardent advocates of wearing your boots in the rain are the grandmothers and great aunts who greet everyone on the family tree with a smooch on the lips. From an infectious-disease doctor's point of view, this is a very bad custom. A kiss on the cheek (or top of the head) is much safer, and a passionate bear hug is the safest of all expressions of affection. The romantic European gesture of men kissing women's hands is, well, dumb, at least in the eyes of a germ doctor. Infectious mononucleosis ("mono") (see chapter 3) is but the most famous disease transmitted by kissing (hence the moniker "kissing disease"). However, all mouth flora, from rhinoviruses that cause the common cold and strep bacteria that cause strep throat to the viruses and bacteria that cause meningitis, can be transmitted by kissing (see chapter 2).

Sex. Well, what's the next logical topic after hugs and kisses? Talking to adolescents about sex is difficult and well beyond the scope of this book (maybe beyond the scope of *any* book). Having said that, sexually transmitted diseases can be the best weapon in your arsenal when you approach the discussion of sex. Your teenager's fear of AIDS, hepatitis, herpes, syphilis,

Table 9.3 ADA recommendations for toothbrush care

1. Do not share toothbrushes (see my comments about sharing in the text).
2. Rinse toothbrushes well with tap water after use.
3. Store toothbrushes upright to air dry.
4. Separate the toothbrush from others in the rack.
5. Don't cover toothbrushes, and don't store them in closed containers (covering them retains moisture, and as you now can recite in your sleep, germs like moisture).
6. Discard and replace toothbrushes every 3 to 4 months.

gonorrhea, genital warts, trichomonas, and chlamydia should be fully exploited (see chapter 3). Remind them that venereal diseases can cause sterility. Find gross and disgusting pictures of chancres and drips and warty growths on the internet or in medical texts in the library; show them to your kids, and then teach them about abstinence or condoms. Teach them that abstinence is a reliable preventive, that condoms are imperfect but much better than nothing, and that there's a lot of sex in their future lives if they can get through this stage without hurting themselves. Teach them that unprotected oral sex can transmit many venereal infections, just not quite as efficiently as unprotected intercourse (see chapter 2).

Blood

Someone else's blood is potentially a dangerous thing. Some of the germs that can be transmitted by blood may cause life-threatening diseases, like AIDS and hepatitis (see chapters 2 and 3). Kids are frequently around other kids who are bleeding—bloody noses, scraped knees and elbows, or loose teeth. Some kids still play the old "blood brothers" game, where they poke each other and rub their bleeding fingers together as a bonding ritual. This is a really bad idea.

Kids need to learn that blood must be taken seriously and usually dealt with by an adult; when a friend gets hurt and is bleeding, teach your kids to run for help rather than tending to the injury themselves. Kids also need to be taught, unfortunately, never to play with discarded needles they may find in trash cans or on the ground in playgrounds and parks.

Other Nuisances and Nuances in Personal Hygiene

Bad Breath. Bad breath, also known as halitosis, although contributed to by oral bacteria, is usually not an infectious disease and is dangerous only for its impact on interpersonal relationships. Despite the enormous marketing of commercial products for bad breath, there are no studies demonstrating effectiveness for any of them in reducing *infections* from mouth germs; only two studies show even a weak benefit in treating the bad breath, and those used tongue scrapers, not washes! The bottom line is that kids don't need mouthwashes or tongue scrapers; regular (twice-a-day) tooth brushing is sufficient. In rare cases, bad breath may be due to severe gingivitis, an inflammation of the gums that does require specific therapy for dental health (usually professional tooth cleaning, followed by conscientious attention to brushing and flossing). Bacteria contribute to the development of gingivitis, which explains the benefit of antibacterial toothpastes and mouthwashes for this condition (see above). The dentist can help determine the possibility of gingivitis as the cause of your child's bad breath.

Body Odor. Bacteria contribute to body odor, which can be cured by improving personal hygiene. Perspiration, although itself odorless, facilitates bacterial growth (remember, germs like moisture), and the combination, particularly after the onset of puberty, can be potent. Showering daily, as well as after physical activity, will eliminate most cases of body odor; morning showers may be more effective because they reduce the bacterial load that accumulates overnight. Body odor is not an infection and is not contagious. The bacteria involved are on the body surface, colonizers rather than invaders (see chapters 1 and 2).

The focal point of most body odor is under the arms, where the warm and moist environment favors bacterial growth. Two classes of products are available for underarm treatment—antiperspirants, which block the sweat glands, and deodorants, which are antibacterial. Antiperspirants are classified as drugs by the FDA because they interfere with a natural process—sweating. Typically, these products contain an aluminum compound, which binds with sweat to form a gel-like plug in the sweat glands; the plug is shed with the rest of our superficial skin layers over time. In contrast, because deodorants act on the surface bacteria only, they are classified as cosmetics. The antibacterial ingredients of deodorants range from alcohol to triclosan. Although no studies have been done to formally prove benefit in treating body odor, enough practical experience has accumulated to validate the use of deodorants for kids, during puberty and beyond, with body odor. Social acceptance, rather than medical necessity, dictates that recommendation. Antiperspirants should be reserved for kids with truly excessive sweating.

Athlete's Foot and Plantar Warts. The fungus that causes athlete's foot (see chapter 1), like all other germs, thrives in moist environments; it comes from other people who have the same infection (see chapters 2 and 3). Plantar warts are caused by papillomaviruses (see chapters 1 and 3) and also come from other people (see chapter 2). When taking a shower in the school gym, at the swimming pool, or at the health club, your kids should wear flip-flop sandals to avoid walking with their chafed and chapped feet on puddles of fungi and viruses.

Bathroom Habits. Kids should be taught at the outset of potty training to wipe themselves from front to back—this is especially important for girls to reduce the risks of urinary tract infections (see chapter 3). Bacteria from the gastrointestinal tract, emerging from the anus, are the cause of almost all urinary tract infections—wiping back to front is thus a bad idea.

Tampons. Adolescent girls should be taught early about the association of tampon use with the risk of toxic shock syndrome (TSS) (see chapter 3). Although rare, TSS can be life threatening. Steps to reduce the risk of TSS include frequently changing tampons (four or five times each day, but no less frequently than every 8 hours) or alternating tampon use with an external pad and using the least absorbent tampon that adequately controls menstrual flow. Girls should also be taught the symptoms and signs of TSS (see chapter 3) and that if any of those signs develop, they should immediately remove their tampon and contact their doctor.

Table 9.4 Top 10 rules for personal hygiene

1. Golden Rule of Infectious Diseases: don't spread to others what you wouldn't want others to spread to you. When you're sick, take measures to prevent passing your infection on to others.
2. Hand washing is the single most important preventive measure against infections, both giving and getting them. Kids should be taught the most important times of the day to wash and that the more time they spend washing, the better.
3. Germs like moisture; hand drying is almost as important as hand washing.
4. There is no evidence that antiseptic- or antibiotic-containing cleaning products are more effective than simple soap and water in preventing infections in the home, school, or day care center.
5. Notwithstanding everything that you've tried to teach your kids in the past, sharing is bad. Teach your kids not to share anything that goes in the mouth or on the face.
6. Antibacterial toothpastes and mouthwashes help treat and prevent gingivitis (inflammation of the gums), a noncontagious disease. There is no evidence, however, that those products reduce infections spread by oral secretions.
7. Almost anything your kids share with a sick sibling or friend can transmit the infection (see "Household Hygiene" in the text).
8. Blood is dangerous. If a friend has a bloody nose, a scraped knee, or a loose tooth, kids should find an adult to help rather than touching the blood themselves.
9. Hugs are better than kisses; kisses on the cheek or head are better than kisses on the mouth.
10. Sexually transmitted diseases can be a useful tool for teaching your adolescents about abstinence and/or safe sex.

In summary, the most important rules for personal hygiene are listed in Table 9.4.

Household Hygiene

Although we are trained to think of public places—restrooms, restaurants, day care centers, and schools—as likely sources of infection, and they are (see below), our perception of home as a clean refuge is fantasy. Taking a culture swab to surfaces at home has been the subject of many a child's school science fair project, and the results are always startling. In particular, the kitchen, bathroom, play room, sick room, and even the laundry can be vast reservoirs of lurking germs. Items that you think of as allies in the war on bugs—dish rags, sponges, mops, and even bottled water—may in fact be double agents, spreading germs faster than they eliminate them.

Although households are packed with germs that have the potential of causing infection, how strong is the evidence that infections actually are caught at home? Very strong! Studies over many years prove that a wide selection of infections, ranging from banal to fatal, are spread in the household, even if all the responsible germs don't originate there.

A brief room-by-room microscopic home tour follows.

The Kitchen

Of the more than 75 million cases of infection-related food poisoning (see chapters 2 and 3) that occur in the United States each year, one-fifth occur in the home, probably the kitchen. In particular, 90% of gastroenteritis cases caused by salmonella infections occur at home. In a house with one salmonella-infected individual, two-thirds of other household members will become infected (some of that may be due to bathroom spread [see below]). The vast majority of *Escherichia coli* O157:H7 infections that caused an outbreak of hemorrhagic gastroenteritis and hemolytic uremic syndrome (see chapter 3) in New Jersey were due to hamburgers prepared and eaten at home! Campylobacter food infection (see chapter 3) in one family member spreads to 15% of other household members (again, a bathroom contribution is probable here [see below]).

One of the recurring themes in this book is that germs like moisture. The highest counts of bacteria in the kitchen are found in the sink and sink drain, dish rags, mops, and sponges. When dish rags are also used to wipe down countertops, the germs from the sink are spread to new surfaces. *Salmonella* species have been found in about 15% of sponges and dish rags in homes like yours. Raw and undercooked meat, poultry, and fish harbor dangerous germs, like *E. coli* O157:H7, salmonella, campylobacter, and pseudomonas, which get on the hands of the chef (you). When cutting boards, faucets, and countertops are then handled by the same food preparer (you), those dangerous germs stick to those new hiding places.

Videotapes taken during studies of typical households routinely show some remarkable lapses in common germ sense. The towel used to wipe up juice from raw meat is then used to dry off washed hands; during food preparation, hands repeatedly go directly from food to face and back again; a towel falls to the floor, is walked on, and is picked up and used to wipe down the countertops—and then is used to cover cooked meat waiting for dinner! The telltale study tapes also show that meatloaf, chicken, and fish were undercooked 35%, 42%, and 17% of the time, respectively.

The Bathroom

Bathroom wetlands are also fertile areas for bacterial, viral, and fungal growth. Toilets, toilet seats, and the rim under the recess of the toilet bowl all harbor salmonella in households where a family member is infected; other fecal flora also stick to the seats and porcelain. Rotaviruses, the leading viral cause of gastroenteritis in young children (see chapter 3), are excreted to the tune of 100 billion virus particles in every gram of feces from an infected child—and the virus can survive for days to weeks on moist environmental surfaces. It takes as few as 10 to 100 of those billions of viruses to cause infection in the next child. Surfaces found to be harboring rotaviruses in the homes of infected children extend well beyond the toilet bowl and include faucets, toys, diaper-changing and disposal areas, and sinks—both bathroom and kitchen!

Although it is always difficult to prove precisely in which room real infections in real children actually originate, one thing is clear: the germs that cause household outbreaks of gastroenteritis and hepatitis A spread from the feces of an infected family member to other family members (see chapter 2). The bathroom is the second most likely place to find fecal germs (the most likely, of course, is on the hands of infected kids who didn't wash after leaving the bathroom [see chapter 2]). Bathroom surfaces are dirty and infectious for the same reasons—dirty and infectious hands touching them.

The act of flushing the toilet represents a violent stirring of the germs in the bowl water, as well as those adherent to the sides of the bowl. Studies have shown that the first flush after a bowel movement reduces the germ load in the water and on the sides of the bowl by 200- to 300-fold, but the first flush also vaporizes 1,000 to 3,000 bacteria and viruses per cubic meter of air space. These germs float for a while and then land on nearby surfaces. How horrible! How disgusting! How meaningless? There has never been a study to show that germs in the air of the bathroom or even on surfaces cause disease—as long as hands are washed when people leave the facility. It is true that when kids touch contaminated surfaces and then touch their mouths, infections with gastroenteritis germs will occur (see chapter 2), but there is no need for your kids to hold their breath in the bathroom (at least, not for infectious reasons!); just teach them to thoroughly wash after every visit (see above). Remember, gastroenteritis germs and other fecal flora are not acquired by breathing them (see chapter 2), only by ingesting them, and that usually happens from contaminated hands. Also, as noted above, it's probably a good idea to keep toothbrushes as far away as possible (e.g., in a medicine cabinet) from the toilet.

The Bedroom/Sick Room

Respiratory viruses are spread by direct (hand) contact with infected secretions and by airborne infected droplets (see chapter 2). In a household study of respiratory syncytial virus (see chapter 1), this cause of upper and lower respiratory tract infection spread to 71% of adults holding and playing with an infected infant and to 40% of adults merely touching surfaces in the baby's room; no adults became ill if they just sat in the baby's room next to the crib without touching anything.

Rhinoviruses, the leading cause of the common cold (see chapter 3), were studied in a real-world infection experiment done in Virginia. Mothers in households in which a child had a rhinovirus cold were randomly assigned to use either an iodine finger dip (iodine kills bugs that come in contact with it) done every few hours or a colored placebo (no infection fighting) finger dip. The mothers in the iodine group had one-third the rate of colds of the placebo dip mothers.

A unique, and still fortunately uncommon, bedroom hazard is bed bugs (see chapter 2). Recently resurgent, this infestation was virtually nonexistent in the United States for the last half of the 20th century. Good laundry practices are only partially protective (see below), as the bugs hide in the cracks and crevices of beds waiting for their early morning feeding hour—and the meal is your kids' blood!

The Laundry Room

Another moisture haven, the laundry room, also nurtures germs. The trends toward using lower-temperature wash cycles, smaller water loads, and permanent press clothes appear to facilitate germ survival on clothing. All bacteria appear to survive a typical wash cycle; adding a rinse reduces but does not eliminate the germ load. In contrast, using hot water with bleach and complete drying cycles reduces viral and bacterial contamination of clothes by 98 to 100%.

Dirty clothes are, well . . . dirty. They house the germs that are shed, sneezed, sweated, and secreted from your kids' bodies. Until recently, however, there was little evidence that how you clean your kids' clothes has any impact on their risk of infections. Now there is a suggestion that there may be an effect of your laundry technique (see the next section).

What's a Parent To Do? Interventions To Make Your Home a Safer Place

Cleaning (versus Disinfecting). Much as with soaps, shampoos, toothpastes, and other personal-hygiene products, household cleaning products come in many flavors, each boasting unique infection prevention attributes. Technically, and by Environmental Protection Agency (EPA) definitions, there is a difference between cleaning and disinfecting; the former refers to mechanically removing dirt, whereas the latter means the incapacitation of germs. Common detergents and water are used for cleaning; chemical compounds with antimicrobial properties, such as those containing bleach (hypochlorite), ammonia, alcohol, antibiotics, and phenols, are disinfecting. The National Institutes of Health Household Products Database lists more than 40 different commercial bathroom and kitchen disinfectants containing one or more of those antimicrobial chemicals.

What are the data regarding the effectiveness of cleaning and disinfecting products in reducing germ load and infection incidence? As with personal-hygiene products, most studies involve measuring germ load reductions under experimental conditions, where known numbers of bugs are subjected to various treatments and then counted again. In these laboratory experiments, bleach is particularly effective against *Staphylococcus aureus* and gastrointestinal germs, like salmonella and *E. coli* (see chapter 1). Ammonia and phenol products, although less effective than bleach, also substantially reduce bacterial counts, whereas cleaning solutions containing baking soda, vinegar, and water alone do not. Rotavirus and rhinovirus reductions are greatest with a phenol-ethanol combination product but are almost as significant with bleach. Ammonia and tap water are not very effective against these viruses.

Studies done in the home have shown that effectiveness in germ reduction is a function of both the product being used and the presence of a defined cleaning schedule. Kitchen contamination with salmonella or campylobacter was not affected much by common detergents (cleaning solutions) and water alone, but adding bleach (disinfectant) resulted in a significant germ load reduction. Surfaces contaminated with meat have been shown to be most effectively sterilized by using a disinfectant, followed by paper towel wiping.

Establishing and adhering to a consistent schedule for cleaning each room of the house is as important as how the rooms are cleaned. A reasonable regimen is to clean kitchen sinks and countertops daily; bathroom sinks, countertops, and flush handles three or four times each week; and toilets and kitchen and bathroom floors weekly. The bedroom and playroom of a sick child should be cleaned twice a day.

Care and common sense in cleaning are essential; for example, kitchen sponges should be either discarded and replaced every day or two or sterilized. Putting the sponges in the dishwasher cleans away many of the germs, but a 2007 study that generated a lot of interest showed that microwaving kitchen or bathroom sponges for 2 to 4 minutes dramatically reduced the number of viable (living and infectious) germs on the sponges. Now, here comes the commonsense part—the same day that the news stories broke about the microwave sponge study, fires, burning-rubber smells, and microwave crashes were reported around the country. The investigators involved in the study had to quickly clarify that sponges must be wet before being microwaved and contain no metal components and that whoever put the sponge in should watch for any signs of trouble during the whole 2-minute microwave cycle.

Here are some more commonsense suggestions for cleaning: use kitchen dish rags for only 1 day before laundering them; don't use the dish rags that you've used to wash or wipe dishes to clean counters or stove tops, and vice versa; and don't store mops in a wet bucket—remember, germs like moisture; thoroughly wash mops in bleach-containing disinfectant or in the washing machine with bleach-containing detergent after each use.

Knowing where the germs are in each room of your house will help you find them and remove them. Focus your cleaning and disinfecting efforts on the hot spots.

- In the kitchen, contaminated surfaces include the sink, faucets, countertops, cutting boards, and floors.
- In the bathroom, scrub the toilet bowl (including the rim under the recessed area), toilet seat, flush handle, sink, faucets, bathtub, shower floor, floors, and doorknob.
- In the baby's room, diaper-changing areas, diaper pail, and crib railings need special disinfection attention.
- In a sick child's room, disinfect all surfaces that may be in contact with the child or the child's secretions.
- In the laundry room, disinfect the sink, areas under dripping clothes, and the floor.

There have been no studies that have shown that a particular product, or even a particular class of commercial products (e.g., simple cleaners versus disinfectants), changes the number or severity of actual infections in households. A recent study, in fact, did not show a beneficial effect of disinfecting or cleaning routines on actual household infection rates, but it did find some other interesting factors that may have an impact (see the next section).

Other Interventions: Laundry, Bottled Water, Exterminators, Etc. Nearly 1,200 members of 238 inner-city households were surveyed (most of the household members were children) in a recent study seeking to determine the effects of a variety of miscellaneous interventions on the incidences of eight common symptoms of infection—diarrhea, vomiting, fever, sore throat, cough, runny nose, skin sores, and conjunctivitis. More than 50 different household hygiene factors were assessed in each household, including food preparation, general cleaning, laundry, sharing of towels and toiletries, cleaning and disinfecting products used, and household members' beliefs about how germs are spread.

The incidences of infection symptoms per month were 9 to 12% among individual household members and 32 to 40% among households. There were several standout findings. Among households that used bleach in their laundry at the outset of the study, only half reported infection symptoms during the study month compared with three-quarters of households not using laundry bleach. Households drinking bottled water had *double* the infection rate (yes, you read that correctly) of those drinking tap water. The reason for this counterintuitive finding was not apparent, but the investigators hypothesized that family members drinking from the same bottle in the refrigerator may have shared germs (remember, sharing is bad!). Using hot water for white laundry reduced the household infection rate by 30%, and households whose members believed that germs were more likely to come from the kitchen than from toys, other people, soiled laundry, or the bathroom were 50% less likely to have infections. The investigators felt that those believing households were more likely to have been fastidious in their cleaning habits in general. Finally, and not surprisingly, households with members in poor health to begin with had more infection symptoms.

Laundering sheets will eliminate bed bugs, but only those on the sheets. Then, new ones creep out from the mattress and from cracks and crevices in the bed frame. The only way to rid a room (or house) of bed bugs once they appear is with the exterminator's help. Where do bed bugs come from? From someone else's bed in another house or hotel, from which they travel in suitcases or on clothing and find their way to your kids' bedroom (see chapters 1 to 3). They're difficult to see, and unless someone tells you their house has bed bugs, there's not much you can do to screen visitors' belongings. All you can do is exterminate the bugs once you find them—and be glad they haven't yet become an everyday worry in most households.

Pets. As many as half of all homes in the United States have at least one pet. The many ways that the many species of pets can transmit the many germs they carry were detailed in chapter 2 (in many pages). The steps that you can take to reduce the risk of infections in your kids from their pets are simple and straightforward and are listed in Table 9.5. More advice on protecting kids from animals' germs can be found under "Petting Zoos" below.

Houseflies. This is a chapter in the reconciliation portion of the book, where we look at Mom's and Grandma's ideas of infection prevention and compare them with science. Remember Grandma's shoo-fly pie, the delicious treat that sat on the windowsill and had to be vigilantly guarded from the flies it attracted? Was Grandma's compulsive loathing of flies warranted? In light of the typhoid fever epidemics at the turn of the 20th century in which flies were found to be a significant vector of disease (see chapter 2), it's certainly understandable where the loathing originated.

There is no question that flies carry germs; those data are presented in chapter 2. Today, though, evidence for a serious risk of infection transmitted by houseflies is weak, thanks to improved sanitation standards for human and animal waste management. However, puppy's poop is still available for flies in your home, as is excrement from other pets; if you live on a farm, the choices for the flies are even greater. Recall from chapter 2 how plentiful the potentially dangerous germs are in animal excrement. Flies land on waste, pick up germs, and then land on your food. If there are enough germs from enough flies, the theoretical risk of transmission by this route could be realized. Simple intervention measures for controlling houseflies include garbage cans with tight-fitting lids, quick and thorough cleanup of pet waste areas, covering food on countertops and picnic tables, good window and door screens, and, finally, a flyswatter.

Antibiotic-Impregnated Surfaces. As noted above, triclosan has found widespread use in commercial products for personal and household hygiene. Under the trade name Microban, this antibiotic has also been incor-

Table 9.5 Protecting your kids from pets' germs

- Make sure your pets are fully vaccinated.
- Kids must wash their hands thoroughly after playing with pets, cleaning up after pets, or handling objects that pets have come in contact with (litter boxes, bird cages, dog kennels, etc.).
- Don't let your kids eat or drink while playing with a pet; they should always wash their hands before eating or drinking.
- Play with pets should *not* be rough and tumble; this reduces the chances for scratches, nips, and bites.
- Your kids should wear masks to clean bird or rodent cages and wear long rubber gloves to clean aquariums.
- Don't let your kids wash their hands or any animal toys or utensils (e.g., water or food bowls) in the kitchen sink after playing with or attending to pets.

porated into the manufacturing process for numerous inanimate household items, including toys. A list of home care product categories, each of which contains multiple commercially available products, as provided on the manufacturer's website, is reproduced in Table 9.6.

Kids' clothing has also been developed with other types of infection-fighting substances woven into the fabrics. However, despite the diverse menu of antibiotic-impregnated inanimate products on the market, there are no studies demonstrating their effectiveness in preventing the spread of household infections. The effect of an antibiotic on a bacterium depends on the binding of the antibiotic to the germ. It is not at all clear how an antibiotic chemically integrated into an inanimate object can effectively bind to a germ that lands on that object.

Household Air and Water Quality. The quality of the water and air in your home, as it affects infections, is discussed, along with community air and water quality, below.

To bring this section on household hygiene to a close, the summary rules in Table 9.7 will make your home less likely to be a magnet for germs that cause infection.

Community Hygiene

This section reviews community determinants of your child's risk for infections. In addition to focusing on issues of water and air quality, I will review the community's role in pandemic control, for example, in the event of a bird

Table 9.6 Classes of antibiotic-containing inanimate household objects

Air filtration
Appliances
Bathroom fixtures
Bathroom safety
Bedding
Cleaning products
Flooring
Food storage
Garden
Humidifiers
Heating, ventilation, and air conditioning
Kitchen fixtures
Office products
Paint and wallcoverings
Personal care
Pet products
Sealants and grouts
Spas
Towels
Vacuum cleaners
Water filtration/storage
Windows and doors

Table 9.7 Top 10 rules for household hygiene

1. Nothing says food poisoning as convincingly as "steak tartare", "I'll have mine rare," "Let's do sushi," or "It's getting late for dinner; the meat's probably done; let's eat." Cook all meats, poultry, and fish thoroughly.
2. Use bleach-containing products to disinfect surfaces that are likely to be heavily contaminated with germs. Disinfecting is better than simple cleaning in reducing germ counts, and even though the evidence for actual reduction in household infections (versus reductions in germ counts) doesn't exist, there is little extra effort, risk, or cost involved in the use of bleach-containing products.
3. Establish and adhere to a regular household cleaning schedule. Clean heavily contaminated areas, like the kitchen, bathroom, and a child's sick room, more frequently.
4. Use common sense in cleaning and disinfecting your home. Sponges, dish rags, mops, and other cleaning tools can spread infection even more effectively than they eliminate germs if not used properly.
5. Use *hot* water whenever possible for all cleaning and washing, including dishes, laundry, floors, and other surfaces.
6. Follow precautions for reducing the risk of household infection spread by pets and houseflies.
7. Wet is bad; dry is good. Always thoroughly dry everything that you've washed.
8. Redouble your disinfecting efforts in all areas of the home when a household member is sick. It *is* possible (really, I'm serious) to prevent the whole household from getting sick when one child brings home a germ (see chapter 2).
9. Impose a gentle quarantine on contacts between an ill household member and others in the home. Cruel as it sounds, everyone is better off if they keep their distance from someone with an infection. Hand washing should be ferocious after all direct and indirect (touching objects that your loved one has touched) contact. Stay away from used tissues. Have separate towels for everyone and, more than ever, *no sharing anything!*
10. Remember, no single cleaning or disinfecting product has ever been proven to be more effective than any other product of the same class; claims of benefits of cleaning or antibiotic-impregnated products in reducing infections are not subject to review by any government agency.

flu outbreak. Finally, I'll also refresh your memory about the "hot zones" for community-acquired infection of kids: day care centers, schools, petting zoos, doctors' offices, and college dormitories. The risks posed by those unique locales are presented in chapter 2; here, I will review the measures you can take to protect your kids before, during, and after their exposure to the hot zones.

Water Quality

The quality of the water kids drink is an important factor in their health. Community standards for water purification, for example, eliminated early 20th century scourges like cholera from our cities and neighborhoods. Chlorination of public swimming pools limited the spread of poliovirus. Still, challenges remain.

Tap Water. Our drinking water comes from rivers, lakes, aquifers, and reservoirs. Strict control over water purification from the time the water leaves these sources until it reaches our homes is enforced by rules from the EPA. In 1993, the largest recognized *Cryptosporidium parvum* (see chapters 1 to 3) gastroenteritis outbreak occurred in Milwaukee. *Cryptosporidium* is a relatively newly recognized parasite that causes gastroenteritis in otherwise healthy people, but it can cause a fatal infection in people with compromised immune systems. In the Milwaukee outbreak, drinking contaminated tap water caused illness in 400,000 people, 40 of whom died. In response, the EPA revised national water purification standards with the goal of reducing *Cryptosporidium* counts to zero. The EPA estimates that the new regulations will reduce the incidence of *Cryptosporidium*-associated gastrointestinal illness by 90,000 to 1.5 million cases per year, with an associated reduction of between 20 and 300 deaths due to the disease. The additional *Cryptosporidium* treatment requirements will also reduce exposure to other microbial pathogens, such as *Giardia,* that travel with *Cryptosporidium.* Although earlier versions of these regulations were already in effect, the newest and most stringent rules became effective in the fall of 2006.

With increased stringency of water purification procedures, however, comes increased concern about the possible safety risks of the disinfecting processes being used. Coincident with the new rules for microbial decontamination, the EPA has also issued new guidelines for limiting the risks of the purification process itself. Chlorine, for example, is a highly effective water disinfectant, but it can react with the water to form by-products, such as trihalomethanes, haloacetic acids, chlorite, and bromate. These by-products, if consumed in excess over many years, may lead to increased risks of cancer and other diseases. The new EPA standards protect public health by limiting exposure to these disinfectant by-products.

Well Water. Private wells, the source of drinking water for 15% of Americans, are not regulated by the EPA, although local and state agencies may have rules specific for their residents. The EPA does, however, provide private water source safety advice, including testing protocols, ways to spot signs of a potential problem, and sources of possible contamination, at their website (http://www.epa.gov/safewater/privatewells/index2.html).

Bottled Water. The Beverage Marketing Corporation reported that in 2003, bottled water emerged as the second-largest commercial beverage product, surpassing milk and beer and trailing only soft drinks. The market continued to grow in 2004, when the total U.S. volume of product sold surpassed 6.8 billion gallons, an 8.6% advance over the 2003 volume. That translates into an average of 24 gallons of bottled water consumed *per person* in the United States in 2004 and a $10 billion-a-year industry.

Bottled water is regulated as a food by the FDA. Not all bottled water is the same, as a quick read of the labels will tell you. The FDA has provided the definitions in Table 9.8 for accurate labeling instructions.

Table 9.8 FDA definitions for bottled water

- Artesian water: water from a well tapping a confined aquifer in which the water level stands at some height above the top of the aquifer
- Mineral water: water containing not less than 250 parts per million total dissolved solids that originates from a geologically and physically protected underground water source. Mineral water is characterized by constant levels and relative proportions of minerals and trace elements at the source. No minerals may be added to mineral water.
- Purified water: water that is produced by distillation, deionization, reverse osmosis, or other suitable processes and that meets the definition of purified water in the *U.S. Pharmacopeia,* 23rd revision, 1 January 1995. As appropriate, it also may be called demineralized water, deionized water, distilled water, and reverse osmosis water.
- Sparkling bottled water: water that, after treatment and possible replacement of carbon dioxide, contains the same amount of carbon dioxide that it had at emergence from the source
- Spring water: water derived from an underground formation from which water flows naturally to the surface of the earth at an identified location. Spring water may be collected at the spring or through a bore hole tapping the underground formation feeding the spring, but there are additional requirements for use of a bore hole.

About a quarter of bottled waters are produced from reprocessed tap water and are therefore initially subject to the disinfecting processes that the EPA requires of tap water. The remaining sources of bottled water, because they are not in contact with surface water, are felt by the FDA to be less likely to be contaminated with germs, like *Cryptosporidium* and *Giardia,* but bottled waters also are not required by the FDA to be disinfected (although some are) or to be tested for *Cryptosporidium* or *Giardia.* The rules are stricter for bottled waters sold nationally than for those that stay in the state in which they are bottled, but all rules for testing bacterial contamination levels are more lax for bottled waters than for EPA-regulated tap water. The bottom line, from an infection point of view, is that tap water is probably more reliably safe than bottled water!

That having been said, proven outbreaks or infections due to bottled water are rare. The Natural Resources Defense Council reported that two cholera outbreaks, neither in the United States, have been associated with bottled water; the most recent of those was in the U.S. territory of Saipan, where FDA standards should have applied. Typhoid and traveler's diarrhea outbreaks have also been linked to bottled water, according to the Natural Resources Defense Council. Earlier in this chapter, it was noted that drinking bottled water was associated with a higher rate of infection in inner-city households, but the authors could not be sure of the reason—they hypothesized that family members drinking out of the same bottle may have been the explanation. You'll still have to make your own decisions about other, noninfectious safety concerns for your kids regarding tap water, such as possible chemical or toxin contaminations; those issues are beyond the scope of this book.

Water Filters and Purifiers. Whereas filters clean water by passing it through small pores, removing particles too large to pass, water purifiers treat the water to remove contaminants. Small-pore filters can remove germs that are larger than the pores—viruses easily escape through most filters. Purifiers, using reverse osmosis or ultraviolet light, kill the germs. In hospital settings, filters, in particular, have been very useful and have found widespread use in removing germs like *Legionella* (see chapters 2 and 3) from the water system. Despite the many purifier and filter products available commercially for home use, none has been shown to reduce the incidence of any household-acquired infection. Public purification of tap water both filters and treats water before we drink it. Home filtration or purification may have other benefits, for example, the removal of noninfectious contaminants, like chemicals or toxins; those studies and the health impacts of home water treatment are very controversial. From an infection point of view, unless you have an immune-compromised child or family member in the home for whom exposure to even minute quantities of otherwise harmless germs could be dangerous, further treatment of tap water is unnecessary. If your home is served by well water, check with local officials about testing your water for germs and the potential need for an in-home treatment to reduce infection.

Air Quality

Outdoor Air. Although of great concern for many health reasons, outdoor air has not been associated with infectious-disease risk. The only exception to this is when the outdoor air is dusty from a construction site or other excavation; in those settings, soil fungi pose a risk to otherwise healthy individuals and a particular risk for patients with abnormal immune systems (see chapter 2 and below). Other than that unusual setting, bringing fresh outdoor air into indoor environments is one way of *reducing* the infectious risks of indoor air.

Indoor Air. There are numerous infectious complications of poor indoor air quality. Among the germs most commonly associated with indoor air infections are the molds. Recall from chapter 1 that molds are fungi, and together with their cousins, yeasts and mushrooms, they comprise approximately one-quarter of the earth's biomass, i.e., they are everywhere. As a result, the health effects of molds are either great or greatly exaggerated, depending on the point of view of the observer. Molds are hearty and need only a simple food source and water (germs like moisture—is this beginning to sound familiar?) to grow.

Molds can live off organic (living or formerly living) matter, such as wood, paper, textiles, and leather, and can survive without active growth on moist inorganic surfaces, like concrete, glass, and applied paint. Molds grow by digesting the organic material, gradually destroying the surfaces they grow on. Molds appear as surface discoloration, green, brown, gray, or black, and release lightweight spores, which travel through the air. Water-damaged homes are particularly susceptible and in extreme cases, such as homes severely affected by Hurricane Katrina, may become uninhabitable.

The original source of many molds is actually outdoors, particularly where leaves, shrubs, and bushes come close to indoor air intake portals; while the molds are outdoors, they are diluted by the sheer quantity of air around them and cause few human effects. Indoors, however, molds can grow and achieve a relatively prominent concentration, affecting those most sensitive to them. The California Department of Health has listed the potential risk factors, shown in Table 9.9, for high concentrations of indoor home mold.

The human effects of molds can be grouped into three categories: inflammation, allergy, and infection. While our emphasis in this book is on the last, a word or two about the other categories is warranted in light of the great attention that "sick building syndrome" has received in the media.

Molds produce toxins, often carried in spores, which can be inhaled from the air and have been associated with eye and mucous-membrane irritation. When larger quantities of toxins are inhaled, a toxic (noninfectious) pneumonia may result. The symptoms are difficult to distinguish from those that may be due to allergic reactions to mold components. Molds also produce allergens, products that can trigger asthmatic or allergic (e.g., hay fever) attacks in sensitive individuals. As many as 20% of asthmatic patients appear to react to fungal allergens. Less well established as mold-related allergic reactions are symptoms such as headache, vomiting, diarrhea, dizziness, and skin irritation allegedly resulting from these toxins.

Actual *infections* resulting from inhalation of molds are particularly hazardous for people with abnormal immune systems, who can develop severe, often fatal fungal lung infections (see chapters 1 and 2). Aspergillus is a fungus that causes two types of lung disease—invasive aspergillosis, a severe pneumonia (and occasionally a whole-body infection) in immune-compromised patients, and allergic aspergillosis, an asthma-like condition in kids prone to reactive airways (like those with asthma, sinus disease, hay fever, and cystic fibrosis). Usually an occupational hazard, this infection has occasionally been reported from exposure to molds in the home or hospital setting. Dust at construction and other excavation sites has been implicated as a reservoir for aspergillus and a risk factor for infection with the fungus. Aspergillus is not spread from person to person. Another soil fungus spread through the air, cryptococcus, causes severe meningitis and disseminated (whole-body) infection in immunocompromised patients.

Table 9.9 Risk factors for high levels of indoor mold contamination

- Flooding
- Leaky roof
- Sprinkler spray hitting the house
- Plumbing leaks
- Overflow from sinks or sewers
- Damp basement or crawl space
- Steam from shower or cooking
- Humidifiers
- Wet clothes drying indoors or clothes dryers exhausting indoors

Certain other fungi and molds have predilections for specific geographic distributions and weather patterns. Histoplasmosis and coccidiomycosis (see chapters 1 and 2) are very similar diseases causing mild lung infections in healthy hosts and severe, potentially fatal infections in patients with abnormal immune systems. Histoplasma fungi prefer the Midwestern and Southeastern United States and the Ohio and Mississippi River valleys, whereas coccidioides fungi prefer the arid Southwest. These are soil fungi that are distributed through the air by spores of the germ and, like aspergillus, are not contagious from person to person.

A very few nonfungal germs, like viruses and bacteria, can also be spread airborne in indoor air (from one infected person to another [see chapter 2]), but these bugs are more likely to be picked up by other routes (see chapter 2).

Air Filters. Air treatment of indoor environments with air cleaners or filters, in the workplace and at home, may reduce mold levels and help certain asthma sufferers whose disease is triggered by fungal allergens. Prevention of fungal infections has been documented in hospitals equipped with expensive high-efficiency particulate air (HEPA) filters. These filters are defined by their effectiveness at removing tiny particles from the air. In high-risk hospital areas, such as intensive-care units, organ and bone marrow transplant wards, and oncology floors, reduction of even small numbers of airborne germs has proven to be life saving.

HEPA air-filtering units are also marketed for home use. There is no evidence to date that these devices, or other air purification methods, reduce household, day care, school, or workplace transmission of infections.

Airplane Air. The unique situation of lengthy airplane travel during which passengers share a limited airspace highlights the potential for airborne infection (the double entendre of "airborne transmission" is unavoidable here) and opportunities for its control. The recycling of airplane air (see below) may make germs that normally are not readily spread through the air (as opposed to by droplets at close range [see chapter 2]) more transmissible. During the severe acute respiratory syndrome (SARS) (see chapter 3) scare of a few years ago, the WHO determined that several cases of the disease appeared to be acquired during air travel from passengers initially infected in Southeast Asia. The WHO issued recommendations for prevention of SARS transmission on airlines that included careful prescreening of passengers for SARS-compatible symptoms prior to boarding and the use of face masks, a designated bathroom for the symptomatic patient, and careful hand washing in the event that SARS-compatible symptoms developed during flight.

More recently, air travel was implicated for potentially facilitating an outbreak of mumps infection (see chapter 3) in April 2006, the first significant U.S. mumps outbreak in more than 20 years. Mumps, like other respiratory infections, is spread by small droplets coughed or sneezed from one infected individual to another (see chapters 2 and 3); this particular outbreak had its roots on college campuses, perhaps spread by students traveling by plane.

Also of particular concern for airplane spread because of their small-droplet route and their potential severity are chicken pox and measles virus infections and tuberculosis, a bacterial infection (see chapters 2 and 3). More dramatic infections, such as Ebola and Lassa fevers and meningococcal meningitis (see chapter 3), are spread by larger droplets, which tend to float in the air less and sink to the floor more, making them less transmissible (see chapter 2). A potential role for air travel in spreading a future influenza pandemic (see chapter 3 and below) has also been theorized.

The air source in airplanes is compressed outside air, which is brought in through the hot jet engines and then cooled before being circulated in the plane. Airplane manufacturers have designed air systems that use up to 50% recirculated air to decrease the costs of cooling the cabin. Cabin air is exchanged through HEPA filters every 3 to 4 minutes—this compares with 5- to 12-minute exchanges of room air at home or in office buildings and schools. Studies of bacterial contamination in airplane air actually find *fewer* floating particles than in the airport terminal or in city buses. Additionally, the top-to-bottom laminar flow of air in the cabin reduces the spread of germs from row to row, forcing airborne particles to the floor of the aircraft. Nevertheless, transmission of tuberculosis has been documented on airplanes and appears to be a function of how infected the index patient is, how much he/she is coughing, the distance that exposed passengers are seated from the index case, and the duration of the flight. The same is probably true for the recent mumps outbreak. The concern regarding airplane air transmission of dangerous infections was highlighted in May 2007 by the international air travel of an American with a resistant form of tuberculosis. Fortunately, the infected individual was not infectious (see chapter 3) and no other passengers developed tuberculosis. Lawsuits for emotional distress are pending as of this writing.

Although generally safe, airplane air remains a slightly greater risk than other sources of airborne infection because of the closed and confined airspace. As the SARS outbreak and resistant tuberculosis scare demonstrate, our highly mobile society brings us within close contact with individuals with both domestic and exotic infections.

The Community's Role in Pandemic Control—e.g., Bird Flu

As you recall from chapter 3, the 20th century saw three influenza pandemics, the worst of which, in 1918, resulted in millions of deaths worldwide and nearly 700,000 deaths in the United States alone. Fears of an impending bird flu pandemic (see chapter 3) have mobilized governments around the world to formulate contingency plans for nations and communities in the event of such an outbreak.

In February 2007, the CDC issued interim guidelines for limiting the impact of an outbreak of bird flu. These guidelines are predicated on the realization that the most effective containment weapon, a vaccine that matches the exact strain causing the pandemic, is unlikely to be available early in the pandemic, and therefore, stalling techniques are needed to delay the spread

Table 9.10 CDC interim guidelines for pandemic control[a]

Setting and actions	Required for pandemic severity index scores 1–5:		
	1	2 and 3	4 and 5
Home			
Voluntary isolation at home of ill adults and kids; antiviral therapy as available (see chapter 6)	Yes	Yes	Yes
Voluntary quarantine of well household members in homes with ill adults and kids; antiviral prophylaxis as available (see chapter 6)	No	Maybe	Yes
School—child social distancing			
Close schools and day care centers; cancel school activities; reduce out-of-school social contacts and mixing.	No	Maybe; ≤4 weeks	Yes; ≤12 weeks
Workplace/community—adult social distancing			
Decrease face-to-face meetings; use teleconferencing; stagger work shifts.	No	Maybe	Yes
Increase distance between people; reduce density in office space, public transport.	No	Maybe	Yes
Modify, postpone, or cancel public gatherings, theater performances, stadium events.	No	Maybe	Yes

[a]February 2007. Modified from http://www.pandemicflu.gov/plan/community/commitigation.html#XVI. These recommendations must be viewed as a work in progress; updates are available continuously at the website.

of the infection as long as possible to allow mobilization of vaccine development and distribution procedures. The guidelines also recognize the importance of children and schools in the perpetuation of a pandemic (see chapters 2 and 3 and "Schools" below).

Integral to the CDC's recommendations are measures designed to create social distancing, keeping people away from each other to reduce human-to-human transmission. Because these recommendations include closing schools for up to 3 months and other strict measures, they have generated significant discussion and some controversy.

Table 9.10 is modified from the CDC's pandemic flu website (http://www.pandemicflu.gov/plan/community/commitigation.html#XVI); the recommendations are determined in part by the assessed severity of the pandemic. The methods for determining pandemic severity are loosely patterned after hurricane grading and are fully described on the same website; a category 1 pandemic is the least severe and category 5 the most severe.

"Hot Zones" for Kids' Infections—Prevention Strategies

Chapter 2 describes the unique infection risks of certain kid hangouts: day care centers, schools, petting zoos, doctors' offices, and college dorms. In this section, I provide guidelines for protecting your kids before, during, and after their visits to these infection amplifiers.

Day Care Centers

You may have concluded from chapter 2 that there is no such thing as a germ-free day care center. You're correct, but there are things you can do to limit the infection risk for your kids beginning before enrolling in the center (Table 9.11) and continuing while your child is in attendance (Table 9.12).

Schools

The main infection risks in schools, where most kids are potty trained, are from respiratory germs (see chapter 2). Prevention of the spread of respiratory infections requires attention to the personal-hygiene recommendations earlier in this chapter. Kids must be taught good hand-washing habits. The rules for hand washing by school-age kids have to be made in your home—schools, unlike day care centers, cannot enforce hand washing before lunch or after bathroom breaks. Reread "Personal Hygiene" above with your kids to remind them of the 10 most important times of the day to wash their hands—many of these apply to school day activities.

Occasionally, schools are also the sites of outbreaks of more serious infections, like meningitis (see chapter 3). The proximity of kids' desks, as well as their frequent hand contacts with each other, allows the spread by direct contact of meningitis germs from the nose or mouth of one child to the eyes, noses, or mouths of others. School outbreaks of *Neisseria meningitidis* meningitis cause particular concern because of the severity of that infection (see chapters 2 and 3). Prompt notification of health authorities, by teachers and other school personnel, of the first case of meningitis in a classroom can help limit the spread. As a parent, you can help make sure that a serious infection like meningitis is reported—check with your school administrator to verify reporting. Classroom contacts of a child with a case of meningitis sometimes require antibiotics or vaccination against *N. meningitidis* (see chapter 7), but the decisions regarding who needs to receive antibiotics and which children should be immunized (if not already immunized against that germ) rest with local or state health authorities; hence the need for their prompt notification.

Startling statistics from national surveys find that 60 to 70% of high school students are sexually active by the 12th grade—and 40% by 9th grade! Particularly adventuresome kids are having sex *in* school. Prevention of sexually transmitted diseases is discussed in "Personal Hygiene" above, but it's clearly never too early to start teaching school-age kids about sex and its risks.

Routine childhood immunizations (see chapter 7) protect your kids against numerous germs that used to spread like wildfire in schools: polio, pertussis, measles, mumps, German measles (rubella), chicken pox, and three bacterial causes of meningitis (*Haemophilus influenzae, Streptococcus pneumoniae,* and *N. meningitidis*). Before they start school, and all the way through the school years, make sure your kids are fully immunized.

Table 9.11 Screening day care centers before enrolling your child: 10 questions to ask (and the 10 answers to hope for)

1. **What are the policies regarding excluding kids and staff members for illness?**
   There are long lists of inclusion and exclusion criteria published by the American Academy of Pediatrics, the CDC, and many state and county health departments to minimize the risk of spreading day care germs. The important answer that you want to get to this question is that the center has a policy, that it is written, and that you can review it (and perhaps take it to your child's doctor to review).
2. **When can kids come back after being home sick?**
   Return policies should be part of the inclusion/exclusion criteria of the center.
3. **Is there a separate room for kids who have colds or other mild infections?**
   Often kids arrive at the center in the morning and are only noticed to be sick later that day (sometimes desperate parents conceal the illness—I know, it's hard to believe). Ideally, a center should have a separate room or area for kids with minor infections to protect the other kids in the center. Some centers advertise this feature, allowing parents to bring their sick kids to the center without having to hide the sniffles or diarrhea.
4. **Do kids need to have up-to-date immunizations to attend?**
   Kids should not be permitted in the center if they do not have their age-appropriate immunizations. Only about half the states in the United States require immunizations for day care entry (see chapter 7), but you should try to find a center in your state that requires vaccines even if the state itself does not.
5. **Do kids need a physical by a doctor before enrolling?**
   As with summer camp, some day care centers require a doctor's examination and approval prior to day care enrollment. This feature of a center attests to its fastidiousness and care in protecting kids, but it is not a common trait among centers.
6. **Are health records kept by the center?**
   Centers should ideally keep a log of infections, at least the major ones, like chicken pox, measles, meningitis, and hepatitis, in the event that they need to help local health departments trace an outbreak.
7. **What are the hand-washing rules for staff?**
   These rules should be written *and* posted prominently. Personnel should wash their hands with soap and water, or an alcohol hand sanitizer, after every diaper change, potty visit, and nose drip wipe-up event. Washing should also be required before snack preparation and serving. There should be sinks near the diaper-changing areas, and *separate* sinks should be available for food preparation and washing eating utensils.
8. **What are the hand-washing rules for kids?**
   Kids must wash their hands before snacks, after the potty, after returning inside from a playground recess, and after touching pets or birds at the center. Policies for protecting kids from pet exposures should be followed (see "Petting Zoos" in the text).
9. **How are toys, sleep mats, play surfaces, and diaper-changing areas cleaned, and how often?**
   All washable toys and surfaces that kids come in contact with should be scrubbed down with a disinfectant (see the text) every evening after the kids leave; this includes nap mattresses and pads. Diaper-changing areas should be cleaned with disinfectant (bleach-containing solution is preferred) *after each use*, or a disposable paper table cover should be discarded after each use. Any surface contaminated with blood should be immediately cleaned with a 10% bleach solution; other blood precautions must be taken to prevent kids from contacting blood or blood-contaminated areas (see "Personal Hygiene" in the text).

(Continued on following page)

Table 9.11 Screening day care centers before enrolling your child: 10 questions to ask (and the 10 answers to hope for) *(Continued)*

10. **How many kids are in the center each day, what are their ages, are potty-trained kids in contact with kids who are not potty trained, and what is the ratio of staff to kids?**
    Ideally, older kids should be kept separate from the younger ones to prevent the fecal-oral spread of germs. This is very difficult to accomplish, especially for smaller centers; the trade-off, however, is that smaller centers have fewer kids to spread germs to your kids. The lower the ratio of staff to kids, the better the chances for good hand washing and other hygienic measures. Kids not yet potty trained should wear clothes over their diapers.

The important role of schools in accelerating pandemics, e.g., of influenza, has been discussed in chapters 2 and 3. Table 9.10 notes the CDC's recommendations for school closures in the event of a bird flu or other pandemic.

Petting Zoos

As you recall from chapter 2, outbreaks of infections among kids visiting petting zoos and other animal exhibits are common; most transmission is fecal-oral, i.e., the animals' feces get into your kids' mouths. The safety tips for parents of kids visiting animal exhibits in Table 9.13 are those I've excerpted and modified from guidelines issued in 2005 by the National Association of State Public Health Veterinarians, Inc., and endorsed by the CDC, for use by animal exhibitors. Infection prevention steps for animal exhibitors are extensive—my modified list for parents is much shorter and is mostly common sense. However, adherence by parents is particularly important in light of a 2007 study, published in the journal *Clinical Infectious Diseases,* which showed poor compliance with infection prevention guidlines by animal exhibitors.

Table 9.12 Prevention measures you can take while your kids are at day care and when they return home

- Make sure your child's immunizations are up to date (see chapter 7).
- Teach older kids (2 years old and above) about hand washing—how to do it, when to do it, and for how long (see "Personal Hygiene" in the text).
- Tell the day care provider if your child is sick (be honest, now!); this protects other kids in the center, and you would appreciate other parents doing the same to protect your kids.
- Wash your child's hands before leaving the center each day; wash your own hands after helping your child, and turn off the faucet with a paper towel (see "Personal Hygiene" in the text). Alternatively, keep disposable alcohol hand sanitizers in the family van (see the text) and have your kids use them as soon as they climb aboard for the ride home.

Table 9.13 Safety tips for preventing kids' infections at petting zoos and other animal exhibits

- Don't carry food, drinking water, pacifiers, sippy cups, toys, or other objects that kids can put in their mouths into animal areas; try to discourage thumb sucking until hands can be thoroughly washed.
- Don't let your kids touch the animals' food or water; only allow your kids to feed animals with feed provided from a container with which the animals have not had contact.
- Closely supervise your children; discourage them from touching animal waste or soiled animal bedding.
- Thoroughly wash your kids' hands (and your own) after leaving animal areas; carrying disposable alcohol hand sanitizer wipes (see the text) is advisable if you can't be certain that hand-washing facilities are provided.
- The gates, pens, and all other areas near where animals are kept are contaminated with animal germs—treat these areas as if they are the animals themselves, and wash accordingly after contact.
- Eat or drink snacks only in areas far removed from the animals and only after thoroughly washing hands.
- Never allow your kids to put their hands in animals' mouths; feeding animals from an open hand must be carefully supervised to prevent bites.
- Kids should have no contact with animal birthing or newly born animals that have not been thoroughly cleaned.
- Don't let your kids drink unpasteurized milk or eat unpasteurized dairy products that may be served at the exhibit.
- Bring bottled water in case you can't be sure of the quality of the water supply at the exhibit.

Doctors' Offices

The waiting room contagion is real; sick kids in doctors' office waiting rooms can make other kids sick (see chapter 2). Having your child hold his breath during the entire waiting room stay at the doctor's office is impractical, and fortunately, very few germs are airborne (see chapter 2). The most effective measure that parents can take when sitting in the waiting room is a complete hand quarantine—do not let your child touch anything. Every toy, book, and puzzle in the waiting room is covered with the droplets of the previous 30 patients that day. Bring your own entertainment for your child (toys, crayons, etc.), and keep your kids in your lap or at least off the floor of the waiting room. You may feel and even look germ-phobic to the other parents in the waiting room, but you'll never see them again anyway, so what difference does it make, right? If your kids must play with objects or other kids in the waiting room, or if your kids escape your white-knuckled grasp and go to ground, make sure you wash their hands with the ubiquitous liquid alcohol hand sanitizers (see above) mounted on the wall in every examination room.

College Dormitories

Forget about it. If you haven't taught your kids good, safe personal-hygiene practices by the time they go to college, it's too late. The old infectious-disease doctor's favorite riddle (What's the most common infection caught on camping trips to the mountains? Gonorrhea.) applies to college dormitories, as well, although gonorrhea may be the least of the sexually transmitted diseases they have to worry about (see chapter 3). College dormitories may also be the first encounter your kids have with athlete's foot, plantar warts, mumps, food poisoning, recurrent staph skin infections, bed bugs, and mono.

The one thing you can do is make sure they have all of their immunizations up to date before they leave for school. Most colleges now have pre-entry immunization requirements, but you should verify with your doctor that your kids are fully boosted and have had the vaccines especially important in college (see chapter 7): those protecting against meningitis, cervical cancer (papillomaviruses), measles, and mumps.

The Supermarket?

Over the years, the classic high school science fair experiments of culturing various surfaces for germs have crossed over into the medical literature. The most recent study showed that shopping cart handles have more germs than public restrooms; yes, those same cart handles that your kids hold and even lick and teethe on while you're picking out breakfast cereals. Some supermarkets are even offering sanitizer wipe dispensers at the shopping cart bays; Arkansas is said to be considering legislating mandatory wipes for shopping carts.

The shopping cart handle is a metaphor for all inanimate objects—they all carry germs, and those that have more human or animal contact carry more germs. Should you wipe every shopping cart handle before putting your kids in the seat? Should your kids keep their hands off the escalator handrail or use their shirt tail to push the elevator button? Of course not. See this chapter's conclusion, "Balancing Prevention with Preoccupation," below.

The "Hygiene Hypothesis"—Are Your Kids Too Clean?

Over the past decades, there has been an increase in the diagnoses of allergic disorders (e.g., asthma), as well as autoimmune diseases (e.g., inflammatory bowel disease and lupus), both of which represent overexuberant immune system reactions (see chapter 4). Conventional wisdom held for many years that the cause of these increases may be related to environmental pollution and toxins (i.e., the junk in the air is causing more wheezing-type diseases, and the junk in our diet is causing more bowel inflammation, for example).

In the late 1990s, after reunification, a German investigator compared the occurrence rates of allergic diseases among East Germans living under impoverished and unhygienic conditions with those of West Germans living in more pristine and generally wealthier environments. Rather than seeing the expected higher allergy and asthma levels in the poor population, she saw the opposite—the cleaner Germans had more allergies and asthma. This led to the hygiene hypothesis, which states the following: a certain critical mass of germs and dirt is required for the healthy maturation of the immune system. If we clean too much, and prevent too many infections, kids will develop aberrant immune system responses that result in more allergies and autoimmune diseases. That is, the immune system needs to be "taught" to respond normally to everyday challenges and to its own body, and that learning requires a certain amount of germs and dirt, without which the immune system goes awry.

What are the scientific data to support this hypothesis? Right now, we have the potential for a classic epiphenomenon (see chapter 12), the existence of two truths that may or may not be related to each other. Truth number 1 is that there are increasing diagnoses of allergies and autoimmune disorders; truth number 2 is that those disorders tend to occur with higher frequency in wealthier socioeconomic environments. In Africa, where hygienic conditions are poor, the incidence of allergic and autoimmune disorders is lower than in the West. Is this genetic or due to the beneficial effects of poor hygiene? Similarly low levels of allergic and autoimmune disorders are diagnosed in Southeast Asia, but that trend reverses itself when Southeast Asians immigrate to Western countries—their children have Western rates of allergic and autoimmune disorders. That seems to dispel a purely genetic explanation.

However, maybe we are simply better at making the diagnoses of those disorders in the West. What factors other than cleanliness are associated with a higher socioeconomic class in the West that could explain the observation? Clearly, there are many differences between Western societies and African and Asian societies that extend well beyond simple hygiene parameters. Also, how plausible is it that, with the extraordinary number of exposures kids get every day (see chapter 2) at day care, school, and in the backyard, there is still a deficit of critical germs and dirt resulting in allergic and autoimmune responses by the immune system?

Although the data are lacking for a true nexus, that is, a proven connection between the two truths noted above, advocates of the hygiene hypothesis have asserted that overuse of antibiotics, including those in household cleaning products, contributes to the excessively clean environments that pose a risk for our kids (although the West Germans with lower rates of allergies and autoimmune disorders in the late 1990s did not have antibiotic-containing soaps yet). Addressing the antibiotic exposure factor, a paper published in March 2006 assessed eight previously published studies of kids who had received antibiotics in the first year of life and assessed whether that exposure predisposed them to asthma later in life. A small statistical association was found, when all eight studies were combined, to suggest there may be a somewhat increased risk of asthma following antibiotics early in life—the authors caution that the quality of the original eight studies was such that a meaningful conclusion cannot yet be drawn about this subject. A single study published in June 2007 in the journal *Chest* implicated antibiotics in the first year of life *and the absence of a dog in the house* during the first year of life as risk factors for developing asthma by age 7 years. See chapter 12 for advice on how to put a leash on those types of reports!

Furthermore, hygiene hypothesis proponents have contended that vaccines may make kids too clean and that natural exposure to vaccine-preventable diseases would be less risky (see the discussion under "The

Hygiene Hypothesis" in chapter 7). That's because those making the arguments against vaccines for being "overhygienic" have never watched a child develop paralytic poliomyelitis or heard the terrifying air hunger whoop of a child with pertussis.

At this time, I recommend that you continue to have your kids wash their hands, take antibiotics when needed (but *only* when needed), and receive all of their childhood immunizations. Maintaining a clean home will reduce the number of infections passed around in your household, infections that keep your kids out of school and you out of work. Those infections can also be dangerous and even life threatening (e.g., food-borne infections). Community sanitation of water will prevent a retreat to the days of epidemics of cholera and cryptosporidiosis; maintaining air quality standards and removing household cigarette smoke exposure (see chapter 11) have been proven (*proven!*) to reduce asthma and other respiratory ailments. We'll wait and watch together to see if the hygiene hypothesis stands the tests of time, careful study, and reproducibility (see chapter 12). In the meantime, and until proven otherwise, clean is still better than dirty.

Balancing Prevention with Preoccupation

Stop worrying that when you turn your head for a moment to take the ketchup jar off the grocery shelf your child will fall ill from the deadly germs on the only part of the cart you neglected to wipe down. When you get home from the supermarket, have your kids wash their hands; do the same after any experience where your kids are likely to come in contact with lots of germs. Follow the personal-hygiene suggestions earlier in this chapter. (Also, stop worrying that if your kids wash their hands after coming home from the supermarket or the little league game they'll be *too* clean and immediately start wheezing; the hygiene hypothesis has a long way to go.)

Yes, it's possible your child will catch a cold from the viruses left on the cart handle by the drooling toddler who last sat there. It's also possible that your child already has had that same cold and is immune or will catch it tomorrow at day care. You can't sterilize the world. It is understandable if, after reading this chapter, you feel compelled to be compulsive about hygiene. Don't be! Spending your waking hours overbathing your kids, overcleaning your home, and overworrying about the world outside is . . . overkill. It will also drive your spouse and your kids nuts. You don't want your kids growing up to wear rubber gloves whenever they go out in public. (And who knows? Maybe there will turn out to be some truth to the hygiene hypothesis and a semiclean home will be discovered to be better for your kids than a superclean home after all.)

No one can get their kids to wash at each of the top 10 most important hand-washing moments every day or to stand at the sink and sing for 20 seconds each time they wash. However, if you succeed with some of the steps on some days and others on other days, you will still notice a dramatic effect on your kids' numbers of colds, numbers of days of missed school, and numbers of antibiotic prescriptions. The same goes for the cleaning routines for your kitchen, bathroom, and laundry; some attention to effective routines and techniques is better than none. When you have a sick child at home, extra efforts *are* warranted and it's well worth the reputation for neurosis that your family will stick you with, but then you should return to normalcy in your cleaning habits when everyone is healthy. They'll all understand, especially when they see that not everyone in the house has to catch the same cold just because one person caught it. In the case of food preparation, good habits that can prevent serious infections aren't more difficult or more time-consuming than what you're doing now—those are easy changes to make.

Rather than becoming compulsive, simply be conscious of the ways infections are spread and the simple steps that can reduce their spread. Be conscious of the circumstances and locales where infection risks are the highest. Consciousness will result in your doing the right things (ultimately even subconsciously) to keep your kids healthier.

There's a lot more about balancing paranoia with prudence and common sense in chapter 12. Relax, it's going to be okay.

10

From the Fridge to the Fringe: Nutrition and Nutritional Supplements

I know, I know. What the world does *not* need now is another vicious cycle, but that precisely describes the relationship between infections and poor nutrition. Nutritional deficiencies increase kids' risk of infection, and infection causes further malnutrition. The common denominator of this reciprocal relationship is clear: both nutrition and infection dramatically affect our kids' immune systems.

Mothers, of course, have known this forever. The legend and the lore surrounding the importance of what kids eat extend well beyond infection protection. Mothers have been convinced that the right foods improve everything from aptitude and attitude to acne and autism and the wrong foods make everything worse. Some of Mom's mythology has been borne out by science—what our kids eat can affect them far beyond lunch hour. The sophistication of Mom's nutritional notions has increased substantially over the past two generations. Where our kids' grandmothers pushed chicken soup and soothing tea, mothers today read the ingredient labels at the supermarket, looking for antioxidants and micronutrients—such as those found in chicken soup and soothing tea.

The pioneering work of Nevin Scrimshaw, founder of the Institute of Nutrition of Central America and Panama, opened the eyes of the scientific and medical worlds to the critical relationship between infections and nutrition nearly 50 years ago. Overlapping the time frame of the "Golden Age of Virology," when hundreds of new viruses and dozens of virus-disease associations were being discovered (see "Weather and Wardrobe" in chapter 11), international societies and health organizations grew to appreciate the potential for interventions in reducing the world's burden of infectious diseases. Children, in particular, were to be the beneficiaries of this enlightenment, as malnutrition hits children's immune systems especially hard.

A recurring theme in this chapter and in chapter 11 is that innumerable variables, acting singly and in combination, can have dramatic effects on *laboratory* measures of immunity. We can now evaluate all the major and not-so-major branches of the human immune system (see chapter 4) with simple blood or skin tests. Well-established norms have been determined for human humoral (antibody) and cellular responses to immunologic challenges, such as infections, allergic reactions, and vaccinations. A first step for many investigators in determining the potential role of any facet of our lives (nutrition, for example) in our risk for infections involves measuring those immunity blood tests in the presence and absence of the suspect factor—for example, chicken soup. Does chicken soup change our antibody levels or immune cell responses? However, there's an important disclaimer: just because there are changes in the *laboratory* measures of our immune system in the presence of factor X, that doesn't mean that factor X will result in increased or decreased incidence of *actual infections*. In this chapter (and chapter 11), I will make the distinction between the theoretical impact of any factor X, as assessed by changes in the laboratory measures of immunity, and the clinical impact as determined by the incidence or severity of actual infections.

The first part of this chapter, like diet itself, is divided into two major categories: macronutrients and micronutrients. Macronutrients are proteins, carbohydrates, and fats. These macronutrients reflect the overall impact of nutrition on infections (and vice versa), because they encompass all the food groups; of these, protein is the dominant determinant of immunologic health. Micronutrients, in contrast, include individual and specific nutritional components, such as vitamins, minerals, trace metals, and the building blocks of the macronutrients: amino acids, nucleotides, and fatty acids. Our survey of micronutrients will also pry open the Pandora's box of herbal supplements. The chapter concludes with a review of specific foods that may or may not impact kids' infection experiences, including breast milk and, yes, chicken soup.

Macronutrients

Protein

The recognition that infection and nutrition are intertwined came in stages, and like many "ah ha's" in science and medicine, resulted from observation at the extremes. The first associations were made in children suffering extremes of malnutrition. Protein-energy malnutrition refers to a broad deficiency of required nutrients, particularly proteins. Dietary proteins are required for all metabolic functions, i.e., for everything our bodies do. Proteins are also required for the formation of antibodies (which are themselves highly specialized proteins) and other vital immune system components. Energy is a term referring to caloric intake—it is possible to consume adequate calories but inadequate protein; the converse is not commonly seen.

It has long been known that the best source of protein for newborns and young infants is breast milk. Infant formula makers have spent decades trying to replicate the essential elements of mother's milk. Beyond the newborn period, sources of protein include dairy, meat, and vegetable products.

The Evidence for a Relationship between Protein-Energy Malnutrition and Infections. First, I'll describe the effects of protein-energy malnutrition on the immune system as measured in the laboratory. Simply put, every element of children's immunity is adversely impacted by malnutrition; the more severe the malnutrition, the greater the measurable changes. Nearly 200 studies of animals have demonstrated similar immune system damage from protein deprivation. The impact of malnutrition on laboratory measures of immunity is unequivocal and dramatic.

Unlike other correlations that follow later in this discussion, the dramatic clinical effects of protein-energy malnutrition accurately reflect the dramatic laboratory changes. Severely protein-energy-malnourished children have a markedly increased overall incidence of actual infections and a significant impairment in their ability to handle specific infections. Scrimshaw's monograph, *Community-Based Longitudinal Nutrition and Health Studies: Classical Examples from Guatemala, Haiti, and Mexico,* summarizes observations and interventional efforts in impoverished communities in Central America. In one 5-year Guatemalan study, providing supplementary food to pregnant women and preschool children resulted in a 19% decrease in the infant death rate in a village in which that rate had remained steady for the prior 9 years. (Newborns and infants did not receive supplementation because of the universality of breast-feeding in these Guatemalan villages well into early childhood.) In that same village, three epidemics of measles swept through during the 5-year study period with only one death—a child not participating in the feeding supplementation study. Neighboring villages, in contrast, had a nearly 7% death rate due to measles during these outbreaks. The average number of illnesses per child per year in the feeding supplementation village and the number of ill days each year were less than in nonsupplemented villages; the best results were seen in kids who participated more than 75% of the time. Diarrheal illnesses were reduced from 111 cases/100 children per year and 255 cases/100 children per year in two control villages (villages not participating in the supplementation program) to 75 cases/100 children in the feeding supplementation village. Death due to respiratory infections was reduced by more than half in the supplementation village. In addition to the observed benefits in reducing infection, the feeding supplementation village kids grew an average of 3 cm taller and 1 kg heavier after the 5-year study than did kids in the control villages. A parallel study provided optimal medical care, with a full-time doctor and nurse and a selection of medicines, to another village—the benefits in the village that received feeding supplementation without additional medical intervention exceeded the benefits in the village with medical advantages but without supplementary feedings!

An elegant study of the reciprocal relationship between infection and nutrition was performed from 1963 to 1972 in an impoverished Mayan village in Guatemala. The village was used as a natural laboratory in which observations of infections and nutrition were made without intervention. Nutrition fell below adequate levels following diarrheal, respiratory, and skin infections; diarrhea resulted in an average 24% reduction in total protein intake, compounding the ill effects of excessive fecal losses of nutrients during gastrointestinal disease. Infection caused prolonged decreases in nutrition—an epidemic of whooping cough (pertussis [see chapter 3]) resulted in reductions of 50% in the consumption of tortillas, lasting a month or more in many of the children studied. Infections while children were still breast-feeding had minimal impact on body weight, but once children were weaned, their body weights fell off of growth curves for months after an acute infectious disease. Children born prematurely or with other underlying diseases suffered the worst consequences of infection-related malnutrition despite normal growth and development during the breast-feeding months after birth. Most dramatically, severe infections precipitated, within a few days to weeks, profound protein-energy malnutrition (kwashiorkor and marasmus) in almost 3 dozen kids; almost all cases occurred during the rainy months, when diarrhea, measles, and other infectious diseases were most common. In this study, and in a similar one done in Costa Rica, infection was identified as the leading cause of malnutrition, growth retardation, and premature death. The vicious cycle of infection and malnutrition is summarized in the Scrimshaw monograph as follows:

> Recurrent infectious disease (and asymptomatic infection) reduce consumption of the village diet. Infections alter digestion and absorption and cause nutrient losses and metabolic alterations. The net result is progressive wasting and eventual stunting. Infectious disease precipitates acute energy-protein malnutrition, causes disability, and ends in premature death. Malnourished individuals have an altered capacity to respond to infection and to heal. The cycle can be interrupted by control and prevention of infections acting on the childhood population, to diminish the negative effects of infection and improve nutrition, growth, and survival.

Malnutrition thus promotes infection by depriving the body of essential ingredients necessary for healthy immune function. Infection, in turn, promotes malnutrition by decreasing food intake, increasing losses of nutrients through the gastrointestinal tract (via diarrhea, for example), reducing the absorption of vital nutrients from the gastrointestinal tract, increasing the use of and need for specific nutritional ingredients, and preventing those ingredients from getting to where they are needed in the body.

Micronutrients

Dietary Supplements

Okay, here goes. Just about every component of mineral, vegetable, and animal matter can now be purchased in powdered form at the health food store and added to your kids' lunch boxes. Because protein-energy malnutrition is rare in the United States and other industrialized nations, the focus of attention by nutritionists—and entrepreneurs—has shifted to the micronutrients. Micronutrients are marketed and sold under the pseudonym "dietary supplements." Dietary supplements are the Wild West of medicine—unregulated territory where it's tough to tell the good guys from the bad guys and the sheriff has gone fishin'.

In contrast to legitimate medicines, those that require prescriptions and those that are sold over the counter with approval by the Food and Drug Administration (FDA) (see chapter 8), dietary supplements can (and do) make outrageous claims of health benefits without having to account to anyone regarding the veracity or safety of those assertions. The FDA, overwhelmed by their regulatory responsibilities for, well, food and drugs, has taken a pass on dietary supplement oversight. For these "nutriceuticals," the government has put the fox in charge of the hen house (my use of yet another metaphor to describe this situation should give you an idea of how upsetting this all is for many of us in the field).

In 1994, Congress (uh-oh!) defined a dietary supplement as follows:

> A product taken by mouth that contains a dietary ingredient intended to supplement the diet. The dietary ingredients in these products may include: vitamins, minerals, herbs or other botanicals, amino acids, and substances such as enzymes, organ tissues, glandulars, and metabolites. Dietary supplements can also be extracts or concentrates, and may be found in many forms such as tablets, capsules, softgels, gelcaps, liquids, or powders. They can also be in other forms, such as a bar, but if they are, information on their label must not represent the product as a conventional food or a sole item of a meal or diet.

A dietary supplement manufacturer is required to do virtually nothing before marketing a product. In contrast to legitimate medicines (both prescription and over the counter [see chapter 8]), Table 10.1 shows what is *not* required before a dietary supplement appears on your grocer's shelves.

It gets worse. The FDA asks manufacturers of dietary supplements to comply with safety documentation only if a new ingredient is to be produced and marketed—this includes anything not already in the food supply. However, there is no listing of already-existing ingredients against which to compare a new product. The government asks the manufacturer to be responsible for determining whether an ingredient is new (and therefore subject to increased scrutiny), and if so, to notify the FDA. I'm not kidding, really!

Table 10.1 Steps *not* required for selling dietary supplements (that are required for legitimate medicines)

Documentation of safety
Documentation of effectiveness (i.e., backup for claims made for benefits)
Registration of the company with the FDA
Registration of the product with the FDA
Establishment of a minimum standard for manufacturing quality
Verification that what's on the label is what's in the bottle
Standardization of a proper dose, serving size, or concentration
Reporting, investigating, or even keeping records of any injuries or illnesses associated with use of the product

In exchange for this carte blanche in marketing dietary supplements, manufacturers are required by law to add the following disclaimer if their products purport to have specific health benefits (e.g., that they boost your immune system or fight infections). "This statement has not been evaluated by the FDA. This product is not intended to diagnose, treat, cure, or prevent any disease." That magnanimous disclaimer should make you feel better about feeding some of this stuff to your kids.

Having now fully exposed my biases, let me backtrack slightly by saying that it is possible that certain dietary supplements are beneficial for kids and may even reduce their risks of infection. A review of products that have received adequate study (few) and/or excessive publicity (many) follows. Implicit in this experience has been the search for one of the holy grails of medicine—a simple immune-boosting supplement that could prevent infections or reduce their severity.

Amino Acids

Proteins are made up of chains of amino acids, i.e., amino acids are the sole building blocks of protein. It is not surprising, therefore, that much attention has been paid to which specific amino acids may be responsible for the aforementioned critical role of protein in maintaining immunity and mitigating infections. Amino acids come in two major flavors: essential amino acids are those required by the body from the outside (dietary), whereas nonessential amino acids are those that our bodies can make on their own (and therefore are not essential to include in our diet).

Glutamine. Glutamine is a nonessential amino acid under normal body conditions but becomes essential under certain conditions where there is breakdown of cells and tissues. Glutamine is used by rapidly dividing cells, including those of the immune system and gastrointestinal tract. Stressful conditions in children and adults result in increased use of glutamine by the body and a decrease in the measurable levels of the amino acid in the blood. Among the stresses that deplete glutamine by as much as 50% are serious infections, like sepsis and malaria (see chapter 3). In the laboratory, white blood cells (the main immune cells of the body [see chapter 4]) grown under glutamine-reduced conditions don't function properly.

Premature infants have reduced levels of glutamine, likely reflecting the numerous stressful conditions that early birth conveys. In a study of 68 very premature babies randomly assigned either to receive or not to receive supplemental glutamine as part of their nutrition, the control group had a nearly fourfold-increased risk of developing sepsis (see chapter 3) compared with the glutamine-supplemented group. Similar results have been found in studies of adult patients in burn units and critical-care units, where marked reductions in complicating infections occur in glutamine-supplemented individuals compared to those in the control groups not receiving extra glutamine.

Extreme exercise, such as marathon running, results in increased infection (see chapter 11) and in depleted levels of glutamine. In a study of 151 elite runners and rowers, half were given glutamine immediately following exercise and then again 2 hours after exercise, while the other half of the group was given a placebo (a sugar pill [see chapter 12]). All the athletes then completed a questionnaire about any infections they may have had during the week after the exercise. Eighty-one percent of the glutamine-supplemented subjects were infection free for that period compared with only 49% of the placebo control group.

The exact way that glutamine works on the immune system isn't clear, and there are likely several ways. In bone marrow transplant patients, glutamine increases the number of circulating lymphocyte white blood cells (important immune cells [see chapter 4]), perhaps reflecting the speeding up of recovery of the patient's immunity from the depletion that occurs with the transplant. Thus, there may be a direct benefit of glutamine on the immune system. Glutamine has also been shown to protect and help heal gastrointestinal tract cells; therefore, improved absorption of other important dietary components may suggest an indirect benefit of glutamine on infection prevention. In adult AIDS patients, glutamine supplementation reduces diarrhea and increases the intestinal absorption of anti-human immunodeficiency virus (HIV) drugs (see chapter 6), reflecting at least gastrointestinal healing and perhaps direct immune benefits, as well.

Should you give your kids extra glutamine to protect them from infections? No. There have been no studies to determine the potential benefit of this amino acid as a supplement in otherwise healthy kids. Recall that glutamine is a nonessential amino acid except under conditions of tissue stress and breakdown. The patient populations that benefit from glutamine with reduced infections are those with significant stress and tissue breakdown: premature babies, adults in critical-care and burn units, bone marrow transplant recipients, AIDS patients, and extreme athletes. Under normal and usual conditions, your kids' bodies make more than enough glutamine to support their immune system functions. Although it may be true that stress, such as marathon running, may decrease glutamine levels and increase the frequency of common respiratory infections (see chapter 11), it is not at all proven that the two occurrences are related—this may be another example of an epiphenomenon, or coincidence (see chapter 12). Furthermore, the common cold and mild fever illnesses do not deplete blood glutamine levels below those which the body can quickly replenish on its own.

Glutamine is plentiful in numerous foods, including chicken, fish, cabbage, beans, beets, and dairy products. There is no risk of glutamine toxicity from eating these foods, whereas glutamine supplements from the health food store are of unknown benefit with risks for side effects. Just have your kids eat a healthy diet if you're worried about their glutamine levels.

Arginine. Like glutamine, arginine is a nonessential amino acid under normal body conditions, but because of its increased use under stress, it becomes essential during severe infections, like sepsis (see chapter 3). A pig model of sepsis shows some benefit of supplementation with arginine during that infection, but human studies are limited and do not yet justify specific treatment with arginine during severe infection. Preliminary studies of elderly patients with wounds (e.g., bed sores) suggest some benefits from arginine in speeding healing; it is not clear from those studies whether the benefits are the result of immune enhancement or improved blood circulation to the wound. The bottom line is that there is no indication for arginine supplementation to prevent or treat infections in your otherwise healthy kids.

Lysine. An essential amino acid, lysine has been shown in limited studies to inhibit the growth of herpes simplex virus (see chapters 1 and 3) in the test tube. Adults with recurrent herpes genital sores or cold sores had fewer outbreaks, and those that did occur were of shorter duration in the lysine-supplemented group than in the placebo group. This study and a couple of others that showed benefit involved small numbers of patients, and other studies have failed to confirm the benefit. Lysine has not been shown to be helpful in preventing or treating other infections, and no studies have been done with kids. Because lysine is plentiful in potatoes, beans, meat, and other staples of kids' usual diets, there is no evidence to suggest that supplemental lysine given to kids with frequent cold sores will be of any additional benefit, even if lysine works against herpesvirus. Skip the lysine supplements.

Vitamins

The term "vitamins" refers to more than a dozen discrete molecules that are required by the body for a variety of basic functions; deficiencies of any of these molecules result in specific disease states, but excesses can also cause serious conditions and must be avoided. There may be numerous reasons for doctors to recommend vitamin supplements for your children, but for normal immunity and infection-fighting ability, they are usually unnecessary—mother's milk and commercial infant formulas are sufficient vitamin sources for most babies, and a well-balanced diet provides adequate vitamins for those purposes in older kids. The following summary, like others in this book, considers only the infection and immune benefits of vitamin supplementation.

Deficiencies of any of the vitamins may result in measured *laboratory* abnormalities of the immune system in animals and, in a few studies, in humans as well. The discussion that follows reviews only the *actual clinical effects* of vitamin deficiencies and supplementation on the development of and recovery from infections.

Vitamin A and Beta-Carotene. Beta-carotene is the precursor to vitamin A, meaning that beta-carotene is converted by the body into vitamin A; supplements of both are available in health food stores. Patients with vitamin A deficiency have defects in their laboratory measures of the immune system, but supplementation with vitamin A has shown negligible effects on these laboratory tests when given to healthy individuals.

Clinically, a clear effect of vitamin A deficiency has been proven. As with other aspects of the vicious cycle of infection and malnutrition, vitamin A deficiency worsens infection and infection depletes vitamin A levels in the body, resulting in even worse deficiency. Vitamin A deficiency in the industrialized world is rare, so most of the data we have come from developing countries. The broadest and most general evidence for a benefit of vitamin A in preventing infections comes from overall death rate studies in parts of the world where vitamin A deficiency is common, such as Asia and Africa. Overall death rates in children 6 months to 5 years of age have been reduced by 30% by community-wide dietary supplementation with vitamin A. These studies involved very large numbers of individuals, and record keeping was difficult. Most deaths were due to infection, as is true in all developing countries. Reductions in deaths due to measles and diarrhea appeared to make the greatest contributions to the overall reduction in death rate. More focused studies of childhood measles and diarrheal disease in developing countries have confirmed that vitamin A supplementation reduces the death rate and leads to more rapid recovery. However, in striking contrast, vitamin A is of no benefit in pneumonia and other respiratory infections of children in developing countries and may even be harmful, resulting in worse outcomes. A single study showed a benefit of vitamin A supplementation in reducing the severity of malaria in children, whereas benefits have not consistently been seen in several studies of patients with HIV infection.

To reiterate, all of these findings have been in parts of the world where vitamin A deficiency is common. There have been no studies to indicate a benefit of supplemental vitamin A in developed countries or in healthy children or adults with normal dietary intake. Toxicity from excess vitamin A can be very severe. Don't give your kids supplementary vitamin A (or beta-carotene).

Vitamin C. A historical fluke has resulted in an inordinate amount of attention being paid to the potential infection prevention benefits of vitamin C. Linus Pauling won the Nobel Prize in Chemistry in 1954 and, as an antiwar activist, was awarded the Nobel Peace Prize in 1962. The fame and high standing, both scientifically and in the lay world, that these awards brought to Pauling gave his next big endeavor great credence. In 1970, Pauling published a monograph hailing the benefits of vitamin C in the prevention and treatment of the common cold. A double Nobel Prize winner was advocating megadoses of vitamin C for otherwise healthy people who had a cold or wanted to avoid one! Since that publication, dozens of clinical trials of supplemental vitamin C have been conducted—Pauling's prestige brought many other investigators into the fray.

Doses of at least 200 mg per day have been studied for the *treatment* of the common cold, i.e., once the symptoms had begun, in seven different clinical trials. No benefit has been seen in reducing the duration or severity of the cold once it has begun. More than 30 clinical trials of cold *prevention* using at least 200 mg of vitamin C daily have been more encouraging. The cumulative results of these many prevention trials indicate that while there is no decrease in the number of colds that vitamin C supplementers get compared with nonsupplementers (who received a placebo, a sugar pill, instead of vitamin C [see chapter 12]), the colds that do develop in those on supplemental vitamin C are of slightly shorter duration (about 13% shorter, so what would normally be a 5-day cold, for example, would last only 4.3 days). The colds in those taking vitamin C are also slightly less severe (as measured by missed days of work or school) than in placebo recipients. That's a lot of vitamin C (at least 200 mg taken every single day) for a very modest reduction in cold severity and length.

Should you give your kids vitamin C every day? Yes, in food. A reasonable approach would be to always feed your kids a diet rich in vitamin C; that's easy, since the vitamin is so ubiquitous. You might consider further supplementing your kids' diet with vitamin C in pill form (250 mg per day), but only when another household member has a cold. During times of household exposure, the risk of your kids catching a cold is greatest and, as a result, so is the potential benefit of supplementation. In other words, rather than giving your kids supplements every day to try and ease one of the 6 to 10 colds that they will catch each year (see chapter 4), save that intervention for when it is most likely to be needed—during a period when there's a cold in the house and the likelihood of it spreading around is high. On other days, just make sure there's enough vitamin C in your kids' diet.

Vitamin D. Vitamin D is generated in your kids' bodies by sunlight. Possible evidence for an immune system-enhancing role for this vitamin is just now starting to emerge. A study in March 2006 found that African-Americans, whose dark skin doesn't absorb the sun's ultraviolet rays well, make less of an important immune protein used for fighting tuberculosis. In December 2006, another study suggested a link between the low levels of vitamin D in our bodies during the winter and the preponderance of respiratory infections, colds, and flu at the same time (it sounds like another potential epiphenomenon, doesn't it? [see chapter 12]). It's much too early in this story to begin supplementing your kids with vitamin D beyond that which they get in milk and fortified juices.

Other Vitamins. Although there has been much laboratory interest in the potential immune system-enhancing effects of vitamin E and vitamin B complex, there is no evidence that either of them, or any other specific vitamins, prevents or treats actual infections in children.

Minerals

Minerals are elements other than the big four (oxygen, nitrogen, hydrogen, and carbon) that are required for life. They also come in two classifications, bulk minerals and trace minerals. Bulk minerals are those that are recommended at greater than 200 mg/day, whereas trace minerals are recommended at less than 200 mg/day. Deficiencies of bulk minerals, e.g., calcium, magnesium, sodium, potassium, and phosphorus, are uncommon in the United States and have not been specifically implicated in infection risk or immune system health. Much attention has been paid to possible benefits of supplemental trace minerals in immune system enhancement and infection prevention.

Iron. Iron deficiency occurs in 2 to 3% of kids under age 11 in the United States and in 10% of teenage girls, making it the most common micronutrient deficiency in this country. Kids with iron deficiency may have anemia (too few red blood cells) and measurable laboratory abnormalities in their immune tests of blood and skin, and supplementation to correct iron deficiency anemia corrects these abnormalities. However, there has been no evidence that iron replacement therapy reduces actual infections in kids who are deficient and absolutely no evidence of benefit in children with normal iron levels. Although correcting iron deficiency is important for several reasons (particularly to reverse the anemia), preventing infection is not among them, and there is at least a theoretical risk that certain bacteria in the gastrointestinal tract will be made more potent with excess iron (many bacteria require iron for *their* growth).

Zinc. The most widely promoted infection-fighting mineral, zinc can be found in numerous dietary supplements that tout immune system enhancement and infection prevention. In particular, zinc is said to be the common cold's worst nightmare. What are the data?

As with other micronutrients, deficiencies of zinc result in abnormal laboratory immune test results; these observations come largely from developing countries, and zinc deficiency of the magnitude to cause such abnormalities is rare in the United States except among the elderly, alcoholics, HIV-infected patients, and people with gastrointestinal absorption problems. Clinically, when supplements are given to zinc-deficient infants and children in developing countries, decreases in diarrhea and respiratory infections are seen.

In the United States, zinc has been evaluated in numerous studies of adults for prevention and treatment of the common cold. The results have been inconclusive. Zinc lozenges taken every 2 to 3 hours have been reported to reduce the duration of the common cold in adults in some studies and to have no effect in other investigations. Side effects, including burning sensations in the mouth, bad aftertaste, and nausea and vomiting, have been notable in numerous studies.

In a single study of children conducted in 1998, 249 kids ranging from elementary school through high school were randomized to receive zinc lozenges or placebo (sugar pills [see chapter 12]) every 2 to 3 hours while awake beginning within 1 day of developing a cold. There were no differences in the severity or duration of the cold, but the zinc group reported many more unpleasant side effects. In a 2006 study reported by investigators in Turkey, giving zinc to healthy kids between the ages of 2 and 10 as a cold prevention reduced the risk of getting a cold, and if the kids did get a cold, it was shorter and less severe than among those kids getting placebo. Confirmation is needed before any recommendation can be made.

Finally, there have been several hundred patients who have reported (to lawyers, the news media, and anyone else who will listen) long-term (perhaps permanent) loss of smell and taste sensations following the use of a very popular intranasal zinc gluconate spray marketed for the common cold. The manufacturers of the product settled a large lawsuit involving 300 patients in January 2006. The product remains on the market and continues to be heavily promoted.

There is currently not enough justification for the use of zinc products in children for the prevention or treatment of the common cold based upon the evidence available to date, and there may be undesirable side effects and risks.

Selenium. You've heard this story before—selenium is a mineral which, in deficient individuals, is associated with measurable laboratory abnormalities in immune function. Animals artificially made selenium deficient develop more severe infections. HIV-infected humans and humans with severe sepsis (see chapter 3) in intensive-care units may become selenium deficient; the latter patients have higher occurrence rates of pneumonia and death. There is no evidence that selenium supplementation in otherwise healthy people, adults or children, provides any protective benefit against infection.

Probiotics and Prebiotics

Probiotics are live, non-disease-causing bacteria; they are "good germs" and have been evaluated in a number of studies for their potential benefits against "bad germs" and for their potential mitigation of antibiotic side effects. Prebiotics, on the other hand, are substances that naturally promote the growth of good germs in our bodies; in other words, prebiotics result in higher concentrations of probiotics without directly feeding germs to the patient. The most common prebiotics are nondigested sugars (fructo-oligosaccharides and galacto-oligosaccharides) that stimulate the good bacteria to grow in higher numbers. Breast milk has naturally high concentrations of prebiotics.

The theory behind probiotics and prebiotics is that by replacing bad germs with safe ones, infections will be reduced in number and severity. The best known probiotic, yogurt, has been shown to have numerous beneficial effects against infections; the live cultures of lactobacillus strains in yogurt are thought to be the reason. The two genera of bacteria most studied as probiotics are *Lactobacillus* and *Bifidobacterium*. Here is the infection-by-infection evidence for benefits of yogurt and other probiotic and prebiotic products.

Infectious Diarrhea

A cumulative analysis of 2 dozen studies involving more than 1,900 patients (1,450 of them kids) has confirmed that the use of probiotics reduces by more than 1 day the duration of diarrhea caused by gastroenteritis germs, like rotavirus (see chapters 1 and 3). Additionally, the percentage of kids and adults with symptoms lasting longer than 3 days is reduced by one-third and the percentage of patients with diarrhea lasting longer than 4 days is reduced by two-thirds. Although a few of these studies were in developing countries, most were in industrialized countries with very low baseline infant death and diarrhea rates, including Western Europe, Scandinavia, Poland, and Israel; as a result, the benefits can be assumed to extend to kids in the United States.

Studies of probiotics to *prevent* diarrhea have been less convincing than those of diarrhea treatment (i.e., use of probiotics once the diarrhea has already begun) and appear to be of greatest benefit in non-breast-fed babies in developing countries, where the occurrence rates of diarrhea are very high. In the United States, there is insufficient evidence for using probiotic supplements all the time in the hope of reducing the occasional gastroenteritis episodes your kids inevitably get.

When your child develops gastroenteritis, add yogurt to his or her diet. If you want something less convenient and more medicinal sounding, *Lactobacillus rhamnosus* strain GG (aka *Lactobacillus* GG) is the most widely studied probiotic strain and the most consistently shown to be effective. It is available in your health food store.

Antibiotic-Associated Diarrhea

Many children develop diarrhea as a consequence of antibiotic treatment for other infections (see chapter 5). There have been several studies that used probiotic preparations concomitantly with antibiotics during treatment for a potpourri of infections requiring antibiotic therapy. Depending on how the cumulative data from these placebo-controlled studies are analyzed, the conclusion is either that there is clear benefit, with reduction in antibiotic-associated diarrhea risk by more than half, or that more study is needed before a benefit can be confirmed. None of the studies showed apparent side effects of feeding good germs to children on antibiotics.

A reasonable plan of action, if your child needs antibiotics for an ear infection, strep throat, or other reason, is to add daily yogurt (containing live cultures) to his or her diet during the antibiotic course and for several days after the treatment is completed.

Respiratory Infections

There is no intuitive reason to believe that feeding good germs to your kids' intestines, where the germs remain until they are excreted, should do anything to reduce infections that are not connected to your kids' guts. However, in a study of nearly 600 kids in 18 day care centers in Finland, the use of probiotic-supplemented milk resulted in 17% fewer episodes of respiratory infections (ear infections, sinusitis, bronchitis, and pneumonia) diagnosed by doctors than with nonsupplemented milk; there was also a 19% reduction in the prescription of antibiotics for those infections. Although there is an esoteric possible immunologic explanation for reduction in respiratory infections by feeding good germs to kids' intestines, more studies are needed to establish a preventive benefit, if any, against these illnesses.

In one study done 10 years ago and never repeated, a probiotic nasal spray was said to reduce the recurrence rate of ear infections. There is no such product available, and there is insufficient evidence that it would be useful if it were available.

Vaginal Infections

There have been very limited studies of probiotics in bacterial vaginosis and vaginal yeast infections (see chapter 3). The theory here is based on the established truism, as indelicate as it may seem, that vaginal germs derive from intestinal germs via the close proximity of the vaginal opening to the anus. Healthy vaginal fluid contains high concentrations of naturally occurring lactobacillus organisms (remember, lactobacilli are also intestinal bacteria), and lower concentrations of those good germs are often found in women with vaginal infections.

Although there are suggestions of benefit from supplementing standard antibiotic or antifungal therapy with yogurt or another form of supplemental lactobacillus, larger and better-designed studies are needed. Again, there is no harm in adding yogurt to your adolescent's diet if she has a vaginal infection. No studies have even hinted that probiotics can replace conventional antimicrobial treatment.

There are no studies to suggest that a diet rich in yogurt will prevent vaginal infections in women who are prone to them, but again, this is a harmless intervention that has some theoretical basis for a belief that it might help.

Infant Formulas

The latest application of probiotics has been their addition to infant formulas; these products are now available in the supermarket. Here's the theory: breast-fed babies have more good bacteria, specifically *Bifidobacteria* species, in their intestines than do babies fed infant formulas; breast-fed babies have fewer infections than formula-fed babies (see "Breast Milk" below); therefore, the bifidobacteria in the intestines of breast-fed babies must be protective against infections and adding those bacteria to infant formulas will reduce infections in formula-fed babies.

By now, you're getting used to recognizing potential epiphenomena (where two truths may or may not be related [see chapter 12]). The fact that breast-fed babies have more good bacteria in their intestines may or may not have anything to do with the lower rate of some infections in breast-fed babies. As you'll read in the discussion of breast milk below, there are many contents of human milk, including immune cells and antibodies, that may contribute to the lower incidence of some infections seen in breast-fed babies. Although some of the good bacteria in breast-fed babies' intestines may come directly from the breast milk, human milk also contains substances that act as prebiotics to stimulate the growth of good bacteria acquired elsewhere by the baby (see "Breast Milk" below).

There are some additional data, however regarding potential benefits of probiotics in infants. Premature babies are prone to a severe intestinal inflammatory condition called necrotizing enterocolitis; germs, both bacterial and viral, have been implicated in the pathway leading to this disease. Necrotizing enterocolitis occurs less frequently in premature babies fed breast milk than in those fed formula (see "Breast Milk" below). Two clinical studies have shown that premature babies fed probiotic-supplemented formula early in life have a lower risk of developing necrotizing enterocolitis, and less severe disease when it does develop, than premature babies who are not given probiotics. This is a very specific subset of infants (premature babies) with a very specific disease (necrotizing enterocolitis); it would be wrong to generalize the conclusions in these studies to otherwise healthy babies.

Finally, the hygiene hypothesis rears its head again (see chapter 9). There are those who believe that the reason more allergic and autoimmune (where the body attacks its own tissues) diseases are diagnosed now than in the past is because babies are exposed to a too-sterile environment. By adding germs to formula, the advocates believe, a healthier immune system will emerge in formula-fed babies.

To date, there are no data to support any actual clinical benefit of infant formula probiotics in reducing infections in otherwise healthy babies or in promoting a healthier immune system; those are theoretical benefits only. There are also theoretical risks to the use of high doses of bacteria, even good bacteria, in very young babies. The ability of the intestines to contain the billions of bacteria within them is not as well developed in babies as it is in older children and adults. Bacteria may be able to escape (the formal term is translocate) from the intestines into the blood and cause serious infections—this occurs with bad bacteria in babies at a higher rate than in older children. There have been no cases of proven sepsis or other severe infections caused by the good-bacteria probiotics given to babies to date.

The bottom line is that both the benefits and the risks of adding good germs to infant formulas are theoretical. There is not enough information to recommend the use of probiotic-supplemented infant formulas for otherwise healthy babies at this time.

Herbal Remedies

As the name suggests, herbal remedies are products derived from plant sources that are purported to have health benefits. Many such offerings claim immune system-boosting abilities, including bee pollen, bromelain, goldenseal, ma huang, noni juice, wheat grass, and barley grass; there are no convincing data that any of these do anything to help immune function or reduce infection. Some have undesirable side effects. Don't give them to your kids.

The four immune system-enhancing herbal remedies that have received at least some legitimate scientific study are echinacea, andrographis, garlic, and ginseng.

Echinacea

Extracts of the plant echinacea, a purple flower, are among the most widely used herbal products in the United States and Europe. Advocates claim that the herb prevents and/or treats the common cold. Because of the widespread use of this product, it is also among the most studied herbal medicines, with nearly 60 published studies. The quality of many of those studies is poor, lacking proper controls or blinding (see chapter 12).

Three acceptable placebo-controlled trials of echinacea for the *prevention* of colds have been done, and none of the three showed benefit—everyday use of the herb does not reduce the number of colds, nor does it reduce the duration or severity of colds.

Nineteen *treatment* studies of acceptable quality have been conducted comparing echinacea to placebo (a sugar pill [see chapter 12]) in people who already had cold symptoms. Echinacea had a beneficial effect, in reducing either the severity or duration of the cold, in nine of the studies and no benefit in six; the one remaining study showed a trend toward a benefit that was not statistically significant (meaning the effect could not be distinguished from chance alone).

One of the studies that showed no benefit was the only one of the placebo comparison studies that has been done with kids. Either echinacea or placebo was given at the onset of cold symptoms to more than 400 children between the ages of 2 and 11 years living in Washington State. Each child was observed in the study for a total of 4 months. A total of more than 700 colds occurred in those 400 kids, a finding that itself reminds us of just how common the common cold is (see chapter 3). There was no difference between the echinacea-treated kids and the placebo-treated kids in the duration or severity of symptoms of the cold or in parents' perception of the severity of their kids' colds. Side effects were more common in the echinacea group, particularly rash; itchiness, headache, and stomach ache also were more common in the echinacea group.

Do not use echinacea to prevent or treat your kids' colds. It doesn't work, and it has side effects.

Andrographis

An Asian shrub, andrographis has been studied in Asian and European venues but has received little attention so far in the United States. Like other herbal and micronutrient substances, andrographis has test tube effects on the immune system, most notably decreasing inflammation. Almost all of the clinical studies done to date have purported to show benefits in the prevention (one study) and treatment (several studies) of the common cold. These investigations have been hampered by poor study design and concerns regarding conflict of interest, as many have been sponsored by the leading commercial manufacturer of the herb. Publication bias may have a role in the lack of studies showing no benefit—there is often little motivation to write and publish results from negative studies (see chapter 12), particularly if they have been sponsored by commercial interests.

With those caveats in mind, andrographis appears to reduce both the duration and severity of common colds in adults. A single trial in 133 children showed benefit over either echinacea or placebo; one-third of the patients received each of the three possible treatments. Unfortunately, there was no blinding of the study, i.e., patients, parents, and doctors all knew which children were getting which treatment. Unblinded studies cannot be relied upon (see chapter 12).

Andrographis also comes with side effects, including at least three reported cases of severe anaphylaxis (overwhelming allergic reaction). Other side effects include headache, fatigue, swollen glands (lymph nodes), nausea, diarrhea, and a metallic taste.

There is not enough evidence to recommend andrographis for kids with colds, and there are risks. Don't use it.

Garlic

Garlic inhibits certain bacteria in the test tube. It has been studied in a single trial conducted by the Garlic Centre in the United Kingdom and was found, when taken daily for 12 weeks, to reduce the occurrence rate of the common cold (remember, though, that colds are caused by viruses, not bacteria [see chapter 3]; the test tube benefits of garlic have been seen with bacteria) in adults by about one-half. Note my comments above regarding sponsored studies. There have been no studies of kids. Season your family's pasta dinners with garlic to taste. Don't feed your kids supplemental garlic.

Ginseng

Two studies of a commercial ginseng product, sponsored by the manufacturer, showed benefits in preventing common colds and reducing the severity of those that occurred. Note my comments above regarding sponsored studies. There have been no studies of kids. Don't use it.

Miracle and Magical Foods

Breast Milk

Protein-energy malnutrition (see above) is virtually unheard of in nursing babies regardless of the nutritional hardships their mothers may face. Indeed, the term kwashiorkor, referring to one of the most serious forms of protein-energy malnutrition, is derived from a word in the language of Ghana meaning the one who is displaced—the child weaned from the breast when a new sibling is born. In the developing world, breast milk has also long been known to be protective against the ravages of gastrointestinal and respiratory infections. However, because the alternative food and water sources for infants are so often tainted in developing countries and because breast-feeding is ubiquitous in those areas, it has been difficult to determine whether breast milk is independently beneficial or merely not as harmful as the other choices. In any case, it has been estimated that optimizing breast-feeding worldwide would save the lives of more than 1 million children who die of gastrointestinal and respiratory infections. As described below, the benefits of breast-feeding in industrialized countries, where infant formulas and the water used to reconstitute them are not themselves a potential source of infection, confirms the independent value of breast milk in protecting against infection.

Analysis of the components of human milk reveals a wide array of immune system-enhancing substances. Antibodies, the essential infection protection proteins targeted against specific germs (see chapter 4), are plentiful in breast milk. The infection experiences of mothers, as reflected in the antibodies circulating in their bodies, are passed on to babies through breast-feeding. In addition to the important immunoglobulin G (IgG), IgM, IgD, and IgE varieties of antibodies (see chapter 4), breast milk uniquely contains large quantities of secretory IgA antibodies. These potent protective proteins are produced in the breast tissue itself by cells that originate in the mother's intestine and either migrate to the breast or otherwise signal breast cells to produce the same antibodies. This remarkable process is called the enteromammary system and is available to an infant only through human milk. Once consumed by the baby, breast milk secretory IgA antibodies coat the baby's gastrointestinal tract, protecting the infant against germs that Mom has been exposed to, and therefore against germs likely to infect the baby. Cow's milk, in contrast, contains only negligible amounts of secretory IgA targeting only germs to which the lactating cow has been exposed, clearly not as useful a protection for human babies.

Beyond antibodies, which provide direct antigerm activity, human breast milk also contains the protein lactoferrin, which binds tightly to iron molecules. Many intestinal bacteria need iron to do their harm, and lactoferrin deprives these dangerous germs of this sustenance, making more room for the good bacteria that babies need to help them digest milk and food. By another mechanism, lactoferrin blocks the binding and penetration of several important viruses, including herpesvirus, cytomegalovirus, and HIV. Finally, lactoferrin stimulates certain parts of the immune system of the newborn while inhibiting other detrimental inflammatory reactions.

Other beneficial protein contents of human milk are lysozyme, which bursts bad bacteria, and casein, which blocks the ability of certain dangerous bacteria to stick to stomach and respiratory cells while promoting the growth of good bacteria in the intestines.

Carbohydrates in human milk also afford protection against bacteria, bacterial toxins, and viruses by binding to them and preventing them, in turn, from binding to babies' intestines and airways. Breast milk also contains carbohydrates that serve as prebiotics (see above), promoting the growth of good bacteria in infants' guts. Fats in human milk are broken down to form smaller molecules, which inactivate germs that contain fat in their own outer walls, such as enveloped viruses (including herpesvirus and HIV) and many bacteria, fungi, and parasites. Still other breast milk compounds help stimulate the development of the infant's immune system. Human milk is also rich in antioxidants, anti-inflammatory enzymes, and vitamins.

The Evidence That Breast Milk Protects against Actual Infections. Okay, there's a lot of good stuff in human breast milk, but what is the evidence for infection prevention? The discussion that follows emphasizes study results from industrialized countries where the simple avoidance of potentially contaminated alternatives to breast milk cannot be credited for reduced infections. In other words, if breast-fed babies in the industrialized world have fewer infections than formula-fed babies, it's really due to the benefits of breast milk and not the avoidance of the dirty water and contaminated formula that plague the developing world. In a nutshell, the evidence for breast milk protection against infection is overwhelming.

Respiratory illnesses, including wheezing and bronchiolitis, during the first 4 months of life occur less frequently in babies breast-fed for the first month. Babies breast-fed for the first 3 months have less vomiting and fewer diarrheal illnesses, as well as many fewer respiratory illnesses, and the infections that breast-fed babies do get are less severe and less often require hospitalization. In a study of more than 1,000 infants followed for the first year of life, babies who were breast-fed for at least 4 months developed only half as many ear infections, the scourge of parent and pediatrician alike (see chapter 3), and half as many recurrent ear infections as bottle-fed babies; similar studies have shown identical findings and demonstrated that, once again, the ear infections that do occur in breast-fed babies are shorter in duration and less severe than in babies on formula. The same has been reported for urinary tract infections and necrotizing enterocolitis, a severe and probably infectious gastrointestinal complication seen particularly in premature infants. Another infection seen with higher incidence among premature babies, bacterial sepsis (see chapter 3), also occurs less frequently among breast-fed premature babies.

Breast milk is the ultimate health food. The benefits of breast milk in preventing infections and reducing the severity of infections when they occur are extraordinary. It is rare in medicine (or in any other discipline) to find such absolutely unanimous agreement.

Cranberry Juice

For decades, cranberry juice has been mothers' favorite elixir for urinary tract infections. The scientific basis to hypothesize a benefit rests in the components of cranberry juice (fructose and tannins) that can inhibit the stickiness of bacteria to the walls of the bladder.

Only recently have a few well-done studies proven what mothers have known all along—cranberry juice (or cranberry extract in tablet form) reduces the presence of bacteria in the urine of women and the occurrence rate of urinary tract infections in women who are prone to them. Unfortunately, all studies have had a large dropout rate, meaning the juice wasn't well liked or well tolerated by many of the participants.

The only study of children to date has been of kids with neurogenic bladders, a condition where kids can't empty their bladders completely because of spinal or other nerve damage. These kids are prone to urinary tract infections, and cranberry juice was not beneficial. No adequate studies in otherwise healthy children or adolescents have been done.

If your daughter is prone to recurrent urinary tract infections and likes cranberry juice, there is no harm in trying a glass a day in the hope of reducing the recurrence rate.

Chicken Soup

Well, you knew we had to discuss it sooner or later. Chicken soup has been prescribed for innumerable ailments, including leprosy and hemorrhoids, by mothers and other healers for centuries. In the 12th century, no lesser authority than the renowned Jewish philosopher and physician to ancient kings, Maimonides, wrote, "The meat taken should be that of hens or roosters and their broth should also be taken because this sort of fowl has virtue in rectifying corrupted humours." Since that time, no lesser authorities than my grandmother and her grandmother have comforted our cold and flu seasons with this wonderful excuse for eating matzo balls. What is the evidence?

Since colds are caused by so many different viruses (see chapter 3), if there is to be a single miracle soup against all of them, it probably targets a common pathway of illness. One common pathway may be hydration—soup improves hydration. Another might be congestion; the delectable-smelling steam from the bowl might clear nasal passages. The cysteine amino acid in chicken soup might thin clogged mucus because cysteine is structurally similar to acetylcysteine, a known mucus thinner.

Finally, there is the immune response, the most common of common pathways in disease (see chapter 4). Investigators in the laboratory have studied the immune effects of chicken soup and found that it inhibits the neutrophil white blood cells (see chapter 4). Since neutrophils lead the inflammatory response to many infections and the inflammatory response contributes to how miserable we feel, the theory has some plausibility as an explanation for any clinical benefits seen with the soup.

As for the clinical evidence for benefits of chicken soup in preventing or treating actual infections, there are none.

My recommendation is to feed it to your kids whenever they are sick—it tastes great, makes the house smell wonderful, and shows your kids that you love them. Serve it to your kids when they're not sick, too, for the same reasons.

Others

Hot tea, oatmeal, milk, and cocoa have soothing warmth with no evidence for benefit, but they can't hurt. As for orange juice, see "Vitamin C" above.

It's Pretty Simple: Good Nutrition Is Good for Your Kids

There is ample evidence that a healthy, well-balanced diet is important for preventing and fighting infections. There is little need, however, to supplement a well-balanced diet that contains proteins and moderate amounts of fats and carbohydrates. Vegetables, dairy products, meats, chicken, and fish contain more than adequate amounts of immune system-enhancing vitamins, minerals, and other micronutrients to protect your kids. There are certain situations in which emphasizing foods rich in specific nutrients may be appropriate. They are described in the sections above, as are the rare instances when taking a supplement may be helpful.

Bon appetit!

11

Miscellaneous Momisms: Sleep, Stress, Weather, Wardrobe, Exercise, and Everything Else

In the never-ending list of eggs and chickens laying claim to which came first, the pairing of mothers' intuition about infection and children's actual experiences with infection is a classic. Do kids actually get more infections when they don't wear their boots in the rain, prompting generations of observant mothers to pass that wisdom on to their kids and to their kids' kids, or did mothers make up that association eons ago because it seemed intuitive to them, thus forever embedding a purely imaginary relationship into our consciousnesses and brainwashing countless generations to come? More simply said, how wise is the conventional wisdom about infections? Mom's hygiene hunches and nutrition notions are discussed in chapters 9 and 10, respectively; this chapter covers the rest.

Inherent in most of this discussion, as was true for chapter 10, is the impact of Mom's wisdom on kids' immune systems. If interventions such as warm clothing in the winter and boots in the rain affect the incidence and/or severity of infections, it will be via perturbations in kids' natural defenses. Once again, referring to the immune system review in chapter 4 will be helpful in understanding the Momisms we're about to discuss. Our scope is necessarily limited to infection prevention benefits—other advantages of sleep, stress reduction, and exercise, undeniable though they may be, are left for other books.

Sleep

Perhaps nowhere is the conundrum of reconciling Momisms with reality better illustrated than in an analysis of the role of sleep in preventing and treating infection. The restorative and healing powers of sleep have been touted since Talmudic times. Professor Sonia Ancoli-Israel, in a review of ancient Hebrew texts, found these testimonials.

> Sleep is like food and medicine to the sick. (*Pirkei de Rabbi Eliezer*)

Sleep is the best medicine. It strengthens the natural forces and diminishes the injurious fluids. (*Sefer Shaashu'im* 9)

Sleep is considered to be good, for when a sick person sleeps, he gets well. (*Talmud Berachot* 57b)

In addition, it is taught that if you visit a patient in the hospital, and the patient is asleep, you should not disturb him. (*Rosh al Hatorah, Vayeira,* reinforcing that nothing else is as important as sleep for the healing process)

Standing by the partially closed bedroom door, lovingly gazing at a sleeping child home sick from school, mothers from Biblical times to today can't help but believe that sleep is just what the doctor ordered. After all, sleep is often the only thing kids want to do when they're sick. How can sleep not be good for sick kids? And if we agree that sleep is good for kids when they're sick, mustn't it be true that lack of adequate sleep can make healthy kids sick?

The Evidence

As you remember from chapter 4, there are many, many measurable laboratory parameters of the human immune system. Laboratory tests can determine the quality and quantity of our infection-fighting cells, the amounts of our specific and nonspecific infection-fighting antibodies, the good cytokines and the bad cytokines. True to Mom's predictions, sleep and sleep deprivation have dramatic effects on these laboratory measures of immune functions. Natural killer cells, antigen uptake by other white blood cells, antibody responses, and cytokine levels are all altered by sleep deprivation. In an ironic twist, some of these immune markers also affect sleep patterns—specifically, certain cytokines directly regulate sleep, both during wellness and during infection. Thus, there is a bidirectional relationship between cytokines and sleep: these chemicals affect our sleep patterns, and our sleep patterns affect the release of the chemicals.

Why do kids (and adults) sleep more when they're sick? The biblical restorative and healing power of sleep is mediated by the aforementioned cytokine chemicals that are released by the immune system. These cytokines regulate both sleep and host defenses—a powerful linkage, indeed, to validate our sense of relief and hope when we see our sick kids sleeping soundly. Other cytokines modulate sleep, that is, although they don't serve a primary role in setting and maintaining sleep-wake cycles, they can alter and disrupt those cycles when released into the body during infection. This explains the tortured sleep of sick kids when they have fever or acute infection—a cytokine-mediated imbalance between deep sleep and shallow sleep (REM [for rapid eye movement] and non-REM sleep). Thus, cytokines are released during infection, causing us to be more sleepy (and perhaps more restless in our sleep), and our increased sleep in turn releases other cytokines to fight our infection. It's a tidy theory.

However, while the laboratory evidence to support the ebb and flow of cytokines during sleep and during infection, as summarized above, is irrefutable, what do clinical studies show regarding the impact of sleep and sleep deprivation on the rate or severity of actual infections? Do our kids really get sicker if they don't get adequate sleep? Do they recover faster if they sleep longer? Is the expression "He's sick; he needs his rest" a medical fact or another of Mom's factoids?

When rats are chronically deprived of sleep, they develop overwhelming and fatal infection. Studies examining the actual clinical effects of sleep and sleep deprivation on human infection are difficult to perform because, unlike experiments with rats, isolating the single variable of sleep from all of the other factors that affect human infection is daunting. In naturally acquired infections, each patient is different, as are the types and numbers of germs causing the infection; it is nearly impossible to control such studies for the prior health of the patients; their nutrition, exercise habits, and exposures to other illnesses; and their unique genetic determinants. It is also nearly impossible to standardize the germs causing the infections to determine if sleepy patients get sicker than well-rested ones.

In order to study the effect of just one of these variables at a time, such as sleep, the ideal experiment would involve volunteers carefully matched for other characteristics and then divided into two groups, one allowed a normal sleep period and one with limited sleep hours. Then, each group must be challenged with the same germ in the same numbers, that is, intentionally infected in the context of a research study.

Data from studies of that design are just now beginning to emerge. In a controlled study of sleep deprivation and response to the flu vaccine, healthy young men were divided into two groups: one group was restricted to 4 hours of sleep a night for 6 nights, and the second group was allowed to sleep their prestudy average of 8 hours. After the fourth night, both groups were given the flu shot. Ten days after the vaccine, the sleep-deprived group had less than half the amount of flu immunity (antibodies measured in the blood) as the normal-sleep group. However, by 3 to 4 weeks after immunization, there were no longer differences between the groups. Although it is known that the level of antibodies in the blood directly predicts the chances of getting the flu, until more such studies are performed, we are left with the logical and laboratory-supported, but still unproven, hypothesis that sleep is important for the prevention of and recovery from infection.

The Bottom Line

The laboratory parameters of immunity that are altered with sleep deprivation, coupled with the very limited clinical studies performed to date, suggest the possibility that even brief episodes of reduced sleep result in increased susceptibility to infections. However, we are far from being able to conclude that a late-night movie, an evening out with friends, or cramming for exams until the wee hours will make your kids sick without other contributing factors (not the least of which, of course, is the concomitant exposure to a germ).

Studies of long-term sleep deprivation relative to actual infection risk have not been performed and would be of little relevance to kids, for whom insomnia is, fortunately, a rare problem.

The increased sleepiness experienced by kids with infections probably tells us that Mom is right about sleep being important for recovery. There are few biological response patterns that have evolved over time that lack some selective advantage for the survival and well-being of the species. Definitive proof, however, is lacking. You can imagine the ethical dilemma in designing an experiment in which already sick patients are randomly assigned to normal sleep versus sleep deprivation.

Stress

Childhood shouldn't be stressful. How many of us would instantly trade our adult hassles and headaches for the carefree childhoods we remember? Still, anyone lifting our kids' backpacks after school, watching them try out for their competitive soccer teams, or dealing with dating, drugs, dermatologists, and driving lessons knows that childhood today is not what it was for Opie in Mayberry. However, is Mom right? Can kids make themselves sick with stress and worry? Does studying for final exams or SATs really make it more likely that your kids will get colds, the flu, or mono? If your kids are run down by the stress of their hectic lives, will a usually short illness become a chronic problem that they can't shake?

In this section, stress refers to psychological stress (as opposed to the physical stress of sleep deprivation, exercise, noninfectious illness, etc.). Humans are evolutionarily hardwired with what is commonly known as the fight-or-flight response to stress. This connection between mind and body allowed our forebears to respond to predators and allows us to respond to gentler threats, such as schoolyard bullies, obnoxious bosses, and public speaking. In addition to the adrenaline released in the fight-or-flight response, our nervous systems and endocrine systems also release immune system mediators. Predictable and reproducible changes in our immunity, as measured by laboratory testing, result from exposure to stress. However, as we explored in the previous section on sleep, do those interesting laboratory observations translate into an actual increase in the incidence or severity of infections for our kids?

The Evidence

Animal studies have confirmed both laboratory and clinical associations between stress and infection—chronic long-term stress suppresses the immune system as measured by laboratory tests, and the suppressed immune system results in increased infection in several animal models. In contrast, short-term stress in animals results in laboratory changes in the immune system but usually not in dramatic changes in infection risk.

Hundreds of clinical studies of humans (very few of them of children) have confirmed the laboratory half of the association—stress clearly affects the immune system as measured by blood tests. Under experimental conditions, where immune functions are measured before and after subjecting study participants to artificial stressful situations, short-term stress (such as fast-paced math challenges or public speaking for 2 hours or less) results in *increased* immunity. There are increased numbers, and therefore increased overall potency, of natural killer cells and large lymphocytes measured in the blood. Other immune elements are also enhanced during short-term stress—there are more neutrophil white blood cells and cytokines released into the circulating blood.

This redistribution of immune elements with short-term stress is similar to that seen in animals, with the presumed teleological explanation being that important protective cells are dispatched to the areas of the body where they may be most helpful during external challenges, such as stress. This may sound familiar; indeed, similar mobilization of our natural immune responses occurs during acute infection (see chapter 4), a biological form of stress. During acute stress, more antibody is released in human saliva, likely another protective mechanism of our adaptive species.

While all of these measures of overall immunity increase with acute stress, the ability of the body to respond to specific infection during those few hours might be decreased—cells responsible for germ-targeted immunity appear to be suppressed. That is, while overall laboratory-measured immunity is enhanced by short-term stress, the ability to respond to any single germ might be impaired, a difficult contradiction to reconcile and an example, perhaps, of the limitations of comparing only laboratory measures of immunity.

Another type of short-term stress experiment more applicable to our kids is termed "naturalistic" stress. These studies have primarily been performed with young adults taking school tests; these individuals agree to have a blood test before and after the school test. The immune responses in this setting are different from those observed in the artificially induced stress discussed above. Here, there is a shift away from cell immunity toward antibody immunity (see chapter 4). So far, this is another observation without explanation or relevance—it is not at all clear to investigators what the purpose or the result of this shift might be.

Finally, in studies again done exclusively with adults, chronic stress, such as living with a handicap, unemployment, or caring for a disabled family member, broadly depresses almost all functional elements of the immune system. In kids, it has long been known that the thymus gland (one of the most important organs of the immune system [see chapter 4]) actually shrinks during periods of prolonged stress.

What about actual infections, though? Do germs take a thrashing during the enhanced immunity of short-term stress? Do they gain a foothold during episodes of chronic stress-induced immune depression? As was seen in the studies of sleep deprivation discussed earlier, clinical correlates of laboratory-observed immune changes are very difficult to come by. Designing the experiments is difficult, as is recruiting enough well-matched study subjects under the correct conditions of stress. As with sleep studies, experimental-challenge studies are just beginning to emerge, and there is a suggestion that infections are increased under conditions of stress. In a study of adults experimentally infected with influenza virus and then quarantined and observed for immune effects, those volunteers whose psychological profiles prior to the study indicated greater long-term chronic stress had more dramatic changes in their immune systems and developed worse flu symptoms. This correlates with at least three other studies that showed decreased response to flu vaccine in individuals with chronic stress (recall a similar result in sleep deprivation experiments noted above). Another proven example of an actual infection that takes advantage of a weakened immune system due to stress is found with our old friend chicken pox. Recall from chapters 3 and 4 that the varicella-zoster virus that causes chicken pox retreats into dormancy after the initial infection (usually in childhood). The virus can then reactivate as shingles later in life as the immune system weakens with old age or concurrent illness. However, another factor that can reactivate chicken pox from its dormancy and trigger shingles is long-term stress—the weakened immune system under stress fails to maintain the varicella-zoster virus' dormancy in the nerve roots.

The Bottom Line

Ample laboratory evidence suggests that stress affects our kids' immune systems (although, admittedly, very little of the evidence actually comes from experiments with kids). There is far less evidence that the effects of stress measurable in the laboratory translate into more infections. Parents can relax a little; it is doubtful that brief episodes of stress, such as the big spelling bee or cramming for final exams, will make your kids susceptible to more frequent or worse infections. On the other hand, longer-lasting stress, such as that associated with the death of a parent, a disruptive family life, or a chronic (noninfectious) illness, may indeed adversely affect how kids handle germs.

Weather and Wardrobe

William Pepper's classic 1911 work, *The Medical Side of Benjamin Franklin,* reminds us of the brilliance of one of our most famous Renaissance men—statesman, scientist, inventor, and philosopher. This particular story is told through the words of John Adams, who shared a room at an inn with Franklin in 1776.

> At Brunswick, but one bed could be procured for Dr. Franklin and me, in a chamber little larger than the bed, without a chimney and with only one small window. The window was open, and I who was then invalid and afraid of the air of the night, shut it close. "Oh", says Franklin, "don't shut the window, we shall be suffocated." I answered I was afraid of the evening air. Dr. Franklin

> replied, "The air within this chamber will soon be, and indeed is now, worse than that without doors. Come open the window and come to bed and I will convince you. I believe you are not acquainted with my theory of colds?" Opening the window, and leaping into bed, I said I had read his letters to Dr. Cooper, in which he advanced that nobody ever got cold by going into a cold church or any other cold air, but the theory was so little consistent with experience that I thought it a paradox. The Doctor than began a harangue upon air and cold, and respiration and perspiration, with which I was so much amused that I soon fell asleep and left him and his philosophy together.

Almost a full century before the seminal works of Koch, Pasteur, and Semmelweis advanced the germ theory (see the introduction to this book), Franklin observed that seamen, spending their days in wet clothes, didn't become ill until they were cooped up with others on shore. He wrote, as quoted in Pepper's book:

> People often catch colds from one another when shut up together in small close rooms, coaches, etc., and when sitting near and conversing so as to breathe in each other's transpiration.

We'll discuss weather and wardrobe together, and not just for their alliterative allure; the linkage, of course, is the possible role of body temperature in contributing to infections. If you were raised like I was, using the word *possible* in the preceding sentence would be, well, impossible. There was no doubt in my mother's mind about the critical roles that warm and dry played in our health. I was born in 1953, and that's the year I began wearing my boots in the rain. It was in the Helen Rotbart School of Disease Theory that I earned my first "M.D."—in microbial denial. All infections, and perhaps all diseases, were caused by a disturbance in the delicate balances between hot and cold, wet and dry, nutritious and not nutritious.

A quick look back at the chronology of virus discovery gives us a clue as to why temperature trumped technology for many of our mothers. Although certain viruses, like rabies virus, were known to exist as far back as the 19th century, it wasn't until the middle of the 20th century that landmark developments in the laboratory identification of viruses allowed them to become household words (at least in some households). The experimental infection of mice with the coxsackie viruses (see chapter 3) in 1948 and the Nobel Prize-winning laboratory advance that grew polioviruses in cell culture in 1949 opened the doors to the discovery of new viruses and, more importantly, new disease associations with viruses. The 1950s and the 1960s became the "Golden Age of Virology," a time when hundreds of new viruses were identified and associated with dozens of diseases that previously were only suspected of being infectious at all. It wasn't until 1956, for example, that the first rhinoviruses (see chapter 3) were successfully identified and grown in the laboratory.

Thus, modern virology emerged at a time when Mom was very busy raising her kids and hand laundering cloth diapers. She can be forgiven for missing a few new scientific discoveries that may have changed her perspective on the relative importance of boots versus bugs. Now, however, with the benefit of another half-century of research on infections, just how wrong was she?

The Evidence

Let's start with the term "the common cold." The linguistic linkage between this most ubiquitous of infections and the climate transcends English to numerous other languages, proving that the nomenclature is not a coincidence; the sniffles are called a cold because of the long-established clinical observation that colds are more frequent when it's cold outside. On the surface, that would suggest that cold climate and cold noses mean more infections, proving Mom right! Or maybe Mom simply fell for the true-true-but-unrelated trap (see chapter 12) and getting chilled has nothing to do with getting sick. The association of colds with the cold could mean, as Benjamin Franklin suggested (see above), that by spending more time indoors in the winter we have more intimate exposures to sick people, or perhaps certain viruses survive better in cooler climates and therefore cause more infections during cold times of the year. Then again, maybe our nasal mucosal surfaces are more raw and susceptible to virus infection when the weather is cold and dry.

Beyond the seasonality of colds, the first biological clues to the possibility that body temperature influences our risk of infection come from observations made by those who discovered how to grow viruses in the laboratory. While most viruses prefer incubation temperatures near normal body temperature (98.6°F), the rhinoviruses prefer a cooler growth environment (91.4°F). The rhinoviruses are the most common of the common cold bugs (see chapter 3), and they grow in the nose, which is a cooler spot in the body than deeper tissues; nasal temperature, particularly in the front part of the nose, may be as low as 86 to 88°F when room temperature air is breathed.

If our kids wore warm wool scarves over their noses during cold weather, would they raise their nasal temperatures and make it more difficult for viruses to grow? The study has never been done. In fact, few studies of weather and wardrobe effects have been attempted. The best study to refute the "wear your boots in the rain" Momism was done nearly 40 years ago and, until recently, has been held as the definitive word, the dogma that fills medical students' textbooks. Researchers studied 49 volunteers (they were prison inmates, an acceptable population from which to recruit "volunteers" in those days) whose blood tests showed no prior exposure to a particular strain of rhinovirus. All of the study participants were then intentionally inoculated, in the nose, with this particular rhinovirus strain. One-third of the subjects were not exposed to cold temperatures at any time in the study, whereas two-thirds were subjected to either a cold room (39°F) or a cool bath (89 to 90°F). Cold exposure came at different times for different subjects, either at the same time the virus was received, during the incubation period, during maximal illness, or during recovery. Various laboratory measures were taken—virus quantities in the nose, blood tests for immune responses, and stress response hormone levels were all measured.

Both groups (those exposed to cold and those not exposed) developed typical common cold symptoms, and there were no differences in the quantities of virus shed or in the formation of antibodies in response to the infection. The severities and durations of the common cold symptoms were similar whether the subjects were exposed to cold or not, except in two groups. Among patients reexposed to cold during recovery, 3 of 7 developed a worsening of the symptoms not seen in volunteers who weren't rechallenged with cold temperatures, and 7 of 13 (54%) patients chilled during the incubation period became ill compared with 4 of 10 (40%) controls. The significance of these differences could not be established because of the small number of patients in each study subset. The only other differences observed were in the laboratory tests, where the usual rise in infection-fighting white blood cells after the development of a rhinovirus infection was not seen in the cold-exposed group—rather, their white blood cell counts *fell*. When those volunteers were exposed to cold temperatures during recovery, their white blood cell counts fell again. Hence, even this study that concluded there was no effect of cold climate on common colds found some adverse effects of cold weather on both clinical symptoms and laboratory measures of immunity.

The flaws of this supposedly definitive study dissociating body temperature from infections are in the numbers of subjects studied under each of the variable conditions—some of the groups had only two or three patients, and the largest variable group studied had only nine patients. As a result, although maternal lore about the impact of chilling on getting sick was dealt a blow, it was not a knockout blow in many peoples' minds—especially mothers' minds.

It took 37 years for the "wear your boots in the rain" Momism to strike back, but in the fall of 2005, investigators in England got over their own cold feet and challenged the entrenched disproval of Mom's conventional wisdom regarding cold feet. The results made international headlines because everyone (except, perhaps, the stodgy old germ theory purists) wanted Mom to win this battle.

The Common Cold Centre of the University of Cardiff in Wales recruits students at the university (a somewhat less captive volunteer group than the prisoners in the 1968 study discussed above) and pays them well to subject themselves to a respiratory virus infection in the context of a clinical trial. Typically, one touted remedy or another is tested in half the group while the other half gets a placebo (a sugar pill [see chapter 12]); pharmaceutical companies hoping to prove the utility of their product usually sponsor the research. In this particular study, however, there was no sponsor; the study was done because of the lead investigator's longstanding interest in nose physiology and the common cold.

The Common Cold Centre study involved nearly four times the number of subjects as the 1968 version. Ninety volunteers sat with their bare feet in a bucket of ice water for 20 minutes, while another 90 controls kept their feet in an empty bucket, still wearing their socks and shoes. None of the volunteers kicked the bucket. (Sorry, I couldn't resist.) But seriously, folks, none of the 180 study participants had symptoms prior to the study or developed symptoms within a few minutes of the intervention, i.e., chilling your feet does not immediately cause you to feel or act sick. However, 26 of the chilled-and-wet volunteers (nearly 30%) developed common cold symptoms within the next 5 days versus only 8 warm-and-dry-footed study subjects (approximately 9%). How can this be, you ask? How can 20 minutes in an icy foot bath cause a nearly threefold increase in common cold symptoms? Will wearing their boots in the rain really protect our kids, as Mom has long argued?

The study authors propose a mechanism to explain the effect they observed, a mechanism that has also been bandied about for many years among boot aficionados. Simply said, the theory is that we are all occasionally infected with respiratory viruses; oftentimes the infection remains subclinical, i.e., we don't feel sick and the virus just goes away in a few days. If, however, we experience a body- or nose-chilling event while the virus is loitering in our nasal passages, our defenses are temporarily lowered—or, alternatively, the virus' growth potency is enhanced because it prefers lower temperatures—and the virus stops being nice and becomes symptomatic. Recall from the 1968 study that there were some measurable effects on the immune system during body chilling even though no significant differences in the infection rate were seen. Another immune effect has been hypothesized. A study done 45 years ago showed that chilling the feet causes constriction of the blood vessels in the nose (go figure!). This reduction in blood flow to the nose, the theory goes, reduces the flow of infection-fighting cells and humors.

It is important to note that the recent British study, like the earlier study that found nearly opposite results, has limitations. First, no virology studies were performed—there is no evidence that the common cold symptoms experienced within 5 days of the ice bath were in fact due to the perking up of latent, lingering, or loitering viruses. Secondly, the study was not blinded (see chapter 12). The subjects with their feet on ice knew they had cold feet, and the others knew that their own feet were toasty. Objective measures of the common cold, such as mucus weight or tissue use, were not employed; rather, subjects were given a standard questionnaire (which has been used in many such studies and is accepted as valid) that asked them to score their own symptoms. However, scoring your own symptoms, even with a valid questionnaire, is subject to bias if you remember which bucket your feet were in a mere 5 days before (see chapter 12).

Finally, animal studies in which the extent of cold-temperature challenge has been greater than that in either of the two aforementioned human studies have confirmed, in numerous species, a clear association between extreme cold exposure and increased incidence and severity of infections.

The Bottom Line

Before we discard or embrace Mom's wisdom about bundling up to fight infections, more studies are needed. Although there are few data in the medical literature to support Mom's musings, there are actually equally few data to refute them. Add to that hundreds of years of folklore and conventional wisdom attesting to a relationship between the elements and our nose. Dress your kids warmly in the winter and keep their noses covered, but don't let them stop washing their hands or avoiding sick contacts, since those protective interventions are beyond reproach (see chapter 9).

Exercise

If you're looking for a justification to avoid exercise, this section may help you—but only if the exercise you're avoiding involves a 26-mile marathon run. As with many things related to the health and well-being of our kids (and ourselves), moderation seems to be the byword in the healthy relationship between exercise and infections. As Mom would say after 3 hours of baseball in the backyard, "Enough's enough; come in and rest up before dinner!"

The Evidence

Animal models across several species have shown that animals subjected to exhausting exercise following experimental infection with any number of different germs have higher rates of death from those infections over the ensuing weeks than do animals that are exercised only moderately or not at all.

As with sleep and stress, the effects of exercise on the laboratory measures of the human immune system have been thoroughly investigated. However, in contrast to the paucity of information regarding real-life infections attributable to sleep and stress, the clinical relevance of the immune system changes with exercise has been much more thoroughly studied. We can thank the very proactive exercising constituency in this country for that clarity. Whether because of the cardiac-protective effects of exercise or because of the compulsive attention to health intrinsic to marathon, triathlon, and other extreme-exercise enthusiasts, it has now been well established that exercise does affect the actual rate and severity of infections.

First there are the laboratory studies. A single episode of moderate exercise enhances human immune functions measured immediately after the exercise period. White blood cells of all types, natural killer cells, and blood cytokine levels were all increased (that is, immunity was enhanced) in 35- to 40-year-old women during the hour or few hours after a 30-minute walk. However, in adults who regularly exercise at a moderate level for several months, minimal if any immune changes are sustained during their resting hours, that is, the immune enhancement seen in the laboratory tests is very transient, limited to the minutes immediately after exercise. In contrast to the momentary enhancement of immunity following moderate exercise, white cell and natural killer immune cell counts measured after extreme exertion, such as running a marathon, are depressed for several hours.

In clinical surveys and questionnaires, adults who regularly and *moderately* exercise report fewer infections, particularly upper respiratory tract infections, than do sedentary adults. Several randomized clinical trials found similar results: daily or almost daily moderate exercise reduces sick days by half compared with nonexercising adults in the same studies. A large population-based study of nearly 550 adults examined the relationship between physical activity and days of upper respiratory tract infections. The results confirmed that moderate to vigorous (but not extreme or exhausting) exercise is protective against these infections; there was a nearly 30% reduction in episodes of respiratory infections in the group with the highest level of regular physical activity. This suggests that the transient laboratory improvements in immune function following moderate exercise accumulate over time and are protective by one or both of two possible mechanisms: (i) with enough minutes after exercise every day the chances of developing infection are reduced by those actual minutes or hours of enhanced immunity alone and/or (ii) because there is broader benefit to brief daily boosts, or priming, of our immunity even if those enhancements can't be measured in the laboratory beyond a few minutes. It's as if exercising gets our immune systems in better shape to respond when challenged by infection, even as exercise gets our muscles and cardiovascular systems in better shape to respond to other challenges.

A paradox emerges when studies are done of highly active adults, such as elite marathon runners. The common perception in surveys of these athletes and their trainers is that extreme exercise results in *higher* rates and *greater* severity of infections, much as is seen with animal models. More than a dozen formal clinical studies have now confirmed this perception—there is a markedly *higher* incidence of upper respiratory infections following marathon running.

Hence, the relationship between exercise and infection is nonlinear, or J-shaped. That is, there is not a direct correlation between the amount of exercise and the number and severity of infections. Up to a point, exercise is protective. At moderate to vigorous levels of regular activity, fewer infections are seen. At extreme levels of exertion, increases in infections occur.

All of the studies discussed above, and all of the studies addressing the relationship between exercise and infections, have been done with adults; there are no data for kids or for kids' activity levels. However, from the adult experience, it can only be concluded that regular physical activity is better for kids than prolonged sitting in front of the TV or computer game box for many reasons, not the least of which is the likelihood of fewer infections.

The Big Question—Can Your Kids Play Soccer When They're Sick?

The immune system-boosting effects of regular, moderate exercise and the immune system-depressing effects of extreme exercise should have great relevance for Mom's mandate about resting when sick—but exactly what is the relevance? It is clear that your kids require a healthy immune system to recover from even the mildest of infections (see chapter 4). Infections tax the immune system, stimulating responses in immune cells and immune chemicals. Very few studies have been done (and none in children) to directly address the effect of exercise *during* infections. In order to properly control such studies, investigators would be required to intentionally infect volunteers with germs (to make sure that each study subject gets the same number and the same type of bugs) and then randomly assign them to exercise or nonexercise groups. Animal studies with this design show that exercise during acute viral infection can make the illness worse and increase death rates due to the infection; however, those experiments typically subject the animals to extremes of exercise, such as forced swimming.

One such human study has been performed. A rhinovirus strain was inoculated into 50 previously healthy volunteers who were tested at the beginning of the study and found to have no detectable antibodies to that strain of virus (meaning there was no evidence for past infection with that rhinovirus strain or for immunity to that virus [see chapter 4]). Thirty-four of the infected adults then exercised every other day for a 10-day period. The exercise was at the mild to moderate level (walking at 70% of heart rate reserve); the other 16 infected study volunteers did not exercise. Every 12 hours, the volunteers' cold symptoms were assessed with a symptom survey, as well as the more objective measure of collecting and weighing the facial tissues used. There were no adverse effects on the course of the common cold symptoms seen in the exercise group compared to the nonexercising controls.

In a study of naturally acquired upper respiratory tract infections, similar results were found. Volunteers who were usually sedentary (not regular exercisers) were enrolled in the study when they developed a cold. In contrast to the inoculation study described above, this study undoubtedly involved multiple different community-acquired virus types. Following development of their colds, 11 volunteers exercised regularly, at the mild to moderate level, for the next 7 days, and 11 others did not. Once again, there were no detectable differences in the severity or duration of the common cold illness between the two groups.

A Somewhat Smaller Question—Does Infection Reduce Exercise Capacity?

The issue of infection and exercise capacity is not as great a concern for parents of young sports stars as it is for elite adolescent and adult athletes. It has long been believed by marathoners and other extreme athletes and trainers that resumption of exercise too soon after an infection can lead to a prolonged period of decreased energy and strength. Individual case studies seem to support this belief; the syndrome of postviral chronic fatigue syndrome (see chapter 3) has been described by numerous investigators who care for athletes who have become chronically debilitated following an acute infection. Most case studies involve exhausting exercise during illness—the marathoner who fears becoming deconditioned with a day or two off during a respiratory infection and therefore keeps on marathoning while ill.

For the everyday couch potato or weekend warrior, however, exercise during a mild infection, such as a common cold, does not seem to increase the risk of prolonged recovery. In another study using rhinovirus inoculation of volunteers, physical performance and exercise capacity were not affected in the group that exercised in the days following infection compared with the control group that did not.

The Bottom Line

The President's Council on Physical Fitness and Sports says that mild to moderate exercise is okay to continue during a common cold but that intensive exercise training should be halted for at least a few days during and after a cold and for 2 to 4 weeks following the flu. Most authorities distinguish between "above-the-neck" symptoms, like the common head cold without fever, and "below-the-neck" symptoms that include cough, chills, body aches, and/or fever. It must be said, however, that these recommendations are best guesses—the only supporting evidence comes from the previously discussed laboratory measurements of the immune effects of exercise and from the studies of rates and severity of upper respiratory infections in populations of exercising and nonexercising adults that are summarized above. There are no studies of humans to validate the avoidance of exercise during episodes of the flu or in febrile infections—common sense, Mom's common sense, again prevails in this arena.

When your kids are sick, keep them out of the soccer (lacrosse, baseball, ice hockey, football, or tennis) game. If their illness is a mild common cold, missing today's game and tomorrow's practice is probably enough down time for the virus to run its course and for the immune system to gain the upper hand against the germ. Use your child's symptoms as a guide; if they're feeling better and their runny noses aren't running anymore, you have a marker for the victory of immunity over the germ. For more significant infections, like the flu, several down days, up to a week or more, may be appropriate; again, your child's energy level and symptoms are your best guide.

Everything Else

Mom has so much to say, it's hard to know when to cut her off and wrap up this section of the book without leaving out someone's favorite "but what about" question. Here are a few leftover Momisms for us to dissect.

Vaporizers and Humidifiers—the Mist Myth

Who can forget the intoxicating smell of Vicks VapoRub permeating our sick room from the hot-steam vaporizer machine in the corner?

First, let it be said about hot-steam vaporizers that there is no justification for their use, and there is significant potential risk. The risk comes from countless kids burned by playing with the device or scalded by accidentally tipping the hot-water container on themselves or their sick siblings. This realization many years ago resulted in the replacement of steam vaporizers in most homes with cool-mist humidifiers. Touted by mothers for generations, enough studies have now been done to finally refute the myth of mist, hot or cold. Mist therapy has now failed to prove effective in controlled clinical trials of the common cold, croup, pneumonia, and sinusitis—the four most widely cited infectious indications for mist therapy in past generations.

Despite the lack of evidence, most pediatricians and reference sources on kids' health recommend cool-mist vaporizers to help moisturize nasal passages and thin secretions. There is probably little or no harm to using this treatment, and it will make you feel like you're doing something that would make your own mother proud. Make sure you clean the vaporizer or humidifier frequently, as they can themselves become colonized with potentially harmful bacteria, like *Pseudomonas aeruginosa* and other waterborne bugs.

Household Cigarette Smoke

This one's not a myth, and there's no way off this bandwagon of bad news regarding household smoke exposure. In addition to increasing the risk of asthma and sudden infant death syndrome, both beyond the scope of this book, exposure of kids to environmental cigarette smoke increases the risk and severity of childhood respiratory and ear infections. The danger of increased infection risk appears with exposure to maternal smoking as early as in the womb, where the risk of fetal growth impairment has long been known.

The mechanism of increased infections in kids exposed to household smoke was once thought to be related to the increased infections seen in smoking parents—more infections in the house mean more opportunity for transmission to the kids. Although still likely a contributing factor, parental infections alone don't explain the breadth of childhood infections increased by household smoke. Bronchiolitis due to respiratory syncytial virus (see chapter 3), pneumonia, bronchitis, and hospitalizations due to respiratory disease have all been proven to increase among children in smoking households. Concomitant evidence shows that lung function in kids exposed to household smoke is impaired—the two observations are very likely linked.

The relationship between environmental tobacco smoke exposure and childhood ear infections (otitis media [OM] [see chapter 3]) has been reviewed by the National Cancer Institute, which published this conclusion: "Overall, the epidemiologic data strongly support a relationship between environmental tobacco smoke exposure in the home and either acute OM or OM with effusion, particularly among children under 2 years of age." More than 15 studies of this association have clearly found increased risk for middle-ear disease in kids living in smoking households. In fact, one of those studies found that exposure in the womb (that is, Mom smoked during pregnancy) was even more important than exposure after birth in contributing to childhood ear infections.

Backyards, Backwoods, Birdbaths, and Bug Spray

Mosquitoes and ticks continue to be the most common U.S. vectors for diseases like encephalitis, Lyme disease, and tick fevers (see chapters 2 and 3), but simple preventive measures can reduce their impact. Your kids don't have to stay indoors all summer to keep them safe from insect bites.

DEET (*N,N*-diethyl-*m*-toluamide) became a household word in 1999 when the first cases of West Nile virus (WNV) infection were reported in the United States. WNV is an arbovirus (see chapter 3), transmitted by mosquitoes, that causes a potentially fatal brain infection (encephalitis). As WNV has run its epidemic course and has now settled into a lower endemic (steady, without big bursts of activity) incidence pattern similar to those of other arboviruses, the panic has subsided, but the wisdom gained remains very relevant. An outbreak in 2006 of eastern equine encephalitis, a familiar mosquito-borne arbovirus infection that has long been absent from magazine covers, reminds us that periodic bursts of increased disease activity often occur after an endemic pattern has lulled us into complacency. WNV will have future bursts, as well.

There are three simple steps for you to take to protect your kids against mosquitoes and the germs they carry (including WNV): insect repellent, barriers, and environmental conditioning. We'll discuss the last two first, because they are the most straightforward. When possible, especially during the mosquito feeding hours between dusk and dawn, kids should wear long sleeves and long pants. Yeah, right. This is not a very practical recommendation for the summertime mosquito season, but your kids will occasionally agree to it on camping trips or other somewhat cooler (that is, less hot; long sleeves and long pants are never cool—as in "cool") times of the summer experience. Mosquito nets on infant strollers and cribs and on sleeping tents and outdoor eating areas are well tolerated by kids, if you can just teach them to close the netting when they leave the enclosure.

Environmental conditioning means getting rid of mosquito breeding grounds in your yard; standing water anywhere is mosquito nirvana. Backyard wading pools, birdbaths, flower pots, dog dishes, old tires, etc., should be emptied or drained each evening.

The use of insect repellents for kids has been the subject of much debate over the years because of potential side effects of the most effective compounds. Those safety concerns have been largely resolved, and prudent use of insect repellents for kids is strongly recommended during mosquito season. Two different chemical compounds have been proven to be the most effective, DEET and picardin. There is little experience with kids using the latter. Other, plant-derived products, such as oil of lemon eucalyptus, have also not been tested adequately with children.

DEET has been extensively studied and proven to be both effective and safe. The higher the concentration of the chemical in the spray or lotion, the longer lasting the mosquito protection. Products containing 10% DEET protect for about 2 hours; products with 24% DEET protect for 5 hours. The American Academy of Pediatrics recommends that DEET products be used at the lowest concentration appropriate for the desired amount of protection and that DEET be applied on children only once per day; the highest concentration of DEET approved for use with kids is 30%.

DEET-containing products should not be used for infants younger than 2 months of age, because skin absorption may result in penetration into the blood and deeper tissues. Insect repellents should not be applied over cuts or wounds (where they may be more easily absorbed into the blood) and should be thoroughly washed off the skin when coming indoors. Other precautions concerning DEET that are unique to kids include not applying the products near the eyes and mouths or to the hands (recognizing that the hands regularly go to the eyes and mouth [see chapter 9]). It is best to apply the product to your own hands first and then rub it onto your child's skin—this is true for both sprays and lotions. Don't use repellent to treat skin that is covered by clothing. Additional instructions for safe use of these products are found on the package labels and should be carefully followed.

Combination products of DEET and sunscreen are not recommended for adults or children, although *separate* DEET and sunscreen products can (and should) be used simultaneously—the application schedules and amounts to be applied differ for each product and indication, resulting in an impractical and potentially harmful combination product.

Tick habitats overlap with those of mosquitoes, but ticks prefer mature, dense woods with heavy undergrowth; most backyards have mosquitoes, but few have ticks. Ticks are the major vectors transmitting Lyme disease, Rocky Mountain spotted fever, tularemia, and other nasties (see chapter 3). The three principles for mosquito bite prevention—insect repellents, barriers, and environmental conditioning—also apply to tick protection, but the details differ a little. Long sleeves and long pants, tucked into socks, are protective; light-colored clothing makes ticks that land on your kids more visible. "Tick checks" should be performed regularly, at least every evening after a day outdoors; you should examine your kids' skin head to toe, and hair should be carefully checked. Older kids who are a bit more sensitive about parents examining them naked from head to toe should be taught to examine themselves after a shower using a handheld mirror. Ticks feed for 12 to 24 hours before transmitting germs, so prompt identification and removal of ticks prevents infection.

Environmental conditioning for tick bite prevention includes avoiding heavy wooded areas or staying on wider paths, both often impractical, especially the part about keeping kids on the trail. If your home is in a wooded area, keep the kids' play spots well mowed and cleared of brush, woodpiles, and weeds. The dog and cat are important parts of the environment in the case of ticks—they bring the insects into the house. Mice also bring ticks indoors.

DEET and permethrin are the two repellents proven most effective against ticks. DEET should be used as described above for mosquitoes. Permethrin is not for use directly on the skin—it is sprayed on clothing, which is then allowed to dry; permethrin is also effective for repelling mosquitoes. Dried permethrin on clothing is tightly bound and is not a risk for absorption through the skin. DEET and permethrin can be used together when heavy infestation with ticks or mosquitoes is a problem.

Condoms

Now for an entirely different type of barrier protection.

A discussion about condoms is probably inappropriate for a chapter titled Momisms, because our mothers, well, just didn't discuss that kind of thing. Our generation had to learn about condoms from our friends in the locker room, but many parents today teach their kids the Trojan Tales. Let's start by making the mandatory disclaimer: abstinence is the only sure protection against sexually transmitted diseases (STDs) (see chapter 3). Next, let's state the obvious: many adolescents don't abstain. Approaches to teaching your teens about personal hygiene, including safe sex, are discussed in chapter 9. This section focuses on what condoms do and do not prevent. The assumption in the ensuing paragraphs is that if your adolescents are using condoms, they are using them consistently and correctly. The validity of that assumption is something I'll let you research on your own.

STDs are transmitted either by direct contact with sores (lesions) or by contact with infected fluid (discharge); a much more complete discussion of these two routes is found in chapter 3. Laboratory studies of condoms show that STD germs cannot penetrate the latex material, so as long as the virus, bacterium, or parasite is confined to the condom, it will not spread to the next individual. Therein lies the rub (sorry)—the infectious sores or fluids aren't always restricted to the condom. Sores are less well contained than discharges because not all sores are on the areas covered by the condom. In contrast, infected male discharges from the penis are expelled into the condom and, in turn, infected female discharges from the vagina are blocked from infecting the penis by the condom (recall, again, the assumption of effective condom use in the previous paragraph).

Transmission of human immunodeficiency virus (HIV), the virus that causes AIDS (see chapter 3), has been most extensively studied and most clearly documented to be blocked by correct and consistent use of condoms. Real-life situations in which an HIV-infected individual and his or her uninfected partner use condoms have unequivocally proven their protective benefits. The routes of transmission of hepatitis B virus and hepatitis C virus are similar to those of HIV (see chapter 2), and it is believed that appropriate condom use will prevent transmission of these viruses, as well, although the studies have not been done.

Other fluid- or discharge-transmitted STDs, such as gonorrhea, chlamydia, and trichomoniasis (trich), are probably also protected against by condoms. Some population-based studies of these germs have found significant protection, while others have not. The disparity lies in the comparative lack of rigor in study design for these germs versus for HIV, where great community concern and anxiety have driven the research approach.

STDs transmitted by sores (ulcers) include syphilis, chancroid, genital herpes, and genital warts (see chapter 3). The same germs that cause genital warts, papillomaviruses, are also associated with cervical cancer, and it is with papillomavirus-associated cervical cancer that the strongest evidence for condom benefit in ulcer STDs is found. Numerous studies have shown a lower incidence of cervical cancer in women whose partners use condoms. For other STD germs, the evidence is again mixed. Some studies have shown reduced transmission of herpes and syphilis with condom use, whereas others have not—once again, the difficulties and limitations of study design and rigor explain these differences.

For STDs other than AIDS and cervical cancer, we are left with common sense and intuition, both of which are compelling, rather than definitive scientific proof for the benefits of condom use. The only potential harm of condom use is a false sense of security—your kids must be taught that no intervention other than abstinence is guaranteed to be protective against *any* STD, including AIDS and cervical cancer.

The Bulb Syringe Torture Technique

Believe it or not, the bulb syringe technique is a very common recommendation of pediatricians and pharmacists (and, I believe, sadistic law enforcement officials and prison guards). Here's how the reasoning goes. Young babies get a cold, their noses get stuffy, and they're miserable. These babies are too young to blow their noses, and the gunk is probably too thick to blow out even if they could. To help relieve the babies' discomfort, we'll squirt salt water into their noses and then stick a turkey baster into their nostrils and suck out the thick goo. I'm serious.

As enlightened parents, this is what my wife and I did when our now 19 year old was a baby (and I was supposed to be even more enlightened than my wife, being the pediatrician that I am). Poor Matt screamed and spat and squirmed and sputtered. But, heck, what did he know? Then, during one particularly fateful common cold in the Rotbart household, our compulsive hand-washing policy must have lapsed because my wife got the baby's cold. I reasoned that I could help decongest her with the same handy-dandy salt water and turkey baster strategy, which she agreed to let me try on her, since we had performed this ritual together on Matt several times. It took a while, but I finally convinced the neighbors not to alert the authorities about the screaming and talked my wife out of the divorce filing. That was the last time I ever used or recommended the bulb syringe for a baby's cold.

Recently, nasal irrigation has had a burst of interest among adults as a sinus infection prevention, especially touted by singers and public speakers for whom a stuffy nose and irritated throat can be incapacitating. Try it once; that should satisfy your curiosity and return you to your senses.

The Evidence. I'm sure one day Matt will tell a therapist about his suppressed memories of being held down and basted by the very people he most trusted. There have been no studies to determine the usefulness of the bulb syringe technique, but it doesn't matter. I don't care if it works or doesn't work in shortening the cold or relieving the symptoms. I don't care if it brings world peace and ends world hunger. Don't do it.

IV. WISDOM OF THE AGES

12

Prudent Paranoia: Common Sense

So much to worry about, so little time. As you make your way through the sometimes terrifying news of each day, bombarded by the infectious threat du jour, it's understandable that by the time you tuck your kids in and kiss them goodnight, you're a bit rattled. What was good and safe for kids yesterday may be bad and hazardous for them today. In your parents' day, they'd give you two aspirin and call the doctor in the morning—today, aspirin puts kids at risk for brain-damaging Reye's syndrome (see chapters 3 and 8). It was just yesteryear that an apple a day kept the doctor away, but today, the organic apple juice in your kids' lunchboxes may be harboring deadly *Escherichia coli* bacteria (see chapter 3). If your kids survive their apple juice, the same devious germs may be lurking where you'd *least* expect them—in the spinach salad you lovingly make at home with all those vitamin-packed, antioxidant-rich, garden-fresh greens. Then again, the *E. coli* bacteria may be where you *most* expect them—at the neighborhood fast-food restaurant. Fresh farm eggs? Salmonella food poisoning. Family hike in the woods? Don't drink the natural spring water! Caribbean cruise? Noroviruses may be along for the ride. Dangers lurk in day care centers, petting zoos, even doctors' offices for goodness sake (see chapters 2 and 9). Where can parents hide their kids?

One way to cope with the continuous onslaught of newspaper Fear Factor is to stop reading—not necessarily a bad option. A better choice, however, might be to read with awareness, not alarm; with prudence, not paranoia; and with the good sense to recognize the nonsense. I hope that this book has helped you do those things. When the next bird flu panic story appears, flip back to that section in chapter 3 and take comfort in the knowledge that not every influenza prediction comes true and that not every outbreak becomes a pandemic. As another dire consequence of childhood vaccines surfaces in the news, turn again to chapter 7 to read the science behind the scare tactics. Are you worried that your kids may be dangerously deficient in their daily herbal tea requirements? Chapter 10 will soothe your nerves. Does this winter's cold and flu season make you worry that your kids are too sick too often and that they surely must be overly susceptible to germs? Reread chapter 4 to bolster your defense of your kids' defenses.

In concluding, there are a few more guidelines I'd like to offer you as you develop your infection prevention philosophy for your own family.

The Meaning of a Single Research Finding

Tens of thousands of doctors and scientists are doing research every day in this country and around the world. Every one of them wants to be the next Louis Pasteur, Edward Jenner, Alexander Fleming, or Jonas Salk. Every one of them wants to win a Nobel Prize or at least a research grant to continue their studies. The currency of research, the path to fame and fortune, is presentation and publication. The results of scientific inquiry are presented at meetings and written into formal reports that appear in journals. Attendees at the meetings and readers of the journals learn of new discoveries and use them to advance their own studies, further pushing the envelope.

New discoveries presented at meetings and published in journals are also invitations—invitations for others to reproduce and confirm the results. Reproducibility of results is the single most vital aspect of the scientific process. A single finding is meaningless unless it can be confirmed by others; there are too many ways for errors to creep into research to trust a single result from a single laboratory or clinic. It is true that everything has to start somewhere—each significant new discovery in medicine or science has to begin with a first report. It's also true that every insignificant finding that will never be reproduced and never be proven to be true also starts with a first report.

Here's the problem. The interface between the laboratory or the clinic discovery and you, the parent, is usually provided by the news media. Reporters for newspapers, radio, television, and the Internet don't know which medical and scientific discoveries are significant. They don't know which breakthroughs will stand the test of time, ultimately reproduced and confirmed by others, nor do the reporters know which promising data will fall by the wayside as yet another dashed hope. Reporters do know buzz words that sell stories. Flesh-eating bacteria! Pandemic! Superbug! Autism linked to vaccines! Vitamin Z cures cancer (in mice)! Every first report of a medical or scientific finding that involves a newsworthy disease makes it into the news, whether the finding itself is newsworthy or not. Newspapers dutifully report each new study of mice with cancer, rats with bird flu, monkeys with AIDS, and humans with miraculous recoveries from incurable diseases after taking megadoses of magical powders. Reporters don't wait for confirmation of research findings before broadcasting the results to the world; news has to sell today, not in 3 years when the results of confirmatory research finally become available. The evening news can't wait for the test of time.

Don't fall for the first report of a medical or scientific discovery. Read it, tuck it away in your brain, and wait to see what becomes of it. Read tomorrow's newspaper, the day after the big discovery, when experts in the field have had a chance to weigh in and put today's new finding in some perspective—usually that perspective includes advice to be cautious about overreacting to the report until it can be confirmed by others. However, be forewarned—you'll have to look hard to find that cautionary perspective in tomorrow's news. Whereas the first report was on the front page, the sobering analysis is buried on page 17, and the refutation of the results that may occur in other laboratories and clinics over the coming years may not be reported at all. Negative results (Vitamin Z doesn't cure cancer! Vaccines don't cause autism! Mice die despite magic powder!) don't sell the news.

The best source of advice and guidance on interpreting health news comes from professional medical societies, which issue regular statements to their constituencies. When the American Academy of Pediatrics or the American Academy of Family Physicians issues a statement on antibiotics, vaccines, or vitamins, that statement is the result of thorough analysis, by experts in the field, of a body of medical and scientific literature. All the first reports and single reports and confirmatory reports and reports refuting other reports are amassed and reviewed for patients and parents. These widely recognized and respected medical societies all have websites with easy-to-access and easy-to-read health updates. The reputation of the societies is on the line—the advice they give and their interpretation of medical developments that affect your kids can be trusted. So, before stocking up on the antioxidant mentioned in today's *New York Times* or avoiding the vaccine decried on today's cable news, consult the experts for perspective.

The Placebo Effect

How susceptible are we to the placebo effect? Feeling better after taking the equivalent of an inert, inactive sugar pill is common and probably accounts for much of medical practice leading up to the 20th century. When we're given an antibiotic for a viral illness, we feel better sooner, even though antibiotics are ineffective against viruses. When we take herbs or homeopathic remedies and feel better, how do we know the effect is real and not in our heads—and if it's in our heads, does that make it any less real, as long as we're really feeling better? As the products proliferate, the elixirs escalate, and the advertisements accumulate, how can we separate fact from hype?

The world is full of anecdotes. Anecdotes are stories about isolated events and collections of related or unrelated vignettes. Anecdotes are not science. Anecdotes should not define medical practice. That's why placebo-controlled studies are performed. Let's say your friend's kids seem to get sick less than your kids do and your friend's kids drink the latest antioxidant–vitamin–amino acid–immune system-boosting energy power drink. That's an anecdote. It doesn't prove that the yucky sludgy liquid has anything to do with the number of colds your friend's kids catch. On the other hand, what if, as part of a well-planned research study, 500 kids are assigned to drink the power drink and 500 other kids—identical in every other way to the power drink group—are assigned to drink the same amount of a non-power drink that looks and tastes just as yucky and sludgy as the power drink? None of the study participants know which drink they've been assigned (that's called a blinded study), and not even the doctors doing the study know (that's a double-blinded study; in a double-blinded study, the assignment of study participants to different study groups is done by a third party not involved in assessing the results). Now, with this placebo-controlled, double-blinded study there's a real opportunity to find a meaningful effect of the power drink, if such an effect exists. The non-power drink is an example of a placebo, a sugar pill given to patients who think they are getting the real medicine and may feel better even if the treatment is bogus. When the parents and doctors of the 1,000 kids in this study report the occurrence rate of infections in the kids, any benefit from the placebo effect will be evident from the reduced infections observed in the non-power drink group. That benefit, resulting from placebo alone, can then be subtracted from any benefit seen in the power drink group to see if the power drink contributed any real effect of its own. If there are equal reductions of infections in both groups, the anecdotal experience of your friend's kids can be explained on the basis of a placebo effect. If, on the other hand, the power drink group has more benefit than the control placebo group, your friend may be on to something.

When reading the exciting results of any new medical breakthrough, look for the line in the story that says miracle medicine X was shown to be twice as effective as a placebo in treating dire disease Y. Without that line referring to the placebo control, the miracle may just be another anecdote.

Even if you do find that placebo control line in the story, wait for other investigators to confirm the results (see "The Meaning of a Single Research Finding" above)—placebo-controlled studies aren't perfect. They may not have included enough patients to provide a truly meaningful and interpretable result. The placebo may not have been properly chosen. (Some placebos actually have their own medically beneficial or harmful effect that must be monitored for; in the power drink example above, had water been chosen as the placebo, electrolyte imbalance and water toxicity would have to be considered in interpreting the results.) The patients may not have been properly matched. (Matched means that, except for the drink, the kids in both groups were identical in their demographics and medical conditions. Proper matching is essential in conducting studies comparing one factor of interest in two different groups of individuals. In the power drink example, the kids in the two study groups should have been matched for age, number of children in the family, size of their classroom or day care center, nutritional status, socioeconomic level, and multiple other factors that can independently influence the number of infections that kids get. Only then can any differences between the two study groups be assumed to be due to the power drink.) Single placebo-controlled trials, although much more valuable than anecdotes, almost never stand on their own as proof of a new medical intervention.

Epiphenomena and the Power of Coincidences

The death rate today among adults who ate pickles as children in 1922 is 10 times the death rate among those who ate pickles as children in 1952 and 100 times the death rate of those who ate pickles in 1982. Those remarkable statistics prove that pickles manufactured today are far less harmful than pickles jarred in 1952, which in turn were far less harmful than 1922-era pickles. Alternatively, the same statistics prove that the kids of 1922 are now octogenarians or nonagenarians, that the kids of 1952 are now in their 50s and 60s, and that death rates are a function of age. Do pickles have anything to do with the death rates, or is this another cucumber coincidence?

The science of epidemiologic research is a precise and statistically driven discipline designed to weed out confusing and confounding coincidences. Things happen, and other things happen at the same time—the two occurrences may both be true and real but may or may not be related to each other. Such epiphenomena are regularly, and breathlessly, reported as fact in the newspapers every day. Just check out the health news section of national newsmagazines.

- A rise in the rate of brain cancers parallels a rise in the rate of cell phone use.
- A flurry of cancers is discovered in kids living near power lines.
- Increasing diagnoses of autism are made amid a rise in the use of certain vaccine products (see chapter 7).
- Breast cancer rates are found to be higher in women who took more antibiotics as children (see chapter 5).
- Chlamydia germs are found in the plaque of diseased heart arteries (see chapter 3), or maybe it's gingivitis that actually predisposes to that heart vessel plaque (see chapter 3).
- Overcleaning and antibiotic-containing soap use increase in parallel with increased diagnoses of allergies and asthma (see chapter 9).

It is undeniable that innumerable medically important cause-and-effect relationships have been definitively proven using epidemiologic studies. Cigarettes cause cancer. Aspirin use by kids during flu and chicken pox infections is associated with Reye's syndrome (see chapter 3). High cholesterol is a risk factor for heart disease. An early version of the rotavirus vaccine increased the risk of the intestinal disease called intussusception (see chapter 7). Sexual promiscuity increases the risk for sexually transmitted diseases (see chapter 3), and sexually transmitted diseases increase the risk for infertility and certain cancers (see chapter 3). These associations of diseases with their causes have been made by meticulously designed studies involving very large numbers of patients and controlling for all the potential confounding variables that can lead to false associations. Perhaps it will someday be proven that cell phones cause brain cancer, but studies to date are conflicting and inconclusive; the same is true for the hygiene hypothesis for allergy and asthma (see chapter 9). Large and meticulously controlled studies have now refuted a causal role for mumps-measles-rubella (MMR) vaccine in autism and autism spectrum disorder, an allegation that was made based on small and poorly controlled epidemiologic and laboratory studies (see chapter 7).

It can get even more confusing. Some reserve the term epiphenomenon for false associations made while overlooking clues to the true cause. For example, if *(if!)* breast cancer is associated with antibiotic use as a child (see chapter 5), maybe it's the recurrent infections and inflammation (which prompted the antibiotic use) that are actually responsible for breast cancer. If *(if!)* gingivitis and chlamydia are both associated with heart vessel plaque (see chapter 3), maybe it's inflammation of any kind, not specifically inflammation due to that germ or that gum disease, that damages the heart.

Coincidences reported as true associations lead to allegations reported as fact. These make great newspaper headlines and prime fodder for lawsuits. The false association of MMR vaccine with autism and the allegation that the vaccine was the cause made the front pages; although that association has long since been disproved, have you run into the headline news coverage reporting the study results that vindicated the safety of the MMR vaccine yet?

When the associations become irrefutable on the basis of the high quality and reproducibility of epidemiologic studies, believe them. Until then, be prudent, not paranoid. It may be prudent to have your kids use handset cords on their cell phones to keep the antennae away from their brains, not because cell phones have been *proven* to cause brain cancer, but because it's an easy intervention to use just in case better studies in the future find a link. There is no proven or even alleged harm from handset use on cell phones. That's very different from subjecting your kids to the real risks of diseases like pertussis, meningitis, hepatitis, and cervical cancer because an unproven association has been alleged with the vaccines that would have prevented those diseases.

Common Things Are Common

During the anthrax bioterrorism scare in 2001, emergency rooms across the country were flooded with tens of thousands of patients who had runny noses, dry coughs, and low-grade fevers. Many thousands were treated with the antibiotic ciprofloxacin in case they had anthrax. Among all of those thousands of patients, only 11 actually had inhalational anthrax (see chapter 3); the rest had colds or sinus infections.

It is true that bad diseases often start out looking like everyday aches, pains, colds, and flu, but almost all kids with symptoms that suggest minor illnesses and infections actually have minor illnesses and infections. As the expression around the hospital goes, when you hear hoofbeats, think horses, not zebras. Your intuition as a parent is your best guide in distinguishing mild from severe. If you're a first-time parent, you'll be in the doctor's office for every sniffle, sneeze, and loose bowel movement—that's the way it should be. Never hesitate to check out even the most benign symptom in your kids if you're worried; that's why the doctor is there, and that's why she sends you a bill. You're not bothering her, and even if you are, too bad. By the time you have your second, third, and fourth kids, and even as your first child gets a little older, the visits to the doctor will be fewer and farther between. With experience, you'll know that virus infections are the most common cause of minor illnesses and that time alone heals most virus infections. Hopefully, you'll soon also realize that antibiotics are unnecessary and potentially harmful when used for the wrong reasons, which include viral infections (see chapter 5).

Of course, occasionally uncommon things occur in the guise of everyday minor illness; uncommon doesn't mean nonexistent. That's where your instinct as a parent is most critical. You know your kids better than anyone else, and the most powerful thing you can tell a doctor about your child is "She's just not acting like herself; I know how she gets when she's sick, and this is different." We listen for those words and take them seriously. Don't get me wrong—it is not your responsibility, nor is it even a good idea, to be your child's doctor; that's the reason that smart doctors don't provide medical care for their own kids. It is your responsibility, though, to be a good consumer regarding your child's health care in the same way that you are about every other product and service you utilize. Yes, common things are common, but when the doctor tells you it's a common thing that's making your child sick and you believe otherwise, it's okay to press the case and even seek out another opinion. I hope you'll find, though, that this level of superassertive parenting in the doctor's office will rarely be necessary—because common things are common—and that cough and fever are probably not anthrax.

Common Sense and the Wisdom of the Ages

Many of Mom's most sage tidbits of health advice turn out to have at least some basis in science, at least plausible if not definitively proven (see chapters 9 to 11). When caught between fact and folklore, use your own common sense. If generations of mothers continue to perpetuate a home remedy and there's no known risk to using it, try it—it may work! Home remedies that haven't survived the test of generations probably don't work.

If a friend tells you that there's a huge conspiracy between the government and the drug companies to cover up the dangers of medicines and vaccines, use your own common sense. Today, we are a healthier society than ever before in history—does that sound like an evil conspiracy?

If the doctor tells you to give your child a medicine that you've never heard of and you're not sure your child needs medicine at all, look it up! The access that parents today have to health information is unprecedented and unlimited. However, unfiltered information can be a double-edged sword. There's a lot of bad information out there, too. Pick your sources carefully. The National Institutes of Health, Centers for Disease Control and Prevention, American Academy of Pediatrics, American Academy of Family Physicians, and most major medical centers and medical schools have online resources for looking up both everyday ailments and esoteric afflictions. Read the blogs, chat rooms, and free encyclopedias if you are curious, but trust only established medical institutions and authorities when making important decisions about your kids' care.

Well, that's about it. Despite the catchy title of this book, it's of course impossible to truly Germ Proof your kids. It is very possible, however, to reduce the number of infections your kids get and to make those they do get less troublesome. You can protect your kids without overprotecting them. Avoid the temptation to wear rubber gloves around the house or to spray the handrails with antiseptic every time your kids go up or down the stairs. Take advantage of the low-tech strategies, like strategic hand washing, and the high-tech advances in childhood immunizations.

Finally, you're doing a great job! You're reading this book because you're concerned about your kids' health and want to do what's best for your family. That love and concern are much more important than the specific medical decisions you make, or which advice you choose to follow, or whether you believe in the germ theory or in Mom's theory. Your love and concern guarantee that you will constantly be assessing your choices, analyzing risks and benefits, and assuring yourself that you're doing everything that should be done for your kids. That's the Golden Rule for parents. The rest is just commentary.

Appendix

Names and Usages of Drugs and Vaccines

Table A1. Brand names, generic names, and usages of drugs that your child's doctor might recommend to prevent or treat infections

Brand name	Generic name	Usage[a]	Chapter reference for more information
Acticin	Permethrin	AI	6
Aftate	Tolnaftate	AF	6
Albenza	Albendazole	AP	6
Alinia	Nitazoxanide	AP	6
Altabax	Retapamulin	AB	5
Amoxcil	Amoxicillin	AB	5
Ancobon	Flucytosine	AF	6
Aralen	Chloroquine	AP	6
Ascarel	Pyrantel pamoate	AP	6
Augmentin	Amoxicillin-clavulanic acid	AB	5
BabyBIG	Hyperimmune botulism IVIG	GG	7
Bactrim	Trimethoprim-sulfamethoxazole	AB	5
Bactroban	Mupirocin	AB	5
Bee-Pen VK	Penicillin VK	AB	5
Biaxin	Clarithromycin	AB	5
Ceclor	Cefaclor	AB	5
Cedax	Ceftibuten	AB	5
Ceftin	Cefuroxime	AB	5
Cipro	Ciprofloxacin	AB	5
Cleocin	Clindamycin	AB	5
Coartem	Artemether	AP	6
Co-Trimoxazole	Trimethoprim-sulfamethoxazole	AB	5
CytoGam	Hyperimmune CMV IVIG	GG	7
Cytovene	Ganciclovir	AV	6
Denavir	Penciclovir	AV	6

(Continued on following page)

Table A1. Brand names, generic names, and usages of drugs that your child's doctor might recommend to prevent or treat infections *(continued)*

Brand name	Generic name	Usage[a]	Chapter reference for more information
Desenex	Miconazole	AF	6
Diflucan	Fluconazole	AF	6
Doryx	Doxycycline	AB	5
Dynacin	Minocycline	AB	5
Dynapen	Dicloxacillin	AB	5
Elimite	Permethrin	AI	6
Epivir	Lamivudine	AV	6
ERYC	Erythromycin	AB	5
Ery-Tab	Erythromycin	AB	5
Eryzole	Erythromycin-sulfasoxazole	AB	5
Eurax	Crotamiton	AI	6
Famvir	Famciclovir	AV	6
Femizole	Miconazole	AF	6
Flagyl	Metronidazole	AB, AP	5, 6
Floxin	Ofloxacin	AB	5
Flumidine	Rimantidine	AV	6
Foscavir	Foscarnet	AV	6
Furodantin	Nitrofurantoin	AB	5
Furoxone	Furazolidone	AP	6
Grifulvin	Griseofulvin	AF	6
Grisactin	Griseofulvin	AF	6
Gyne-lotrimin	Clotrimazole	AF	6
Gynezole	Butaconazole	AF	6
HBIG	Hyperimmune hepatitis B IMIG	GG	7
Humatin	Paromomycin	AP	6
Intron A	Interferon alpha	AV	6
Keflex	Cephalexin	AB	5
Ketek	Telithromycin	AB	5
Kwell	Lindane	AI	6
Lamisil	Terbinafine	AF	6
Lariam	Mefloquine	AP	6
Ledercillin VK	Penicillin VK	AB	5
Levaquin	Levofloxacin	AB	5
Licide	Pyrethrin-piperonyl butoxide	AI	6
Loprox	Ciclopirox	AF	6
Lorabid	Loracarbef	AB	5
Lotrimin	Miconazole	AF	6
Lotrimin	Clotrimazole	AF	6
Macrobid	Nitrofurantoin	AB	5
Malarone	Atovaquone-proguanil	AP	6
Mepron	Atovaquone	AP	6
Micatin	Miconazole	AF	6
Minocin	Minocycline	AB	5
Monistat	Miconazole	AF	6

(Continued on following page)

Table A1. *(continued)*

Brand name	Generic name	Usage[a]	Chapter reference for more information
Mycelex	Clotrimazole	AF	6
Mycolog II	Nystatin-triamcinolone	AF	6
Mycostatin	Nystatin	AF	6
Mytrex	Nystatin-triamcinolone	AF	6
Nix	Permethrin	AI	6
Nizoral	Ketoconazole	AF	6
Nydrazid	Isoniazid	AB	5
Nystat-Rx	Nystatin	AF	6
Omnicef	Cefdinir	AB	5
Ovide	Malathion	AI	6
Palivizumab	Monoclonal respiratory syncytial virus antibody	GG	7
PCE Dispertab	Erythromycin	AB	5
Pediazole	Erythromycin-sulfasoxazole	AB	5
Pen-Vee K	Penicillin VK	AB	5
Pfizerpen	Penicillin G	AB	5
Pin-X	Pyrantel pamoate	AP	6
Prevpac	Amoxicillin	AB	5
Principen	Ampicillin	AB	5
Pronto	Pyrethrin-piperonyl butoxide	AI	6
Proquin XR	Ciprofloxacin	AB	5
Qualaquin	Quinine	AP	6
Quinamm	Quinine	AP	6
Quiphile	Quinine	AP	6
Rebetol	Ribavirin	AV	6
Relenza	Zanamavir	AV	6
Retrovir	Zidovudine	AV	6
RID	Pyrethrin-piperonyl butoxide	AI	6
Rifadin	Rifampin	AB	5
RIG	Hyperimmune rabies IMIG	GG	7
Rocephin	Ceftriaxone	AB	5
Roferon-A	Interferon alpha	AV	6
Septra	Trimethoprim-sulfamethoxazole	AB	5
Sporonox	Itraconazole	AF	6
Stromectol	Ivermectin	AP, AI	6
Sulfatrim	Trimethoprim-sulfamethoxazole	AB	5
Sumycin	Tetracycline	AB	5
Suprax	Cefixime	AB	5
Symmetrel	Amantidine	AV	6
Tamiflu	Oseltamivir	AV	6
Tequin	Gatifloxacin	AB	5

(Continued on following page)

Table A1. Brand names, generic names, and usages of drugs that your child's doctor might recommend to prevent or treat infections *(continued)*

Brand name	Generic name	Usage[a]	Chapter reference for more information
TIG	Hyperimmune tetanus IMIG	GG	7
Tinactin	Tolnaftate	AF	6
Tindamax	Tinidazole	AP	6
Tisit	Pyrethrin-piperonyl butoxide	AI	6
Trimox	Amoxicillin	AB	5
Valcyte	Valganciclovir	AV	6
Valtrex	Valacyclovir	AV	6
Vantin	Cefpodoxime	AB	5
VariZIG	Hyperimmune varicella (chicken pox) IMIG	GG	7
V-Cillin	Penicillin VK	AB	5
Vermox	Mebendazole	AP	6
Vibramycin	Doxycycline	AB	5
Yodoxin	Iodoquinol	AP	6
Zinacef	Cefuroxime	AB	5
Zithromax	Azithromycin	AB	5
Zovirax	Acyclovir	AV	6
Zyvox	Linezolid	AB	5

[a]Abbreviations: AB, antibiotic; AF, antifungal; AI, anti-insect (insecticide); AV, antiviral; GG, gammaglobulin.

Table A2. Names and usages of vaccines that your child's doctor might recommend to prevent infections[a]

Vaccine name	Germs protected against
ActHIB	*Haemophilus influenzae*
Adacel	Tetanus, diphtheria, acellular pertussis
Boostrix	Tetanus, diphtheria, acellular pertussis
Comvax	*Haemophilus influenzae,* hepatitis B
DAPTACEL	Diphtheria, tetanus, acellular pertussis
DT	Diphtheria, tetanus
Engerix-B	Hepatitis B
Flumist	Influenza
Fluvirin	Influenza
Fluzone	Influenza
Gardasil	Papillomavirus (warts, cervical cancer)
Havrix	Hepatitis A
HibTITER	*Haemophilus influenzae*
IPOL	Poliovirus
Menactra	*Neisseria meningitidis*
Menomune	*Neisseria meningitidis*
MMR II	Measles, mumps, rubella
MR	Mumps, rubella
Pediarix	Diphtheria, tetanus, acellular pertussis, hepatitis B, inactivated polio
Pedvax HIB	*Haemophilus influenzae*
Pneumovax	*Streptococcus pneumoniae*
Prevnar	*Streptococcus pneumoniae*
ProQuad	Measles, mumps, rubella, varicella (chicken pox)
Recombivax HB	Hepatitis B
RotaTeq	Rotavirus
Td	Tetanus, diphtheria
TriHIBit	Diphtheria, tetanus, acellular pertussis, *Haemophilus influenzae*
Vaqta	Hepatitis A
Varivax	Varicella virus (chicken pox)

[a]See chapter 7 for more information on vaccines.

Index